普通高等教育“十四五”规划教材

“十三五”江苏省高等学校重点教材
（2020-1-063）

废水处理
原理与技术

WASTE WATER TREATMENT
Principle and Technologies

主 编 李大鹏 李 勇
副主编 潘 杨 陈重军

中国石化出版社

内 容 提 要

本书以市政废水、农村生活污水、工业废水为对象，系统介绍了各种水处理单元技术方法的基本原理、工艺流程、设计计算、管理及应用等。内容包括总论、筛滤与调节、混凝、沉淀与上浮、深层过滤、化学处理、吸附、离子交换、膜分离、其他相转移分离法、循环冷却水处理、废水生物处理理论基础、好氧活性污泥法、好氧生物膜法、厌氧生物处理、生物脱氮除磷、人工生态处理、污泥处理与处置、废水处理厂设计等。

本书适用于环境工程专业研究生和本科生，可作为双语教学或者全英文授课教材或者参考书。

图书在版编目(CIP)数据

废水处理原理与技术：英文 / 李大鹏，李勇主编．
—北京：中国石化出版社，2022.3
普通高等教育“十四五”规划教材
ISBN 978-7-5114-6540-5

Ⅰ．①废… Ⅱ．①李… ②李… Ⅲ．①废水处理-高等学校-教材-英文 Ⅳ．①X703

中国版本图书馆 CIP 数据核字(2022)第 031729 号

中国石化出版社出版发行
地址：北京市东城区安定门外大街 58 号
邮编：100011　电话：(010)57512500
发行部电话：(010)57512575
http://www.sinopec-press.com
E-mail：press@sinopec.com
北京富泰印刷有限责任公司印刷
全国各地新华书店经销
*
710×1000 毫米 16 开本 23.5 印张 389 千字
2022 年 4 月第 1 版　2022 年 4 月第 1 次印刷
定价：68.00 元

Contents

Chapter 1　Overview of waste water

Emphasis: Waste water quality index.
Discharge standard.
Environmental effect of waste water.
Typical domestic sewage treatment methods.
The characteristics of rural sewage and its treatment technology.
Nut: Principle of secondary treatment of waste water.

1.1　What is waste water

Waste water is the flow of used water from a community or industrial process. Most people believe that waste water is only "sewage" but waste water also comes from industrial and commercial sources, and from storm and ground water. Generally, each building or home has a drainage system that connects to a network of larger sewers that carry the waste water to the waste water treatment plant. Sanitary sewers carry only domestic and industrial waste water, while combined sewers carry waste water and storm water runoff. Usually every reasonable effort is made to exclude storm and ground water from the sanitary sewer system. In older sewer systems, however, these efforts are sometimes unsuccessful due to leakage and other design considerations. Typical waste water contains 99.94% water and 0.06% suspended solids (i.e. particulates). The working process of the waste water is showed in the Figure 1-1.

Since dissolved oxygen (O_2) is the key element in the life of a body in water, it is important to measure how much oxygen is used by bacteria to consume an amount of waste water. The measure of the required oxygen is called the biological oxygen demand (BOD). The higher the BOD, the more oxygen will be demanded by the bacteria as they consume the wastes. The removal of BOD is a major goal of waste water treatment.

Occasionally, waste water may contain substances that cannot be consumed by

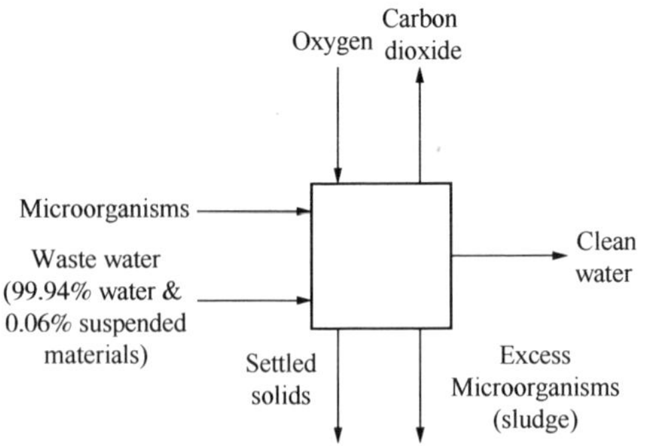

Figure 1-1 Inputs and outputs in waste water treatment

micro-organisms, and therefore are not part of the biological oxygen demand. Examples are pesticides, heavy metals such as chromium, lead, and zinc, and nutrients which micro-organisms cannot consume. These substances may have adverse health effects in downstream water supplies. It is therefore necessary to use complex, auxiliary processes to remove what the bacteria cannot consume.

Waste water contains harmful, pathogenic (i.e. disease-causing) bacteria, viruses and other micro-organisms. The treatment of this water to remove pathogenic organisms through disinfection is critical before the waste water is discharged.

As we noted earlier, municipal waste water is about 99.94% water, thus the concentrations of the various contaminants discussed are very low. The concentrations are expressed in milligrams per litre of water (mg/L). One mg/L of a constituent is equivalent to one part of a constituent (by weight) in one million parts of water or expressed as another often-used term, one part per million (ppm).

Waste water treatment is designed to remove small quantities of pollutants as the survival or death of fish, for example, depends on the presence or absence of 2~3 mg/L of dissolved oxygen in the water. A minimum of 4~5 mg/L of dissolved oxygen is usually needed to sustain good fish populations.

1.2 The types of the waste water

1.2.1 Domestic waste water

Domestic sewage is the waste water discharged from the daily life of residents. It mainly comes from the residential buildings and public buildings, such as the apartments, offices, schools, hospitals, markets, other public places and the toilets in the factories. Therefore, the main pollutants in the domestic sewage are the organic matters (protein, carbohydrate, fat, the urea, ammonia nitrogen) and lots of pathogenic microorganism (parasitic worm eggs and intestinal infections). However, the organic matters are not stable, and it is easy to corrupt, resulting in the stench. The bacteria and the pathogens breed depending on the organic matters, resulting in the infectious diseases. Therefore, the domestic sewage must be treated before drainage.

The development of sewage treatment in China is late. But the demand on the sewage treatment increased due to the economy increase after the reform and opening. The growth of the demand on the sewage treatment in China is far higher than the demand of the whole world after the 1990s.

From the 1990s, the growth of the demand on the sewage treatment in the whole world is up to 6% (average). While in China, the average growth is up to 17.73% during the decade of the 1990s. It is 2.9 times of the average growth of the whole world. It is coming into the 21st century, the demand on the sewage treatment increases very fast. The waste water consumption increased from 1.88 million tons to 4.47 million tons from 2000 to 2004. It is calculated that the average growth of the demand is up to 27%. In addition, the apparent consumption of the sewage was up to 2.25 million tons at 2001, while China has become the world's largest consumer of sewage, more than the USA. China's sewage treatment imports 1 million tons at 1998 and China has become the world's largest importer of sewage at the same year. Making a comparation between 2004 and 1998, the average growth of the import on the sewage treatment is up to 27.14%.

Till 2020, over 10, 113 waste water treatment plants had gotten discharge permit. According to treatment scale, there are 4, 873 waste water treatment plants under 10, 000 m^3/d and there are 3, 147 for waste water treatment plants between 10, 000 m^3/d and 50, 000 m^3/d. Super waste water treatment plants with the scale of million tons are very few.

According to provinces, the waste water treatment plants are mainly distributed in Guangdong, Shandong, Jiangsu and Zhejiang. In 2019, the treatment capacity is 3, 035 ten thousand tons/day, 1, 834 ten thousand tons/day, 1, 746 ten thousand tons/day and 1, 515 ten thousand tons/day, respectively. The numbers of waste water treatment plants are 697, 554, 828 and 173, respectively. If 80% water supply amount can be attributed to waste water based on 2017, the increase of waste water treatment is up to 13,197 ten thousand tons/day in the whole country. The increase is about 57.9%, compared with the treatment scale, only 22, 792 ten thousand tons/day. Those data of waste water treatment plants are showing in Table 1-1.

Table 1-1 Waste water treatment capacity (ten thousand tons/day) in provinces in 2019

Provinces	Numbers of plants	Capacity of 2019	Water supply of 2017	The increase
Jiangsu	828	1, 746	4, 568	1, 908
Guangdong	697	3, 035	5, 834	1, 632
Beijing	134	657	2, 241	1, 136
Zhejiang	373	1, 515	2, 840	757
Shanghai	43	619	1, 642	695
Hubei	391	954	2, 035	674
Hunan	346	928	1, 864	563
Sichuan	1, 125	1, 013	1, 942	541
Shandong	554	1, 834	2, 966	539
Xinjiang	146	322	968	452
Fujian	213	653	1, 347	425
Jilin	103	377	998	421
Jiangxi	233	501	1, 144	414
Heilongjiang	155	379	962	391
Neimenggu	121	272	790	360
Guangxi	285	523	1, 094	352
Henan	338	1, 263	1, 995	333
Anhui	380	999	1, 624	300
Liaoning	246	770	1, 281	255
Shanxi	157	343	688	207
Hebei	344	1, 033	1, 483	153
Gansu	116	288	548	150
Guizhou	392	375	613	115
Ningxia	44	127	299	112
Yunnan	220	455	702	107
Xizang	22	37	178	105
Chongqing	743	529	777	93
Hainan	50	111	247	87
Tianjin	130	442	522	-24
Shanxi	213	601	680	-57
Qinghai	53	91	172	
Total	9, 195	22, 792	45, 044	13, 197

1.2.1.1 The pollutants

(1) Pathogen contamination Mainly from urban sewage, hospital sewage, garbage and surface runoff, etc. The characteristics of pathogenic microorganisms are: the quantity is large; the wide distribution; the longer survival time; the propagation speed; easy to produce resistance, it is difficult to eliminate; after the traditional two stage biochemical sewage treatment and chlorine disinfection, some pathogenic microorganisms and the virus can still survive a lot; such pollutants can enter the body through a variety of ways and, result indeceases.

(2) Aerobic organic pollution The common characteristics of organic matter are these substances can directly get into the water, and be decomposed into simple inorganic substances, such as carbon dioxide and water, dissolved oxygen in the decomposition process need to be consumed in the water, in anoxic conditions, corruption decomposition will occur, water quality will be deteriorated.

(3) Eutrophication pollution It is a phenomenon of water pollution caused by excessive content of nutrients such as nitrogen and phosphorus. The eutrophication of aquatic ecosystems by chemical pollutants caused by two ways: one is the increasing amount of inorganic nutrients in plants in normal condition; another is increasing in decomposer organic matter.

(4) Stench Odor is a common pollution hazard, it also occurs in polluted water. People can smell up to more than 4, 000 kinds of odor, there are dozens of harmful. Harm foul: 1, interfere with the normal respiratory function, digestive dysfunction; mental irritability, decreasing work efficiency, judgment, memory decline; 2, in the long-term work and living environment will cause the stench of olfactory dysfunction, cerebral cortex damaged the central nervous, excited and adjusting function; It can produce hydrogen sulfide, formaldehyde and the other toxic hazard.

1.2.1.2 Waste water quality index

(1) Physical Index

① Temperature. The effluent from the houses has high temperature, which is discharged into the water body to increase the temperature of the water and causes the heat pollution of the water body. The increase of water temperature affects the survival of aquatic organisms and the use of water resources. The solubility of oxygen in water decreases with the increase of water temperature. The dissolved oxygen content decreases. On the other hand, the water temperature rises and accelerates the consumption of oxygen, which leads to the water body's anoxia or the deterioration of water quality. The temperature of surface water varies with season and climate. At 0.1 ~ 30℃, the temperature of underground water is relatively stable.

② Color. Color has euchroic and allochromatic color. Euchroic is caused by the dissolved substance or colloidal substance in the water. Therefore, the color is presented after removing the suspended matter in the water. Allochromatic color consists of the color that is caused by the dissolved substance, the colloid substance and the suspended substance. Generally, pure natural water is clear and transparent, that is, colorless. It only determines euchroic of natural water and drinking water, but the sewage containing metallic compounds or organic compounds and other colored pollutants has various colors.

③ Odor and taste. Odor and taste, like color, are also sensory indicators, which can qualitatively reflect the amount of a certain amount of pollutants. Natural water is odorless and tasteless. When a body of water is polluted, it produces a peculiar smell. The odor of water is derived from compounds such as reductive sulfur and nitrogen compounds, volatile organic compounds and chlorine gas. Different salt can bring different odors in water supply. For example, sodium chloride with salty taste, magnesium sulfate with bitter taste, calcium sulphate with a slight sweetness and so on.

④ Turbidity and transparency. The water is turbidity due to the impurity of the suspension and colloid state. The turbidity of water can be expressed with turbidity. The content of suspended matter in water is one of the basic indexes of water quality, which shows the insoluble suspended and floating substances in water, including inorganic and organic matter. The part of the suspended substance that can be precipitated within 1 to 2 hours is called a settling solid, which can roughly represent the amount of suspended matter in the water body. The precipitated substances in domestic sewage are usually called sludge; the precipitated particles in industrial waste water are called sediment.

(2) Chemical index

① Organic matter. Organic matter such as carbohydrates, proteins, fats and other organic matter in sewage and some industrial waste water is decomposed into simple inorganic substances, carbon dioxide and water under the metabolism of microorganisms. These organic matters need to consume a large amount of oxygen in the process of decomposition, so they are oxygen consuming pollutants. Oxygen consumption is one of the main causes of black and odorous water body.

(3) Inorganic index

① Plant nutrient elements. N and P in sewage are plant nutrient elements. From the perspective of crop growth, plant nutrient elements are valuable substances, but too many N and P are easy to lead to eutrophication. The content of nitrogen and phosphorus in water is closely related to the degree of eutrophication. The role of phosphorus is much greater than that of nitrogen in the eutrophication of sewage.

② pH value. pH value is mainly to indicate the acidity and alkalinity of water samples.

③ Heavy metals. Heavy metals are mainly toxic elements such as Hg, Cd, Pb, Cr, Ni and metal arsenic. They also include toxic heavy metals such as Zn, Cu, Co and Sn.

(4) Biological index

① Total number of bacteria. The total number of bacteria in the water reflects the degree of bacterial contamination in the water body. It is obvious that the source of pollution does not be explained by the total number of bacteria, it must be combined with the number of coliform to determine the source and safety of water pollution.

② Escherichia coli. Water is an important medium for the spread of intestinal diseases, and coliform bacteria is regarded as the most basic feces indicator flora. The value of coliform group can indicate the degree of water pollution in water samples by feces, which indirectly indicates the possibility of enteric pathogens (typhoid, dysentery, cholera, etc.).

(5) Comprehensive index

① Chemical oxygen demand (COD). COD refers to the content of oxidants consumed in water samples treated with a strong oxidizer under certain conditions, represented by the milligram/liter of oxygen. Chemical oxygen demand reflects the degree of pollution of the reducing substances in water.

② Biochemical oxygen demand (BOD). BOD refers to the amount of dissolved oxygen consumed by microorganisms in the biochemical process of certain oxidizable substances, especially in the biochemical process of organic matter, under the specified conditions. Biochemical oxygen demand (BOD) reflects the content of biodegradable organic matter in the water body.

③ Suspended solids (SS). SS, which is the total non-filterable residue dried at 103 to 105℃. The presence of suspended matter in the water will make the water body cloudy, reduce the transparency, and affect the respiration and metabolism of aquatic organisms. In the treatment of water and waste water, suspended solids have specific significance.

④ Total nitrogen (TN). When a certain amount of nitrogen is contained in the water body, it will lead to the vigorous reproduction of plankton and a eutrophic state. Therefore, the total nitrogen is one of the important indexes to measure the water quality.

⑤ Ammonia nitrogen (NH_3-N). NH_3-N, in the form of free ammonia or ammonia salt in the water. The main source of ammonia nitrogen in water is the decomposition product of nitrogen containing organic matter in the sewage. The determination of the content of ammonia nitrogen in water helps to evaluate the pollution and " self purification" of the water body.

⑥ Total phosphorus (TP). Phosphorus is one of the essential elements of biological growth, but the high phosphorus content in water can cause algae overproduction and cause eutrophication.

Typical concentration range of pollutants in waste water is as Table 1-2.

Table 1-2 Typical concentration range of pollutants in waste water

No.	Basic control item	Concentration		
		Low	Middle	High
1	COD/(mg/L)	250	430	800
2	BOD/(mg/L)	110	190	350
3	TOC/(mg/L)	80	140	260
4	Total nitrogen/(mg/L)	20	40	70
5	Organic Nitrogen/(mg/L)	8	15	25
6	NH_4^+-N/(mg/L)	12	25	45
7	Nitrite/(mg/L)	0	0	0
8	Nitrate/(mg/L)	0	0	0
9	Total phosphorus/(mg/L)	4	7	12
10	Organic phosphorus/(mg/L)	1	2	4
11	Inorganic phosphorus/(mg/L)	3	5	10
12	Total solid/(mg/L)	390	720	1, 230
13	Dissolved solid/(mg/L)	270	500	860

continued

No.	Basic control item	Concentration		
		Low	Middle	High
14	Nonvolatile dissolved solid/(mg/L)	160	300	520
15	Volatile dissolved solid/(mg/L)	110	200	340
16	Total suspended solid/(mg/L)	120	210	400
17	Nonvolatile suspended solid/(mg/L)	25	50	85
18	Volatile suspended solid/(mg/L)	95	160	316
19	Settling solid/(mg/L)	5	10	20
20	Oil/(mg/L)	50	90	100
21	Chloride/(mg/L)	30	50	90
22	Sulphate/(mg/L)	20	30	50
23	Volatile organic matter/(mg/L)	<100	100~400	>400
24	Coliform bacteria/(CFU/L)	$10^6 \sim 10^8$	$10^7 \sim 10^9$	$10^7 \sim 10^{10}$
25	Fecal coliforms/(CFU/L)	$10^3 \sim 10^5$	$10^4 \sim 10^6$	$10^5 \sim 10^8$
26	Cryptosporidium/(CFU/L)	0.1~1	0.1~10	0.1~100
27	Lambeau Giardia Lamblia/(CFU/L)	0.1~10	0.1~100	0.1~1000

Annotation: Low concentration, middle concentration and high concentration are referred to 750 L/(cap · day), 460 L/(cap · day) and 240 L/(cap · day) of waste water quantity, respectively.

1.2.1.3 The standard for domestic sewage drainage in China

Standards for the discharge of the domestic sewage are the control standards for the pollutants and toxic factors on the environment due to the discharge of domestic sewage. According to the water quality of the receive water body, the standard was formulated, combing the environmental characteristics, social, economic and technology conditions. The standards consist the national discharge standard, the local discharge standards.

(1) The national discharge standard The national discharge standard is the standard formulated by the national competent administrative department of environmental protection and applied nationwide or within a specific region. 《Comprehensive waste water discharge standard of the People's Republic of China》(GB 18918—2002) is suitable nationwide (Table 1-3~Table 1-5).

Table 1-3 The maximum allowable discharge concentration of basic control items in national standard (mean a day)

No.	Basic control item	First grade		Second grade	Third grade
		A	B		
1	COD/(mg/L)	50	60	100	120
2	BOD/(mg/L)	10	20	30	60
3	SS/(mg/L)	10	20	30	50
4	Animal and vegetable oils/(mg/L)	1	3	5	20
5	Petroleum/(mg/L)	1	3	5	15
6	Anionic surfactant/(mg/L)	0.5	1	2	5
7	Total N (nitrogen)/(mg/L)	15	20	—	—
8	NH_3-N/(mg/L)	5 (8)	8 (15)	25 (30)	—
9	Total P (phosphorus)/(mg/L)	0.5	1	3	5
10	Color	30	30	40	50
11	pH	6~9			
12	Number of fecal coliforms/(CFU/L)	10^3	10^4	10^4	—

Annotation: the value is for the water temperature less than 12 degrees in the bracket, while outside the bracket is for the water temperature higher than 12 degrees.

Table 1-4 The maximum allowable discharge concentration of first kind pollutants in national standard (mean a day)

No.	Item	Standard value
1	Total mercury/(mg/L)	0.001
2	Alkyl mercury/(mg/L)	Not detectable
3	Total cadmium/(mg/L)	0.01
4	Total chromium/(mg/L)	0.1
5	Cr VI/(mg/L)	0.05
6	Total arsenic/(mg/L)	0.1
7	Total lead/(mg/L)	0.1

Table 1–5 The maximum allowable discharge concentration of selective control pollutants in national standard (mean a day)

No.	Item	Standard value/ (mg/L)	No.	Item	Standard value/ (mg/L)
1	Total Ni	0.05	23	Trichloroethylene	0.3
2	Total Be	0.002	24	Tetrachloroethylene	0.1
3	Total Ag	0.1	25	Benzene	0.1
4	Total Cu	0.5	26	Total benzene	0.1
5	Total Zn	1.0	27	*o*–xylene	0.4
6	Total Mn	2.0	28	*p*–xylene	0.4
7	Total Se	0.1	29	*m*–xylene	0.4
8	Benzopyrene	0.00003	30	Ethylbenzene	0.4
9	Volatile penol	0.5	31	Chlorobenzene	0.3
10	Total cyanide	0.5	32	1,4–dichlorobenzene	0.4
11	Sulfide	1.0	33	1,2–dichlorobenzene	1.0
12	Methanal	1.0	34	*p*–nitrchlorobenzene	0.5
13	Anilines	0.5	35	2,4–dinitrochlorobenzene	0.5
14	Total N nitroso compounds	2.0	36	Phenol	0.3
15	Organophosphorus pesticide (P)	0.5	37	*m*–Cresol	1.0
16	Malathion	1.0	38	2,4–dichlorophen	0.6
17	Dimethoate	0.5	39	2,4,6–trichlorophenol	0.6
18	Parathion	0.05	40	Dibutyl phthalate	0.1
19	Parathion–methyl	0.2	41	Dioctyl phthalate	0.1
20	Pentachlorophenol	0.5	42	Acrylonitrile	2.0
21	Trichloromethane	0.3	43	Adsorbable organic halides	1.0
22	Tetrachloromethane	0.03			

(2) The local discharge standard The local discharge standards are approved by the people's governments of provinces, autonomous regions, and municipalities, applicable in specific administrative regions.

For example, Shanghai Municipal Waste Water Integrated Discharge Standard (DB 31/199—2009) is applicable to Shanghai City.

(3) The relationship between the national discharge standard and the local

discharge standard Article 10 in the 'Environmental Protection Law' said: if the items are not specified in the national discharge standard, the people's governments of provinces, autonomous regions, and municipalities can formulate the local discharge standard. If the items are specified in the national discharge standard, the local discharge standard can be stricter than that in the national discharge standard. When two kinds of standards coexist, it performs the local standard.

1.2.1.4 Oversea standards for domestic sewage drainage

Those developed countries, such as USA, European Union (EU) had published discharge standard according to domestic sewage, but some parameters and maximum allowable discharge concentration of pollutants are different from our standards.

(1) USA　Secondary treatment guidelines in USA federal regulations (40CFR133) showed the detailed introductions about sewage drainage. USA parliament gave the requirement that domestic and industrial sewage had to be secondary treated before discharged into natural water bodies. In addition, EPA (Environmental Protection Agency) showed the discharge standard based on BOD, SS and pH (Table 1-6). The plants of USA are mainly used for treatment of domestic waste water. Therefore, those hazardous chemicals and unconventional pollutants should be removed by pretreatment processes before the waste water were discharged into the plants based on USA federal regulations (40CFR133). Additionally, USA built discharge permit regulations that are called national pollutant discharge elimination system (NPDES). NPDES pointed out that any point source releasing pollutants into the water requires an EPA permit. The state can make specific standards based on the actual situations of the basins in the state. Taking protection of the water bodies as an example, the concentrations of the pollutants in the effluent should be lower than that in domestic sewage discharge standard of secondary treatment in USA (Table 1-6). Moreover, the specific standard can increase other parameters that were not required in domestic sewage discharge standard of secondary treatment in USA, such as N and P. Table 1-7 showed the standard values of NPDES in Haiey, Idaho.

Table 1-6　Domestic sewage discharge standard of secondary treatment in USA

Items	BOD/(mg/L)	BOD for Carbon/(mg/L)	TSS/(mg/L)	pH
Average of 30 days	30	25	30	6~9
Average removal of 30 days	85%	85%	85%	—
Average of 7 days	45	40	45	6~9

Table 1-7 The standard values of NPDES in Haiey, Idaho

Items		Average of 30 days	Average of 7 days	Instantaneous maximum
BOD	mg/L	30	45	N/A
	kg/d	43	64	
SS	mg/L	30	45	N/A
	kg/d	43	64	
NH_4^+-N	mg/L	1.9	2.9	3.3
	kg/d	4.1	6.4	7.1
TP	kg/d	6.8	10.4	N/A
Kjeldahl-N	kg/d	25	35	N/A
E. Coli	CFU/100 mL	126	N/A	406
Fecal coliform	CFU/100 mL	N/A	200	N/A

Annotation: N/A is referred to not applicable.

(2) European Union Council directive concerning urban waste water treatment (91/271/EEC) regards discharge standard (Table 1-8). The standard is suitable for treatment and discharge of waste water from cities and specific industry. To prevent the running of the plants from being adversely affected by the import of industrial waste water, there is a general need for pretreatment of industrial waste water. The waste water treatment plants of EU member states must accept the council directive (91/271/EEC), and then EU member states can propose stricter directives than the council directive (91/271/EEC) base on their economy and technology level.

Table 1-8 Domestic sewage discharge standard in EU

<table>
<tr><th rowspan="2">Plant scale (population equivalent)</th><th colspan="2">SS</th><th colspan="2">COD</th><th colspan="2">BOD</th><th colspan="2">TN (sensitive area)</th><th colspan="2">TP (sensitive area)</th></tr>
<tr><th>Content/(mg/L)</th><th>Removal/%</th><th>Content/(mg/L)</th><th>Removal/%</th><th>Content/(mg/L)</th><th>Removal/%</th><th>Content/(mg/L)</th><th>Removal/%</th><th>Content/(mg/L)</th><th>Removal/%</th></tr>
<tr><td>2,000~10,000 m³/d (400~2,000 m³/d)</td><td>60</td><td>70</td><td rowspan="3">125</td><td rowspan="3">75</td><td rowspan="3">25</td><td rowspan="3">70~90</td><td>—</td><td>—</td><td>—</td><td>—</td></tr>
<tr><td>10,000~100,000m³/d (2,000~20,000 m³/d)</td><td rowspan="2">35</td><td rowspan="2">90</td><td>15</td><td rowspan="2">70~80</td><td>2</td><td rowspan="2">80</td></tr>
<tr><td>>100,000m³/d (>20,000 m³/d)</td><td>10</td><td>1</td></tr>
</table>

Annotation: The water sample in Table 1-8 is referred to the mixed sample of 24h concentrations or flow rates.

(3) Japan Sewage directive (2014) regards discharge standard (Table 1-9). In addition, local government can propose stricter directive than national directive according to actual situation.

Table 1-9 Domestic sewage discharge standard in Japan (average of day)

Items	Unit	Content
pH		5.8~8.6
BOD	mg/L	10/15 (separated system/combined system)
		40(combined system)
SS	mg/L	40
TN	mg/L	10/20 (separated system/combined system)
TP	mg/L	0.5/1.0/3.0①
E. Coli	CFU/ mL	3, 000

① BOD is referred to 40 mg/L in combined system to prevent the running of the plants from being adversely affected by the heavy rain.

1.2.1.5 The treatment method for domestic sewage

Generally, the biotechnical processes are used in the treatment of domestic sewage. The main reason is that the biodegradability of the domestic sewage is high, up to 0.5. In addition, there are some processes for the treatment of domestic sewage (Table 1-10).

Table 1-10 The selection of the waste water treatment according to the pollutants

Pollutants	Treatment processes
Suspended solids	Screen, grinding, screen, filter, sedimentation, flotation, centrifugation, coagulation
Biodegradable organic matters	Activated sludge, biofilm, stable ponds, land treatment system.
Nonbiodegradable organic matters	Physical-chemical system, activated carbon, ozone or other advanced oxidation, land treatment system
Pathogene	Disinfection: chlorination, ozone, chlorine dioxide, ultraviolet ray, bromine or iodine, radiation; land treatment system
Nitrogen	Biological nitrification and denitrification, nitrogen blow off, ion exchange, land treatment system
Phosphorus	Chemical treatment: aluminum salt, ferric salt, lime; biological-chemical methods, A/A/O process, land treatment system

continued

Pollutants	Treatment processes
Heavy metals	Coagulation and sedimentation, flotation, ion exchange, electroosmosis, reverse osmosis, activated carbon, ferriter technique
Dissolved inorganic solids	Ion exchange, electroosmosis, reverse osmosis, evaporation
Oil	Oil separator, flotation, coagulation and filtration, coarse graining, electrolysis
Heat	Cooling tower, cooling tank
Acid, alkali	Neutralization, dialysis, evaporation
Radioactive contamination	Coagulation and filtration, ion exchange, evaporation, storage

(1) Physical processes The processes are referred to the separation and recovery of insoluble pollutants (including oil film and oil beads) in waste water by physical methods. Gravity separation method, centrifugal separation and filtration are the main methods. Additionally, the method, based on the heat exchange, is attributed to the physical process.

(2) Chemical processes The processes are referred to the separation and removal of the soluble pollutants or transformation of the pollutants from toxic forms to nontoxic forms. The units based on the chemical reaction of adding chemicals are as followed, coagulation, neutralization and oxidation-reduction, and so on. The units based on the mass transfer effect are as followed, extraction, stripping, blow off, adsorption, ion exchange, electrodialysis and hyperfiltration. It is noticeably, the units based on the mass transfer effect are related to the chemical mechanism and physical mechanism, therefore, the process could be separated from the chemical processes. It is named physical-chemical processes.

(3) Biological processes The processes are referred to thetransformation of the soluble, colloidal, superfine suspended pollutants from toxic forms to nontoxic and stable forms under the metabolism of the microorganisms. According to the microorganisms, biological processes can be separated to aerobic processes and anaerobic processes. Nowadays, the waste water treatment plant mainly uses the aerobic processes. The aerobic processes can be divided into activated sludge and biofilm methods. For the former, it is a treatment unit and has many running modes. For the latter, it consists the biofilter, rotating biological disk, biological contact oxidation pond, biological fluidized bed, and so on. The method of biological ponds is named natural biological processes. Anoxic processes are also named biological reduction

process. It is mainly used for the treatment on the organic waste water with high concentration and sludge.

(4) Biological contact oxidation process The filters were filled into the bioreactor, and they were submerged by the waste water with the oxygen and the water passed the fillers with some flow rate. The biofilms covered the fillers, and the waste water contacted completely with the biofilms. The pollutants can be removed under the metabolism of the microorganisms. The process is in the middle of the activated sludge and the biofilter.

1.2.2 Industrial waste water

Generally, industrial waste water is referred to the waste water during the industrial processes. The main pollutants are the industrial production materials, process products, by-products, and those pollutants during the processes. The types of the waste water are rich and the constituents are complex. For example, the mercury is the main pollutant in the electrolysis brine waste water. The lead and the cadmium are the main pollutants in the heavy metals smelt waste water. The cyanide and the chromium are the main pollutants in the electroplate waste water. The phenol is the main pollutant in the petroleum refining waste water. Therefore, the industrial waste water consists many toxic materials and harms the environment and the health.

1.2.2.1 The types of the industrial waste water

(1) According to the chemical materials The waste water can be called inorganic waste water due to the main inorganic pollutants, while the water can be called organic waste water due to the main organic pollutants. For example, the water can be called inorganic waste water if it comes from the electroplating and mineral processing. On the contrary, the water can be called organic waste water if it comes from the food processing or petroleum refining. But the water can be called mixed sewage if it comes from the printing and dyeing. Therefore, the component in the different waste water from different industry was different completely.

(2) According to the materials in the processing and the product The waste water includes many types, such as papermaking, metallurgy, coking, metal pickling, chemical fertilizer, printing and dyeing, leather making, pesticides, electric power station, and so on.

(3) According to the main pollutants in the waste water The waste water includes many types, such as acid waste water, alkali waste water, cyanide waste water,

chromium waste water, mercury waste water, phenol waste water, aldehyde waste water, oil waste water, sulfide-bearing waste water, organic phosphorus waste water, radioactivity waste water.

1.2.2.2 The typical industrial waste water

The components are very complex in the industrial waste water. Nowadays, the industrial waste water could be divided into some kinds according to the main pollutants.

(1) Pesticide waste water There are many kinds of pesticides, and the quality of pesticide waste water is complicated. Its main features are as followed:

The concentration of pollutants is high, and chemical oxygen demand (COD) can reach tens of thousands of milligrams per litre.

It is highly toxic. In addition to containing pesticides and intermediates, waste water also contains toxic substances such as phenol, arsenic and mercury as well as substances that are difficult to degrade by many microorganisms.

It has a foul odor and is irritating to the human respiratory and mucous membrane.

The quality and quantity of the waste water are not stable. Therefore, the pollution of pesticide waste water to the environment is very serious. The purpose of the treatment on the pesticide waste water is to reduce the concentration of pollutants, improve recycling efficiency, and strive to achieve harmless. The methods for treating pesticide waste water include activated carbon adsorption, wet oxidation, solvent extraction, distillation, and activated sludge. However, the new pesticides that are highly efficient, low-toxic, and low-residue is the development direction of pesticides. Some countries have banned the production of organic chlorine and organic mercury pesticides such as BHC, and actively researched and used microbial pesticides. This is a new way to fundamentally prevent pesticide waste water from pollution on the environment.

(2) Phenolic waste water The phenolic waste water mainly comes from coking plants, gas plants, petrochemical plants, insulating material plants and other industrial sectors. Decontamination of high concentration ammonia nitrogen industrial waste water and production of ethylene, synthetic phenol, polyamide fibers, synthetic dyes, organic pesticides and phenolic resins. Phenolic waste water mainly contains phenolic compounds such as phenol, cresol, xylenol, and nitrocresol. Phenolic compounds are protozoal poisons that can make proteins coagulation. When the concentration of phenol in water reaches 0.1~0.2 mg/L, the smell of the fish is odor and cannot be eaten; if the

mass concentration increased to 1 mg/L, fish spawning will be affected. The amount of phenol in the water reaches 5~10 mg/L, and fish will die in large numbers. Phenol in drinking water can affect human health. Even if the concentration of phenol in the water is only 0.002 mg/L, chlorophenol odor will be produced when chlorine is used for disinfection. Phenolic waste water with a concentration of 1, 000 mg/L is often referred to as high-concentration phenolic waste water. This waste water must be treated after phenol recovery. Phenolic waste water with a mass concentration of less than 1, 000 mg/L is called low-concentration phenolic waste water. This kind of waste water is usually recycled and the phenol is concentrated and recovered before being processed. The methods for recovering phenol include solvent extraction, steam stripping, adsorption, closed loop, and so on. Waste water containing phenol at a concentration of less than 300 mg/L can be discharged or recovered after being treated by biological oxidation, chemical oxidation, and physical chemical oxidation.

(3) Mercury-containing waste water Mercury-containing waste water mainly comes from non-ferrous metal smelters, chemical plants, pesticide plants, paper mills, dye factories, and thermal instrumentation. Methods for removing inorganic mercury from waste water include sulfide precipitation method, chemical coagulation method, activated carbon adsorption, metal reduction method, ion exchange method, and microbial method. Alkaline mercury-containing waste water is usually treated by chemical agglomeration or sulfide precipitation. Acidic mercury-containing waste water can be treated by metal reduction. Low-concentration mercury-containing waste water can be treated by activated carbon adsorption, chemical agglomeration, or activated sludge. Organic mercury waste water is difficult to treat. Organic mercury is usually oxidized to inorganic mercury and then be removed.

The toxicity of various mercury compounds varies greatly. Elemental mercury is basically non-toxic. Mercuric chloride in inorganic mercury is a highly toxic substance. Phenylmercury in organomercury is rapidly decomposed and has little toxicity. Methylmercury is easily absorbed into the human body, and it is not easily degradable. In addition, the excretion is slow. It is particularly prone to accumulate in the brain and is the most toxic. For example, mink disease is caused by methylmercury poisoning.

(4) Oily waste water Oily waste water mainly comes from petroleum, petrochemical, steel, coking, gas generation stations, and mechanical processing industries. The relative density of oil pollutants in the waste water, except heavy tar, is

greater than 1.1, and the remaining relative densities are all less than 1. Oils usually exist in three states in waste water.

Floating oil. oil droplets larger than 100 μm, easy to separate from waste water.

Dispersed oil. The particle size of the oil droplets is between 10 and 100 μm, suspended in water.

Emulsified oil. oil droplet size is less than 10 μm, difficult to be separated from waste water.

Because the oil concentrations are different in waste water discharged from different industrial sectors. For example, the waste water is generated in the oil refining process, the oil content is about 150 to 1, 000 mg/L, but the tar content in coking waste water is about 500 to 800 mg/L. The tar content in waste water discharged from the gas generation station is 2, 000 to 3, 000 mg/L. Therefore, the treatment of oily waste water should first use oil separator to recover slick or heavy oil. The efficiency is up to 60% to 80%, and the oil content in the effluent is about 100 to 200 mg/L. Emulsified oil and dispersed oil in waste water are difficult to remove, so emulsification should be avoided or lightened. One of the methods is to reduce the emulsification of the oil in the waste water during the production process. The second method is to use air floatation and demulsification.

(5) Heavy metal waste water Heavy metal waste water mainly comes from the waste water discharged from mining, smelting, electrolysis, electroplating, pesticides, pharmaceuticals, paints, and pigments. The types, contents, and existing forms of heavy metals in waste water vary with different manufacturers. Because heavy metals cannot be decomposed, we only transported their place of existence and change their physical and chemical forms. For example, after the chemical precipitation treatment, heavy metals in the waste water are converted from dissolved ions to insoluble substrates and precipitated, and then they are transferred from the water to the sludge. After ion exchange treatment, the heavy metal ions in the waste water are transferred to the ion exchange resin, and after regeneration, they are transferred from the ion exchange resin to the regenerated waste liquid. Therefore, the principle of heavy metal removal is to reform the production process. We can decrease the use of the toxic heavy metals or no use. The second is to use reasonable process, scientific management and operation to reduce the amount of heavy metals and the amount of waste water, and to minimize the amount of discharged waste water. Heavy metal waste water should be disposed of locally at the site of production and not mixed with other waste water to avoid complicating the

treatment. It should not be directly discharged into urban sewers without treatment to avoid of the expansion of heavy metal pollution. The treatment of heavy metal waste water can generally be divided into two categories. The first is to convert the dissolved heavy metals in the waste water into insoluble metal compounds or elements and remove them from the waste water by sedimentation and floating. Possible methods include neutralization precipitation method, sulfide precipitation method, floatation separation method, electrolytic precipitation (or floatation) method, diaphragm electrolysis method, and the like. The second method is to concentrate and separate heavy metals in waste water without changing their chemical forms. Reverse osmosis, electrodialysis, evaporation, and ion exchange can be applied. These methods should be used alone or in combination depending on the quality of the waste water, the amount of water, etc.

(6) Cyanide-containing waste water Cyanide-containing waste water mainly comes from the electroplating, gas, coking, metallurgy, metal processing, chemical fiber, plastics, pesticides, chemical and other departments. Cyanide-containing waste water is a kind of industrial waste water with relatively high toxicity. It is unstable in water and easily decomposed. Both inorganic cyanide and organic cyanide are highly toxic substances, and human ingestion can cause acute poisoning. The lethal dose of cyanide to the human body is 0.18 g and that of potassium cyanide is 0.12 g. The mass concentration of cyanide in fish killing is 0.04 to 0.1 mg/L. The treatment measures for cyanic waste water include:

① Reforming the process to reduce or eliminate the effluent containing cyanide-containing waste water. If non-cyanide electroplating is used, industrial waste water from electroplating workshops can be eliminated.

② Waste water containing high amounts of cyanide should be recycled. Waste water containing low levels of cyanide should be purified before disposal.

Recovery methods include acidification aeration-alkali absorption method, steam desorption method, and the like. Treatment methods include alkaline chlorination, electrolytic oxidation, pressurized hydrolysis, biochemistry, biological iron, ferrous sulfate, and air stripping. Among them, the alkaline chlorination method is widely used, and the ferrous sulfate method is not complete and unstable. The air blow-off method not only pollutes the atmosphere but also fails to meet the discharge standards for the effluent. Therefore, it is less frequently used.

(7) Food industry waste water The food industry has a wide range of raw materials and a wide variety of products. The amount of water discharged from waste

water and the quality of water are very different. The main pollutants in the waste water are:

① Solid materials floating in the waste water, such as vegetable leaves, peels, minced meat, poultry feathers, etc.

② Substances suspended in the waste water are grease, protein, starch, colloidal substances, etc.

③ Dissolved in the waste water of acid and alkali, salt and sugar, etc.

④ Silt and other organic matter entrained with raw materials, etc.

⑤ Poison of pathogenic bacteria.

The waste water from the food industry is characterized by a high content of organic substances and suspended substances, which is easily corrupted, and generally has no major toxicity. The main danger is the eutrophication of the water body, which causes the death of aquatic animals and fish, promotes the production of odor from the organic matter deposited on the bottom, deteriorates water quality, and pollutes the environment. In addition to proper treatment of water quality according to the characteristics of water quality, waste water treatment in the food industry should generally adopt biological treatment. If the quality of the effluent water is very high or the content of organic matter in the waste water is high, a two-stage aeration tank or a two-stage biological filter, or a multi-level rotating biological contactor, or a combination of two biological treatment devices may be used. Anaerobic-aerobic series biological treatment systems can also be chosen.

(8) Waste water from papermaking industry Papermaking waste water mainly comes from the two production processes of pulping and papermaking in the papermaking industry. Pulping is to separate the fiber from the plant raw material into a slurry and bleach it. Papermaking is to dilute, shape, squeeze, and dry pulp to make paper. Both processes discharge large amounts of waste water. The pollution in the pulping is the biggest. Waste water discharged during pulp washing is dark brown, which is called black water. The concentration of pollutants in black water is very high, BOD is as high as 5~40 g/L, and contains a lot of fibers, inorganic salts and pigments. The waste water discharged from the bleaching process also contains a large amount of acid-base substances. The waste water discharged from the paper machine is called white water, which contains a lot of fibers and fillers and compounds added during the production process. Waste water treatment in the paper industry should focus on increasing the rate of recycled water, reducing potable water use and waste water discharge, and it should

also actively explore various reliable, economical, and treatment methods that can make full use of the useful resources in waste water. For example, the flotation method can recover fibrous solids in white water, and the recovery rate reaches 95%, and then the clear water can be reused. The combustion method can recover sodium hydroxide, sodium sulfide, sodium sulfate and other sodium salts combined with organic compounds in black water. Neutralization method is to adjust the pH value of the waste water; coagulation, sedimentation or flotation method can remove suspended solids in waste water; chemical precipitation method can decolorize; biological treatment can remove BOD, more effective for kraft paper waste water; wet oxidation is suitable for the treatment of the sulfurous acid pulp waste water. In addition, reverse osmosis, ultrafiltration, electrodialysis, and other treatment methods are used at home and abroad.

(9) Printing and dyeing industry waste water The printing and dyeing industry has a large amount of water consumption, and usually consumes 100 to 200 tons of water per ton of textiles. 80% to 90% of the waste water is printing and dyeing waste water. Recycling and harmless treatment are the commonly used methods. Recycling can be divided into: ① The waste water can be respectively recycled according to the water quality characteristics, such as the diversion of the bleaching scouring waste water and the dyeing printing waste water. The former can be reused and then reduce discharge. ② The evaporation was often used for the alkali liquor recycling. ③ The dye recovery.

For example, the indanthrene dye can be acidified into cryptic acid. Because the acid is colloidal particles, and then suspended in raffinate. Therefore, the acid can be recycled after sedimentation and filtration. Another treatment is harmless. It can be divided into:

① Physical treatment. The method includes precipitation and adsorption. The precipitation mainly removes the suspended solids in the waste water. the adsorption mainly removes the dissolved contaminants in the waste water and decolorizes them.

② Chemical treatment. The method includes neutralization, coagulation, and oxidation. The neutralization is to adjust the acidity and alkalinity of waste water, but also to reduce the color of the waste water. The coagulation is to remove dispersed dyes and colloidal substances in the waste water. The oxidation is to oxidize the reducing substances in the waste water to precipitate the sulfur dyes and vat dyes.

③ Biological treatment. There are activated sludge, biological rotating disc, biological drum and biological contact oxidation. To improve the quality of effluent water, it meets discharge standards or recycling requirements. The methods are often combined.

(10) Dye production waste water Waste water from dye production contains acids, alkalis, salts, halogens, hydrocarbons, amines, nitro compounds, dyes and their intermediates. In addition, some also contain pyridine, cyanide, phenol, benzidine, and heavy metals, such as mercury, cadmium, and chromium. The composition of the waste water is complicated. It is toxic and difficult to clean. Therefore, the appropriate methods for the treatment of dye production waste water should be selected according to the characteristics of the waste water and its discharge requirements. For example, the removal of solid impurities and inorganic substances can be performed by coagulation and filtration. The removal of organic substances and toxic substances mainly include chemical oxidation, biological methods and reverse osmosis. The decolorization can generally be performed by a process consisting of coagulation and adsorption. The removal of heavy metals can be ion exchange method.

(11) Chemical industry waste water Chemical industry waste water mainly comes from the production waste water discharged from the petrochemical industry, coal chemical industry, acid and alkali industry, fertilizer industry, plastics industry, pharmaceutical industry, dye industry, and rubber industry. The main measures for the prevention and control of chemical waste water pollution are as followed. Reforming the production process and equipment, reducing the pollutants, preventing the waste water from being discharged, and conducting comprehensive utilization and recovery. The treatment has to choose the suitable methods according to the water quality and requirements. The primary treatment is to separate suspended solids, colloids, slicks, or heavy oils from the water. The water quality and water regulation, natural precipitation, floating and grease and other methods can be chosen. The secondary treatment is main to remove the available biodegradable organic matters and part of the colloidal matters, reduce the biochemical oxygen demand in the effluent, and part of the chemical oxygen demand. Generally, the biological treatment is used. After the biological treatment, a considerable amount of COD is remained, sometimes with high color, smell and taste. Therefore, the advanced treatment is further required for further purification. The tertiary treatment is to remove organic pollutants and soluble inorganic pollutants that are nonbiodegradable in the waste water. In addition, activated carbon adsorption and ozone oxidation, ion exchange and membrane separation techniques can also be used. Various

types of chemical industrial waste water can use different treatment methods according to different water quality, quantity of water, and the requirements of treated drainage quality.

(12) Acid and alkaline waste water Acidic waste water mainly comes from steel plants, chemical plants, dye factories, electroplating plants, and mines, which contain various harmful substances or heavy metal salts. The acid mass fractions vary widely, ranging from less than 1% to greater than 10%. Alkaline waste water mainly comes from printing and dyeing factories, tanneries, paper mills, oil refineries and so on. Some of them contain organic matters or inorganic matters. Some of the matters have a mass fraction of more than 5%, and some are less than 1%. Acid-alkali waste water, in addition to containing acid and alkali, often contains acid salts, basic salts and other inorganic and organic matters. The acid-alkali waste water is highly corrosive and needs to be properly managed before it can be discharged. The principle for the treatment of acid-alkali waste water is as followed.

① High-concentration acid-alkali waste water should be prioritized for recycling.

According to water quality, water volume and different process requirements, plant or regional dispatching should be carried out. It is to try the best to reuse the waste water. If the repeated use is difficult, or the concentration is low, the amount of water is larger, the concentrated methods can be used to recover acid and alkali.

② Low-concentration acid-alkali waste water, such as washing water in pickling tanks, and rinse water in caustic tanks, should be neutralized.

For the neutralization treatment, the acid waste water can be considered for the neutralization of the alkali waste water. However, neutralizer treatment can be used without these conditions.

(13) Beneficiation waste water Beneficiation waste water has the characteristics of large amount of water, high suspended solids content, and large number of harmful substances. Its harmful substances are heavy metal ions and beneficiation agents. Heavy metal ions include copper, zinc, lead, nickel, antimony, cadmium, arsenic, and rare elements. The flotation agents added in the beneficiation process are as follows.

① Collectors, such as RocssMe, black drug [$(RO)_2PSSMe$], white drug [$CS(NHC_6H_5)_2$];

② Inhibitors, such as cyanide (KCN, NaCN), water glass (Na_2SiO_3);

③ Foaming agents, such as turpentine, cresol ($C_6H_4CH_3OH$);

④ Active punishments, such as copper sulfate ($CuSO_4$), heavy metal salts;

⑤ Vulcanization agents, such as sodium sulfide;

⑥ Mineral paddle modifiers, such as sulfuric acid and lime. Beneficiation waste water mainly passes through the tailings dam and the suspended solids can effectively be removed in the waste water.

The content of heavy metals and flotation chemicals can also be reduced. If it does not meet the discharge requirements, the advanced treatment should be further processed. The commonly used treatment methods are as follows.

① Lithium neutralization and roasting dolomite adsorption can be used to remove heavy metals.

② The ore adsorption, activated carbon adsorption can be chosen for the main removal of the flotation agent.

③ Chemical oxidation was used for the cyanide-containing waste water treatment.

(14) Metallurgical waste water Big water consumption, various types, complex and variable water quality are the main characteristics for the metallurgical waste water. According to the sources and characteristics of the waste water, the waste water consists the cooling water, pickling waste water, washing waste water (dust removal, gas or flue gas), scouring waste water, coking waste water, and the waste water that is condensed, separated or spilled during the production. The development of the treatment on metallurgical waste water are as followed.

① The new technology and the new process with the characteristics of no water or less water and no pollution or less pollution for the production should be commonly used, such as the use of dry quenching of coke, coking coal preheating, direct desulfurization and cyanide removal from coke oven gas.

② Comprehensive utilization technology should be developed, such as the recovery of useful substances and the heat from the waste water in the waste gas, reducing the fuel loss.

③ According to different water quality requirements, comprehensive balance and stream use should be considered. Another measure is to improve the stability of water quality, and constantly improve the utilization rate of the water cycle.

1.2.2.3 The quality and quantity characteristics of the industrial waste water

The quality and quantity characteristics of the industrial waste water were very big due to the production process and the production mode. For example, the inorganic matters are the main pollutants in the power station and the mine, while the organic matters are the main pollutants in the papermaking and food. For the latter, the content of the BOD_5 is generally over 2, 000 mg/L, even up to 30, 000 mg/L. Although in the same production process, the quality maybe different. Take the steel-smelting furnace as an example, during the different smelting, the value of pH is between 4~13, and the content of suspended solids is between 250~25, 000 mg/L.

The other characteristic is that the industrial waste water consists many types of matters, and the formation is different, except of the indirect cooling water. For example, the fluorine is hydrogen fluoride or fluorine ions in the glass industry and electroplating waste water, but in the phosphate fertilizer plant, the fluorine is silicon tetrafluoride. Another example is nickel. The nickel could be ionic state or complexing form. It is unquestionable that it is difficult to clean the industrial waste water due to these characteristics.

As forthe other characteristics, the quantity of the industrial waste water depends on the uses. Smelting, papermaking, petroleum refining, power station uses the more water and then the waste water are big. Take the steelmaking workshop as an example, the waste water is up to 200~250 tons per ton steel. However, the real drainage waste water is related to the cycle rate. The real drainage waste water is only 2 tons per ton steel if the cycle rate is high in the steel plant.

1.2.2.4 The discharge standards for the industrial waste water

Today, China has many discharge standards for the industrial waste water due to the variable industrial waste water, while the different standards are suitable different waste water. For Example:

① Discharge standards of water pollutants for dyeing and finishing of textile industry (GB 4287—2012)

② Emission standard of pollutants for electroplating(GB 21900—2008)

③ Effluent standard of pollutants for pesticides industry(GB 21523—2008)

④ Discharge standard of water pollutants for sugar industry (GB 21909—2008)

⑤ Discharge standard for pollutants from synthetic fat acid industry(GB 3547—83)

⑥ Discharge standard for pollutants from synthetic detergent industry(GB 3548—83)

⑦ Discharge standard of water pollutants for leather and fur making industry(GB 30486—2013)

⑧ Discharge standard for pollutants from petroleum exploitation industry (GB 3550—83)

⑨ Emission standard for pollutants from petroleum refining industry(GB 31570—2015)

⑩ Discharge standard for pollutants from film processing(GB 3553—83)

⑪ Discharge standards for pollutants from chromium salt industry(GB 4280—84)

⑫ Emission standard of pollutants for petroleum chemistry industry (GB 31571—2015)

⑬ Emission standard of pollutants for sulfuric acid industry (GB 26132—2010)

⑭ Discharge standards for pollutants from phosphorus industry(GB 4283—84)

⑮ Emission standards for pollutants from light metal industry(GB 4912—85)

⑯ Emission standards for pollutants from heavy non-ferrous metal industry (GB 4913—85)

⑰ Emission standards for pollutants from asphalt industry(GB 4916—85)

⑱ Effluent standard for pollutants from scrubbing goods train(GB 5469—85)

⑲ Discharge standard of water pollutants for fermentation alcohol and distilled spirits industry (GB 27631—2011)

⑳ Discharge standard of water pollutants for medical organization (GB 18466—2005)

㉑ Discharge standard of pollutants for livestock and poultry breeding (GB 18596—2001)

㉒ Discharge standards of water pollutants for pharmaceutical industry (GB 21907—2008)

1.2.3 Rural waste water

Rural sewage is mainly composed of domestic sewage and agricultural waste water. Sewage component is fixed, mainly contains carbohydrates, protein, fat, amino acids and other organic matter, more suitable for the growth of bacteria. Bacteria and viruses become the survival. But the sewage generally does not contain toxic, and has a certain effect, can be used for irrigation of farmland. The composition of agricultural waste water is various, different seasons, different places, different development goals of villages and towns, the waste water needs to use different methods of treatment. The rural waste water is discharged into the natural water bodies without the treatment (Figure 1-2).

1.2.3.1 The status of rural waste water

At present, China's annual total discharge of waste water reaches 60 billion m^3, of

Figure 1-2 Current situation of domestic sewage discharge in rural areas

which township waste water reaches 20 billion m^3 and rural waste water reaches 8 billion m^3. There are more than 600,000 administrative villages and 2.5 million natural villages in China, 96% of which have no drainage channels and waste water treatment systems. Domestic waste water is discharged at will, and the resulting environmental pollution is not only a potential safety hazard in rural water sources, but also the crisis of freshwater resources has been exacerbated and the irrigation of cultivated land has not been effectively guaranteed, which has jeopardized the survival and development of farmers. Therefore, rural waste water has become an important source of pollution affecting the environmental quality of water bodies. Currently, 75% of natural and artificial water bodies in China are eutrophic.

Generally, the rural waste water consists three main sources toilet water, washing water, and kitchen water.

The rural waste water treatment system is divided into a centralized waste water treatment system and a decentralized waste wäter treatment system in China. The centralized waste water treatment system is established through the establishment of a centralized pipe network collection system and a large-scale waste water treatment plant, on which an advanced treatment is performed and then discharged into a natural water body or reused. It is commonly used for domestic sewage in rural that are very close to the city. The decentralized waste water treatment system refers to the treatment of rural domestic sewage nearby, which is a low-cost, sustainable treatment system. It is generally applicable to the vast rural and remote areas where people live more widely. Most of the rural waste water that has been treated currently uses a decentralized treatment model. Considering the characteristics of scattered, small, and large quantities of rural waste water in China, the domestic waste water treatment model needs to consider such factors as stable treatment process, low investment and operating costs, and convenient maintenance and management. At present, there are mainly the following types of widely used decentralized processing technologies: treatment technology of stable pond, artificial wetland treatment system, land treatment technology, biofilm technology and integrated device processing technology.

1.2.3.2 The characteristics of rural waste water quality

The rural waste water mainly consists the three kinds, the kitchen waste water, the washing waste water and the rest room waste water. Generally, there is no collection systems for the waste water, resulting in that the waste water goes to the canals, ponds, reservoirs, lakes and so on. The main characteristic of these waste water is the high content of organic matters (Table 1-11 and Table 1-12).

① The rural population is small, widely distributed and scattered, and most rural areas have no sewage discharge pipe network.

② The concentration of domestic sewage in rural areas is low and variable.

③ The nature of most of the rural domestic sewage is not much different. The water basically does not contain heavy metals and toxic and hazardous substances (but as people's living standards increase, some domestic sewage may contain heavy metals and toxic and hazardous substances), containing a certain amount of nitrogen, phosphorus, water quality fluctuations, strong biochemical.

④ The water quality is different at different times.

⑤ The effluent discharged from the toilet has poor water quality but can be used as a fertilizer in septic tanks.

Table 1-11 The main source and the main component in the rural waste water

Source	Component	Characteristics
Kitchen	Washing rice Washing vegetables Washing dishes	Kitchen garbage, rice bran, oil - water mixture. Increasing the suspended solids, COD, BOD.
Washing	Washing and rinsing Bathing Laundry	The percentage of the washing water is over 50% to the total rural waste water. Consisting the chemical substances, such as N, P.
Rest room	Toilet and urinal flushing	The content of COD, N, P in the rest room waste water is higher than that in kitchen and washing waste water. It is the main pollutants source for the rural waste water.

Table1-12 The characteristics of the rural waste water

Source	Amount /%	BOD/ [kg/(cap · a)]	COD/ [kg/(cap · a)]	NH_3-N/ [kg/(cap · a)]	TP/ [kg/(cap · a)]
Kitchen	16	11.0	16.0	0.3	0.07
Washing	46	1.8	3.7	0.4	0.10
Rest room	26	9.1	27.5	4.4	0.70
Others	12	—	—	—	—
Total	100	21.9	47.2	5.1	0.87

1.2.3.3 The discharge characteristics of the rural waste water

The discharge of rural waste water is irregular, with large instantaneous changes, and the discharge of sewage is in a discontinuous state. Rural waste water is usually sprinkled directly onto the ground, discharged into the river nearby or merged into the river through a drainage ditch. The direct discharge and indiscriminate drainage of rural waste water not only causes serious environmental pollution, but also has a major impact on the rural ecological environment and the health of residents.

① The pipe network for collecting rural waste water is not perfect. In most rural areas in China, there is no sound waste water collection networks, and the rural waste water is discharged into natural water bodies. The waste water characteristics consist small runoff, scattered distribution and extensive coverage. In most areas, rural waste water is discharged directly and there is no treatment facility.

② The daily discharge of rural waste water is regular. Waste water in rural areas is greatly affected by the level of water used by rural households, which shows a cyclical change. The daily change rate of rural domestic waste water discharge is relatively large, and there are three peak periods in one day, namely morning, noon, and night. In the morning, it is time for rural residents to wash, cook and use toilets. The sewage generated during this period is mainly kitchen waste water, wash water and flushing water. The peak amount of sewage water at noon mainly comes from kitchen waste water and laundry waste water. The peak of sewage water in the evening is mainly the discharge of kitchen waste water and bath water.

③ The seasonal discharge of rural waste water is regular. There are more domestic sewage discharges in rural areas in summer and relatively less in winter. At the same time, due to more rainfall in summer, the concentration of pollutants in rural domestic sewage is low; On the contrary, in winter, as a dry season in a year, the concentration of pollutants in rural domestic sewage is relatively high. The impact of on-site discharge of rural domestic sewage on the wet season is relatively low, while in the dry season in winter, the concentration of pollutants in domestic sewage is relatively high, and the nearby discharged sewage pollutes the surrounding ponds, rivers and other water bodies more seriously. Therefore, in the investigation of the status of rural domestic sewage, it is necessary to conduct a separate analysis, there are certain differences in the degree of impact of sewage in different periods.

The discharge characteristics of the rural waste water are showed in the Table 1-13.

Table 1-13 The discharge characteristics of the rural waste water

Area	Amount/ [L/(cap · d)]	BOD/ (mg/L)	COD/ (mg/L)	NH_3-N/ (mg/L)	TP/ (mg/L)	SS/ (mg/L)	Discharge mode	Discharge area
Jiangsu	30	80~200	120~400	0.5~17	1~4	70~220	Sewers	Sewage treatment station
Tianjin	42	100~220	115~360	0.1~10	0.8~3.5	90~200	Open or blind ditch	Rivers, ponds, farms, Sewage treatment station
Sichuan	90	167~285	60~320	3~42	0.5~13	100~500	Direct drainage	Nearby water body or soils
Guangdong	71	150~270	180~350	0.3~15	1~3	60~100	Sewers	Sewage treatment station

1.2.3.4 The quantity characteristics of the rural waste water

Generally, the rural water are the rivers, lakes, ponds and wells, potable water. While the water resource are the wells, potable water, the rivers, lakes and ponds can be used as the irrigation and washing. Nowadays, the villages are dispersed. In addition, the economy level, the natural conditions, the technology and the water mode are different completely, so that the quantity characteristics of the rural waste water are different too. In general, the main characteristics are as followed.

(1) The total waste water amount is huge and increases year by year It was reported that the total rural waste water was up to 9.86 billion tons. Accordingly, the total amount of COD was up to 9.04 million tons, the total amount of the total nitrogen was up to 1.2 million tons, and the total amount of the total phosphorus was up to 0.1 million tons. In the future, the total waste water amount will increase year by year, due

to the promotion of the urban-rural integration and the development of the economy in the village and the increase of the living level.

(2) The difference is obvious due to the areas The quality and the quantity of the rural waste water are different due to the geographical conditions, the economy, the living mode and the seasons. Generally, in the flat area, the distribution of the villages is centralized relatively, therefore, the amount of the waste water is centralized relatively. On the contrary, for the hills, the distribution of the villages is not centralized, so that there are no the collection systems for the waste water. The waste water is drained into the nearby water bodies according to the geographical conditions. For the areas with low economy, the sanitary wares are few, so the amount of the waste water is low. It is mainly the washing water and the kitchen waste water, suggesting that the content of the N and P is low, relatively. For the areas with high economy, the sanitary wares are common and complete, so the amount of the waste water is higher than that in the areas with low economy. In addition, the content of the N and P is higher.

(3) Seasons change For the rural waste water, the quantity changes very big in a day or in the different seasons in a year.there are three peak periods in one day, namely morning, noon, and night. For the other time, the quantity of the waste water is low relatively. Even then at night there is no waste water in the sewers. In the dry season, the pollution is heavy due to the rural waste water, while the pollution is light in the wet season.

1.2.3.5 The problems existing in the treatment of rural waste water

It is well known that the rural waste water has been the big problem in China. The main reason is the amount of the waste water is big and the lack of the policies. The problems were as followed.

① Lack of effective pipe network collection and processing systems. The residents of rural areas lived in a more dispersed manner, causing a lack of concentrated sources of domestic waste water. Whether it is the collection of pipe networks laid in rural areas, or single-family or small-area waste water pipe networks are extremely lacking or imperfect, and the construction of sewage treatment facilities is not perfect. The direct discharge of most of the waste water through the ditches is one of the major causes of the deterioration of the water quality of local rivers, and it is also an important factor that causes the pollution of the local water environment and the eutrophication of lakes.

② Lack of targeted waste water treatment technologies and treatment methods.

Since the 1980s, some unpowered or micro-powered integratedwaste water treatment equipment has been gradually used to disperse rural waste water. These equipment require little investment and to a certain extent solved the collection, treatment and discharge of scattered waste water. However, this kind of waste water treatment equipment adopts a single process with low processing efficiency, and there are many problems, especially the removal efficiency of nitrogen and phosphorus is very low. The eutrophication of rivers and lakes is still serious.

③ Lack of discharge standards that meet the needs of rural waste water treatment. At present, China has no national emission standards for the treatment of rural domestic waste water. The existing waste water treatment facilities basically refer to *Discharge standard of pollutants for municipal waste water treatment plant* (*GB 18918—2002*), *Integrated waste water discharge standard* (*GB 8978—1996*), and *The reuse of urban recycling water - Quality of farmland irrigation water* (*GB 20922—2007*). Compared with urban areas, rural waste water has its own characteristics. That is, the key problem of rural waste water pollution is due to eutrophication caused by nitrogen and phosphorus. The management level and economic level in rural areas are generally weak, so applying the above standards is not reasonable. To this end, local standards for some provinces have been formulated, and Shanxi, Ningxia, Hebei, and Zhejiang have issued standards for the discharge of waste water from villages and towns. For example, the Ningxia local standard *Discharge standard of pollutants for rural domestic waste water* (*DB 64/T700—2011*); Zhejiang Local Standard *Discharge standard of pollutants for rural sewage treatment facilities* (*DB 33/973—2015*); Hebei Local Standard *Emission standard of rural domestic sewage* (*DB 13/2171—2015*); Shanxi Local Standard *Discharge standard of pollutants for rural sewage treatment facilities in shanxi province* (*DB 14/726—2013*). In accordance with the characteristics of the local rural environment, the functions of water areas, and the use of tailwaters, various regions have made different provisions for pollutant emission limits.

④ Lack of effective management measures. In the past, there were few laws and regulations concerning the environmental pollution in rural areas in the country. The management of rural waste water treatment and the setting of responsible parties were rather vague, and the management systems in each place were quite different. This made it difficult to carry out rural waste water treatment. Problems that exist include the impractical size of the treatment facility and high operating costs, which can lead to unsustainable decommissioning.

1.2.3.6 Typical technologies of rural waste water treatment

At present, the typical treatment technologies for rural domestic sewage are: septic tanks, oxidation ponds, artificial wetland, rotating biological contactor, and integrated treatment technologies (Figure 1-3).

Figure 1-3 A rural sewage treatment station in Caojiabang

(1) Septic tanks A septic tank is a small-scale processing structure that separates and precipitates domestic sewage and performs anaerobic digestion of the sewage (Figure 1-4). The septic tank can remove suspended solids, organic matter and pathogenic microorganisms in the sewage, and at the same time can effectively reduce the organic pollution load of subsequent treatment units. After entering the septic tank for 12 to 24 hours, the sewage can remove 50% to 60% of the suspended matter. The precipitated sludge is decomposed by anaerobic fermentation for more than 3 months to decompose the organic matter in the sludge into stable inorganic substances, and the easily corruptible raw sludge is converted into a stable mature sludge, which can be used as a fertilizer. Septic tanks have the advantages of simple structure, low cost, simple maintenance and management. Its disadvantage is that it needs to clean up the sediment sludge regularly and it is easy to pollute groundwater.

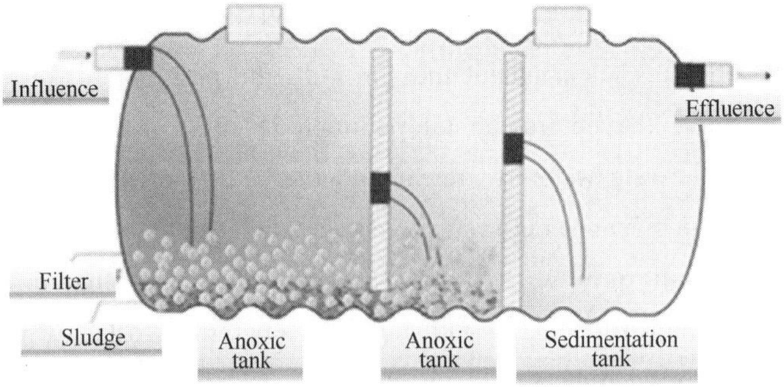

Figure 1-4 A septic tank of a toilet

(2) Oxidation ponds Oxidation pond technology has been widely used in the treatment of rural domestic sewage because of its low cost and convenient maintenance (Figure 1-5). Oxidation ponds are divided into aerobic ponds, facultative ponds, aeration ponds, and anaerobic ponds. Its decontamination principle is that after waste water enters the pond, decomposition reactions, nitrification reactions, and photosynthetic reactions occur under the action of bacteria, algae, and other organisms to remove organic pollutants. Oxidation pond treatment technology has a good treatment effect on rural domestic sewage, can achieve better effluent quality, and some also have the function of nitrogen and phosphorus removal, the removal rate of BOD_5 is as high as 80%. For the oxidation ponds, there are many problems such as long hydraulic retention time, large floor space, serious mud deposits and odors, and sewage purification effects affected by natural conditions. Many types of new ponds have been developed, such as active algae ponds, aquatic plant ponds, and high-efficiency algae ponds, suspended artificial medium ponds, ultra-deep anaerobic ponds, mobile aeration ponds, etc. It has a good treatment effect on rural domestic sewage. The average removal rates of COD, NH_4^+-N and PO_4^{3-} are 70%, 90% and 50%, respectively.

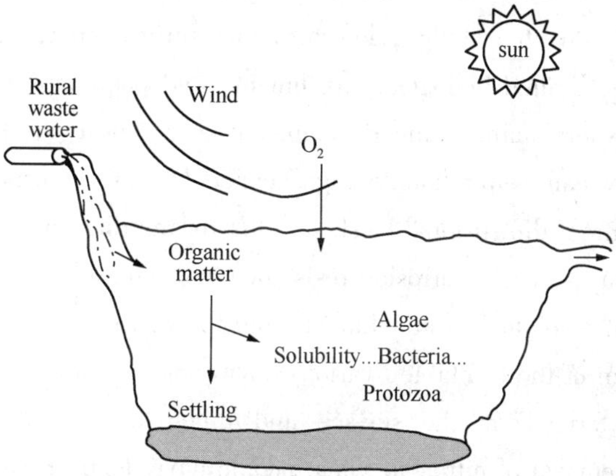

Figure 1-5 The sketch diagram of the oxidation pond

(3) Artificial wetland Artificial wetland treatment system consists of five parts: filler layer, microorganisms, aquatic plants, spine and spineless animals, and water bodies. Purification of waste water is achieved through filtration, adsorption, precipitation, ion exchange, plant absorption, and microbial decomposition. It has many advantages such as high efficiency, low consumption, investment saving, wide application range, and so on. The pollutants targeted by the constructed wetland sewage

treatment system are mainly nitrogen, phosphorus, suspended solids, organic matter, heavy metals, etc(Figure 1-6).

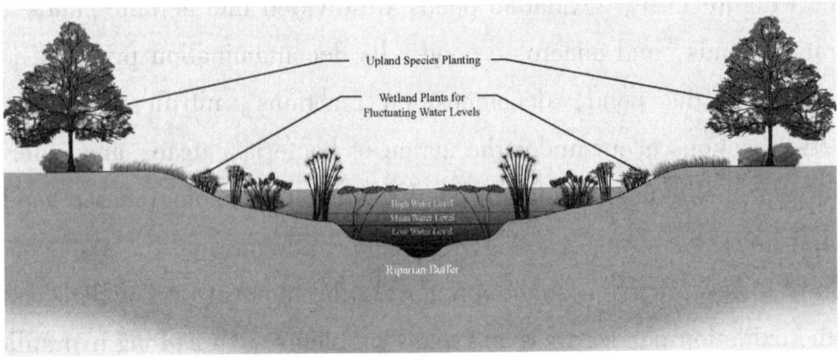

Figure 1-6 The concept map of the artificial wetland

(4) Rotating biological contactor(RBC) The rotating biological contactor is a biological treatment structure composed of a water tank and a rotating disk partially immersed in sewage. The microbial film growing on the surface of the tray repeatedly contacts the oxygen in the sewage and air in the tank, so that the sewage is purified. However, the traditional rotating biological contactor has the disadvantages of easy detachment of activated sludge, low turntable surface area, poor strength, non-corrosion resistance, limited biological attachment, and poor impact load resistance. In view of the above shortcomings, the development of a new type of rotating biological contactor technology can better handle high-concentration domestic sewage. The new technology uses a three-dimensional mesh structure of the dial, with high strength, high specific surface area, superior corrosion resistance and other characteristics, can greatly increase the amount of biological attachment, and the biofilm is not easy to fall off. Due to the special design of the turntable, various environments such as aerobic, anoxic and anaerobic can be formed on the surface and inner layer of the turntable, which contributes to the removal of nitrogen. Some people have done research on the treatment of high-concentration domestic sewage with a new type of rotating biological contactor. The results show that the new type of rotating biological contactor has the advantages of high organic load and large amount of biological attachment. It is suitable for the treatment of high-concentration organic waste water, and its organic load can reach 1.3~1.6 kgCOD/(m^3 · d), the removal rate of organic matter can reach more than 90%(Figure 1-7).

(5) Integrated equipment for sewage treatment The integrated sewage treatment equipment is based on biochemical reaction. It combines several units with different func-

tions, such as pretreatment, biochemistry, precipitation, disinfection and sludge reflux, into a single unit, forming a sewage treatment assembly.

Its equipment is simple to install. It can run unattended after a simple debugging. Compared with the traditional sewage treatment plant, the integrated equipment is small in volume, which can save land effectively.

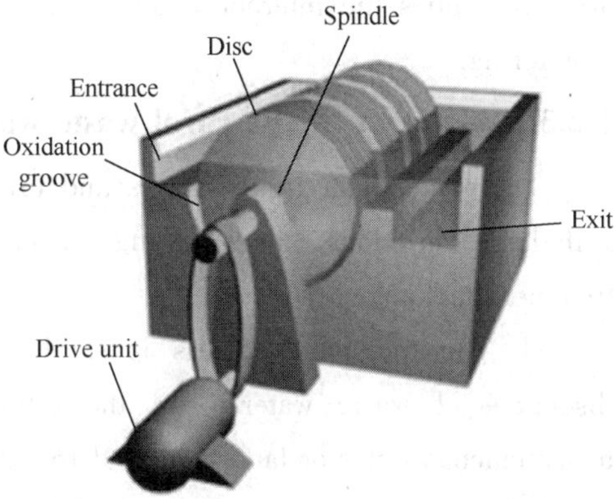

Figure 1-7 The rotating biological contactor

The integrated sewage treatment equipment can not only mix aeration and sedimentation operations according to time or space order, but also be built together. The integrated sewage treatment equipment is a good choice for the rural treatment and even for the small scale sewage treatment(Figure 1-8).

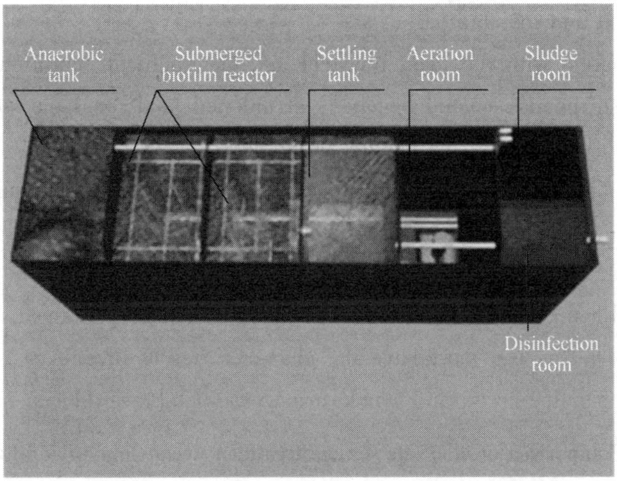

Figure 1-8 The typical integrated equipment

Integrated membrane bioreactor (MBR) is a new type of sewage treatment device developed by combining membrane separation technology and biological treatment process. It has advantages of good treatment effect, low energy consumption, compact structure and less excess sludge. In the Beijing Olympic Games, 2008, the Hong Kong, China equestrian competition venue has adopted the "integrated membrane bioreactor" to treat the horse manure from the horse concentration point, cleaning sewage and the staffs' domestic sewage. With a high energy efficiency, the equipment covers a small area. It implies the internet water quality and running procedures of automatic online monitoring.

1.2.3.7 How to choose rural waste water treatment technology

The waste water from villages and towns is scattered, with small quantities, and with large variations, the following aspects should be fully considered when selecting treatment technologies:

① The treatment process is stable, and it should be able to achieve stable discharge of waste water, and the effluent can be directly reused for domestic miscellaneous water or landscape and irrigation water.

② The one-time investment and construction cost of technology is relatively low and should be within the financial capacity of the town, township and village.

③ Running cost is low, chemical agents are not used, and electricity consumption is low. The cost of running the equipment must match the capacity of the residents in the village and township area. After training the local village technicians, they can be properly operated and maintained.

④ Combination with the local physical and geographical conditions, such as the use of local waste ponds, coated beaches, abandoned land, and pay attention to save the area. In addition, it does not occupy farmland.

⑤ The operation and management are relatively simple, and the equipment does not require a high level of user operation. Therefore, the equipment requires a high level of automatic control. With the weak technical and management capabilities in rural areas, the management and maintenance of the treatment facilities can be performed. Therefore, rural sewage treatment technology must have the characteristics of investment, high processing efficiency, low operating costs, and simple management(Table 1-14).

Table 1-14 Comparison of domestic sewage treatment technology in rural areas of China

Items	Underground aerobic treatment tank	Underground anaerobic treatment tank or constructed wetland	Stabilization Pond	Buried sewage treatment	Decentralized anaerobic biogas sewage purification tank without power
Treatment efficiency	The removal rate of COD is 85.0%	The removal rate of COD is 87.5%	COD, NH_3-N, P can reach the standard	COD, SS, BOD can reach the standard	COD, BOD can reach the standard but it was unapparent to N and P

continued

Items	Underground aerobic treatment tank	Underground anaerobic treatment tank or constructed wetland	Stabilization Pond	Buried sewage treatment	Decentralized anaerobic biogas sewage purification tank without power
Construction Cost	600~800 yuan/(m^3 · d)	600~800 yuan/(m^3 · d)	500 yuan/(m^3 · d)	Build in situ	1,000 yuan/(m^3 · d)
Operation cost	0.40 yuan/m^3	0.05 yuan/m^3	0.05 yuan/m^3	0.10 yuan/m^3	Almost not exist
Application scope	A relatively concentrated area, hotel, school, scenic spot, market town, etc.	Centralized or decentralized treatment of rural sewage	An area in which a reservoir is available	Centralized or decentralized treatment of sewage	Decentralized domestic sewage treatment

1.2.3.8 Design standards and costs of rural waste water

The analysis results of the waste water treatment design standards and construction costs of the 99 rural domestic waste water treatment facilities that have been built in a certain city in southern Jiangsu show that the sewage production volume of these sewage treatment facilities has a design standard of 231 to 371 L/(household · d). Of these, 38% of the waste water treatment facilities have a design standard of 231 to 250 L/(household · d), and 52% of the facilities have a design standard of 291 to 310 L/(household · d) (Table 1-15). In the Taihu Lake region, many rural household waste water discharge characteristics have been studied. The results showed that the average daily drainage of these households was 46.6 L/cap. If the calculation is based on one family of three people, the above research shows that the amount of rural domestic waste water generated in the Taihu Lake Basin should be less than 150 L/(household · d), so the waste water treatment design standard of existing rural domestic waste water treatment facilities is obviously too high. Compared with the amount of waste water produced, the above-mentioned construction cost standards for rural domestic waste water treatment facilities are relatively consistent, mostly between 16, 000 to 21, 000 yuan/(t · d) or 4, 800 to 6, 000 yuan/household. Among them, 68% of the construction cost of the facility is between 15, 001 and 20, 000 yuan/(t · d), and 97% of the construction cost of the facility is above 4,000/household. However, the study found that the construction costs for small-scale urban sewage treatment plants with a daily treatment capacity of less than 10, 000 tons were mostly between RMB 3, 000 and RMB 4, 000/(t · d). It is only a quarter of the cost of the construction of most of the rural domestic sewage treatment facilities in Table 1-15. It is indicating that the waste of resources in the construction of rural sewage treatment projects is serious.

Table 1-15 Design criteria of existing rural sewage treatment facilities

Design standards/ [L/(household · d)]	Percentage/ %	Construction cost/ [yuan/(t · d)]	Percentage/ %	Construction cost/ [yuan/household]	Percentage/ %
231~250	38	<10, 000	2	<3, 000	2
251~290	5	10, 001~15, 000	5	3, 001~4, 000	1
291~310	52	15, 001~20, 000	68	4, 001~5, 000	68
>310	5	>20, 000	25	>5, 000	29

1.3 The effects of the waste water

1.3.1 Chemistry of water impurities

1.3.1.1 Suspended solids and dissolved solids

As we have seen, water contaminants can be classified into four broad categories: suspended solids or particles, dissolved solids or salts, dissolved gases, and organic matter.

(1) Suspended solids Suspended solids or particles are compounds that do not dissolve. They include sand, silt, and precipitated salts. They range in size from visible grains of fine gravel to microscopic particles of metal oxides and clays. These particles may be removed from water through mechanical processes such as filtration. Some particles are large enough to settle out on their own, or with a little help from a device called a separator. Other minute particles may not settle on their own, even though they are not in solution. These are called colloids. Mechanical filtration alone cannot remove these colloidal particles. Rather, chemical coagulation is usually required which involves combining or clumping colloidal particles together to form them into filterable larger particles.

(2) Dissolved solids or salts Salt is the general term that applies to ionic compounds, that is, compounds consisting of a metallic and a non-metallic component. When in solution, they cannot be separated from water by mechanical

means. Dissolved salts can be removed, however, by various chemical processes, which cause ions to precipitate as insoluble compounds. Chemical processes can also be used to attract and hold undesirable ions while replacing them with other "less undesirable" ions. Lastly, dissolved salts can also be removed by specialized processes involving membranes or heat which essentially remove the water from the salts (rather than removing the salts from the water).

(3) Dissolved gases Dissolved gases enter water through contact with the atmosphere or underground deposits that generate gases. In water exposed to the air you will find the main atmospheric gases—nitrogen (80%), oxygen (18%), and carbon dioxide (3%)—in roughly the same proportions as they appear in the atmosphere (78%, 21%, and 0.03%). Carbon dioxide forms carbonic acid in solution, but at a concentration in most surface waters of 2 to 3 per cent. This concentration is usually of little significance in water treatment.

Dissolved oxygen in surface water on the other hand makes up about eighteen per cent of the dissolved gases and can have a significant effect on the corrosiveness of water. While the last atmospheric gas, nitrogen, is the most abundant at approximately eighty percent, nitrogen gas is virtually inert and is of no concern in water treatment applications.

In ground water, higher concentrations of carbon dioxide can be encountered—high enough to make the water acidic and thus corrosive. This is due to confinement and pressure beneath the earth. Once exposed to atmospheric pressure, however, carbon dioxide and other noxious gases such as methane and hydrogen sulfide (the cause of the "rotten egg" smell sometimes found in well water) dissipate quickly into the atmosphere.

(4) Organic matter Organic compounds are the class of substances that are based on a carbon skeleton. The chemical characteristics of carbon allow it to form virtually limitless combinations with other elements. Most organic molecules are composed of oxygen, hydrogen, nitrogen, and sulfur with carbon as the backbone of the molecule. Some organic molecules are soluble, such as the common, naturally-occurring tannic, humic and fulvic acids which are all products of vegetable decay, or pollutants such as pesticides and herbicides. Other organic molecules, such as petroleum products and living organisms such as bacteria, protozoans, viruses, algae and their remains, are insoluble.

(5) Turbidity and sediment Turbidity is the cloudiness in water caused by

suspended particles. When turbid water is viewed through clear glass, the water lacks the clarity and brilliance of clean water. The suspended particles may be inorganic or organic and range in size from clearly visible specks to particles too small to see with the naked eye.

Turbidity is measured in nephelometric turbidity units (NTU), an arbitrary standard that measures the light scattering tendency of suspended particles. (The Greek word nepheloees means cloudy.) Water with a turbidity of 20 NTU could appear clear to the eye, while water with a turbidity of 75 NTU would likely appear slightly cloudy. Most drinking water standards require a turbidity of less than 1.0 NTU.

Suspended particles in turbid water can settle in unwanted areas causing inefficiency in heat transfer equipment, and fouling or plugging of small orifices. Common sediments found in hydronic (hot water heating) systems are iron oxides which may deposit on heat exchange surfaces, causing poor heat transfer, and which may stain or discolor surfaces. In industry, these particles can interfere with sensitive chemical, electronic or robotic processes and may require a unscheduled shutdown of the entire manufacturing system.

1.3.1.2 Water Hardness

Among soluble contaminants, the most common water problem is hardness, which is caused by calcium and magnesium salts dissolved in water. These salts arise from water in the hydrologic cycle coming in contacting with such mineral deposits as limestone, gypsum and dolomite. These hardness ions are soluble to begin with, but through reactions with other ions or compounds, or by being heated, they may form insoluble deposits. There are two forms of hardness in water: Carbonate or Temporary Hardness, and Non-Carbonate or Permanent Hardness.

Carbonate or Temporary Hardness is that hardness which precipitates when water is boiled. It is caused by calcium and magnesium ions which are present and combined with carbonate and bicarbonate ions in water. The chemical compounds involved are:

Calcium Carbonate $CaCO_3$

Calcium Bicarbonate $Ca(HCO_3)_2$

Magnesium Carbonate $MgCO_3$

Magnesium Bicarbonate $Mg(HCO_3)_2$

Non-Carbonate or Permanent Hardness is that hardness which will not precipitate when the water is boiled. It is caused by calcium and magnesium ions combined with ions other than carbonate and bicarbonate ions to form chemical compounds. Such chemical compounds are:

Calcium Sulfate $CaSO_4$

Calcium Chloride $CaCl_2$

Calcium Nitrate $Ca(NO_3)_2$

Calcium Hydroxide $Ca(OH)_2$

Magnesium Sulfate $MgSO_4$

Magnesium Chloride $MgCl_2$

Magnesium Nitrate $Mg(NO_3)_2$

Magnesium Hydroxide $Mg(OH)_2$

The term hardness applies to anything that precipitates in contact with soap, so iron and manganese, and other polyvalent ions are technically hardness minerals. Next to hardness caused by calcium and magnesium, iron and manganese are the most common problem causers and are normally found at much lower concentrations than are calcium or magnesium.

Although hardness does not pose any health-related problems, it does cause many economic and aesthetic problems. Hardness causes excessive consumption on cleaning agents, detergents and other chemicals, including water treatment products. Scaling due to hardness and plugging of piping and heat exchange surfaces translates to lower water treatment efficiencies, higher maintenance costs, reduced equipment life, increased equipment failure, and possibly increased water consumption in various systems. As we will discuss in the next section, corrosion may also occur beneath scale deposits. This type of corrosion is called under-deposit corrosion.

1.3.2 Metallic corrosion

Corrosion is an electrochemical process, similar in principle to the chemistry of a battery. It inevitably consists of two chemical "half-cells" (Figure 1-9):

An anode, at which metal atoms lose electrons to go into solution as positive ions (cations) in a reaction called oxidation.

A cathode, at which positive ions that are in solution, such as hydrogen (H^+) gain electrons in a reaction called reduction.

The solution referred to above must be capable of conducting electrical current by the movement of the anions and cations; this solution is called an electrolyte.

In the figure, iron metal atoms undergo oxidation by losing 2 electrons per atom; the resulting iron ions, called ferrous ions, go into solution; the half-cell reaction is:

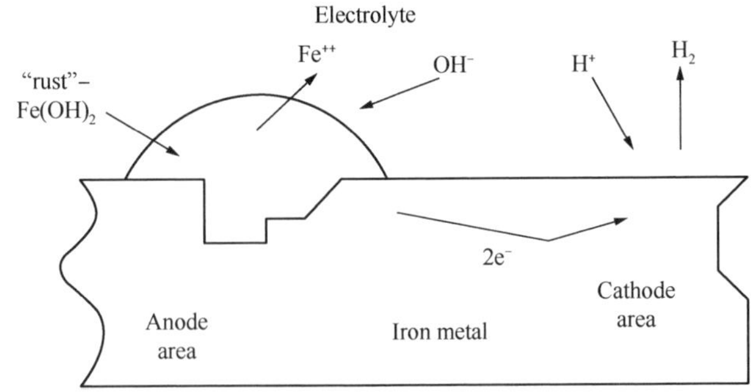

Figure 1-9 A corrosion cell

$$Fe \rightarrow Fe^{++} + 2e^-$$

It is important also to understand that corrosion can take many different forms depending on the metal, the nature of the electrolyte, and various conditions such as flow rates of liquids, the presence of stress in the metal, and so on. There are eight main forms of corrosion classified by appearance.

(1) Uniform attack Uniform attack is the most common form of corrosion; in this type, the metal becomes thinner over its entire surface until it eventually fails. This type of attack is generally of no concern in hydronic systems as inhibitors reduce corrosion rates to acceptable levels. It is usually found when inspecting piping and heat exchange equipment does not present any operational problems. Unfortunately, most corrosion problems constitute localized attack at susceptible metal areas which may be detrimental to the lifespan of heat exchange equipment.

(2) Galvanic corrosion Galvanic corrosion occurs where two different metals are in contact due to their different tendencies to lose electrons; as a general, we avoid contact between dissimilar metals for this reason. This type of activity will be observed by an increased attack usually close to the junction of the two metals. In certain instances, dielectric couplers may be employed which act as an insulator to prevent electron flow between the two metals. Once the anodic and cathodic metals are electrically separated from each other, galvanic corrosion will not take place.

(3) Crevice corrosion Crevice corrosion occurs where there are shielded areas on metal surfaces such that small volumes of stagnant liquid can occur and differences in the concentrations of oxygen and/or other aggressive ions arise; this can occur where there are holes in the metal, gaskets, lap joints, even surface deposits of dirt, bolts, and rivet heads. The prevention of crevice corrosion is both a problem of design—to

avoid the crevices in the first place—and cleaning to avoid deposition of dirt. Generally, this type of attack is not common, except for materials such as stainless steel and titanium which depend on an air-formed oxide film for protection.

(4) Pitting Pitting is a very problematic form of corrosion because it can occur without being detected such that the first evidence that something has happened is the leak that appears in the pipe or tank. This type of attack is the result of one or more highly active anodic sites which penetrate deep into the metal surface. This penetration at the anode may result in serious damage or even tube failure within heat exchange equipment or associated piping. In many instances, pitting leads to system shutdown or may disrupt an entire plant process. The causes of this type of attack may be unequal ionic concentrations, oxygen differentials, high temperature areas, metallurgical defects, cuts, scratches or crevices on the metal surface. Some stainless steels alloys are particularly susceptible to pitting if they are used in the wrong environment.

(5) Intergranular corrosion Intergranular corrosion refers to localized attack along the boundaries of the actual crystals or grains that are part of the metal structure. Again, some alloys, such as the 300 series stainless steels and brass alloys, are susceptible to this form of corrosion. The surfaces of metals that are experiencing this form of corrosion can feel "sugary" due to the small particles that fall away from the metal as corrosion proceeds.

(6) Selective leaching Selective leaching is the preferential removal of one of the constituents of an alloy; for example, the loss of zinc in brass alloys (brass is mainly a zinc-copper alloy) can occur in some environments, leaving a copper coloured material at the point of attack. Many variations of selective leaching exist, but the three common forms are dezincification, graphitization and dealuminification.

① Dezincification: the leaching of zinc from brass leaves the metal a reddish, copper colour instead of a yellow colour associated with brass. A common example of this is the dezincification of brass impellers in recirculation pumps due to caustic attack. Dezincification on galvanized surfaces may also occur at pH values above 8.2 with newly-constructed cooling towers. This will be discussed further in the section on cooling water treatment.

② Graphitization: cast iron may experience a loss of iron which weakens the metal leaving graphite and iron oxides behind. Although the metal may appear undisturbed, it can be penetrated with a sharp object. This phenomenon is a rare occurrence with cast iron piping and associated components.

③ Dealuminification: this form of leaching is related to aluminum bronzes and is generally associated with seawater environments.

(7) Erosion corrosion Erosion corrosion occurs where there is movement of metal upon metal, flow of fluid over the metal surface, or cavitation (the formation and sudden collapses of gas bubbles in a flowing liquid). The mechanical action involved eroded metal at the points of contact.

① Impingement. Waters that are high in suspended and dissolved solids or dissolved and entrained gases often damage metal oxide films causing localized corrosion. The abrasive tendency of this type of turbulent water flow physically tears away metal in horseshoe shaped pits which may be aligned. Severe attacks may occur at the inlet of heat exchanger tube bundles, points of restricted flow, elbows or sharp bends etc. Although many metals may suffer from impingement, the most serious problems occur with copper and its various alloys. If impingement corrosion is evident, water flow rates should be restricted to five feet per second (1.8 m/sec) or possibly less if total dissolved solids or dissolved gas concentrations are high.

② Cavitation corrosion. If the absolute pressure in any part of a pressurized hydronic system approaches the vapour pressure of the system fluid, vapour pockets will form in the low pressure regions. This is called cavitation and occurs in variable or low pressure situations with fluids containing dissolved or entrained gases. The most common area of concern is that of recirculation pumps in closed recirculating systems as vapour pockets will form and collapse in the impeller passages.

This attack of the impeller by pockets of vapour will cause noise that is similar as that of an outboard boat motor prop that is not completely in the water, generally described as a "gurgling sound" or "clapping sound". As liquid surfaces meet during bubble collapse, considerable kinetic energy is released. This has been estimated to create pressures as high as 687,000 kilopascals (100,000 psi). The claps of energy impinge on the metal surface and may break a protective surface film and lead to the initiation of corrosion. If this type of attack continues for a prolonged period of time, pump impellers, piping and associated components may be damaged or completely destroyed. This condition is similar as water hammer that occurs in water pipes. This type of attack may also be found in sharp bends in piping systems or the discharge side of gate or globe valves. Usually this type of destruction appears as deep circular pits.

③ Fretting corrosion. The rapid localized corrosion that occurs on closely-fitting surfaces under a load and subject to a small amplitude slip or movement (i.e. chattering or vibratory motion) is called fretting corrosion. This phenomenon has also been called friction oxidation because the corrosion product is an oxide, or false Brinelling since indentations or pits may be present (similar as those caused by Brinell hardness testing). Almost all alloys are subject to fretting corrosion and its incidence is high in vibrating equipment. It occurs mainly on machine parts such as ball and roller bearings, shafts and gears etc., and in heat exchange equipment where two surfaces are subject to movement under a load such as the tube/tube sheet interface in heating boilers and heat exchangers. Heat exchange equipment in cooling applications rarely suffers fretting corrosion as temperature changes are not as drastic as in the heating of boilers and temperatures are much lower; leading to very little thermal expansion and contraction at the tube/tube sheet interface.

(8) Stress corrosion Stress corrosion, as the name implies, can occur in metal parts that are under mechanical stress. This form of corrosion often occurs as cracks in the structure; however, it is more than just mechanical failure as a corrosion process, involving oxidation and reduction, takes place. Stressed areas of a metal or alloy are anodic to unstressed areas which are cathodic. Stress corrosion is rarely a concern in the building environment unless standard system design practices are not followed. Industrial processes are more likely to encounter these two types of stress corrosion.

Corrosion monitoring as well as routine visual inspections of piping and heat exchange equipment by building or plant personnel and water treatment suppliers are vital to the operational efficiency and longevity of heat exchange equipment.

As should be apparent, other metals besides iron (or steel) are subject to corrosion in specific environments. Copper (Cu) and copper alloys such as brass, for example, are subject to accelerated corrosion when ammonia is present. Even the various "stainless steels" will corrode under certain conditions. What all these situations have in common is that there is an oxidation/reduction pair of reactions involved.

1.3.3 Scale formation

Scale is produced when soluble salts are precipitated and deposited from recirculating water systems. The rate of formation depends mainly on the following factors:

① The amount of scale-forming material (which we have previously termed temporary hardness), iron, and possibly silica, present in the water.

② The total alkalinity or acidity of the water.

③ The temperature of the system water and the " skin" temperature of heat exchange surfaces, the latter being very important in heating and industrial process systems.

To determine whether or not scale formation will occur, we can analytically measure the mineral levels and related conditions of the system water, and we can also accurately predict whether scale deposits may be prevented. A prediction of scale formation and the permissible cycles of concentration may be calculated by determining any one of three scaling indices:

the Langelier Saturation Index (LSI);

the Ryznar Stability Index (RSI);

the Puckorius Scaling Index (PSI).

1.3.4 Fouling

In municipal system water and well water, fouling from suspended matter and micro-organisms is not as common as in once-through systems. Once-through systems commonly involve the use of surface water for cooling purposes only. Fouling can be a serious problem in once-through systems using a surface water source.

Fouling deposits tend to settle in low-flow areas or heat-exchange surfaces and possibly " bake" onto heat-exchange surfaces, depending of course on the skin temperature of the heat-exchange surface. Fouling is usually defined as deposition of non-scale forming materials such as:

① Iron, silt or sand suspended in the system or makeup water.

② Naturally occurring organics in the makeup water.

③ Micro-organisms in the system or makeup water.

④ Particulate matter scrubbed (i.e. washed) from the atmosphere.

⑤ Deposition of chemical additives due to poor control.

⑥ Corrosion products migrating to low flow areas or heat exchange surfaces.

⑦ Organic contamination from process leaks such as oil, grease or process contamination products.

To avoid the fouling, the measures, such as dispersants or surfactants or Flocculants, can be used.

(1) Dispersants To reduce the amount of suspended matter settling in low flow areas or heat exchange surfaces, dispersants may be added to adsorb (attach) onto a specific impurity that is a bothersome deposit. Some of these dispersants are polymers

which are specific to hardness and iron, while others are broad spectrum performers. Dispersants function by attaching themselves to the suspended particles. This attachment or adsorption to the particle involves electrostatic forces which modify the surface of the particle to the extent that the particles actually repel each other and hence disperse widely instead of clumping together in a deposit.

For example, in open recirculating systems most of the suspended matter in the cooling water will have a slight negative charge. A dispersant having a negative charge, namely an anionic polymer, is commonly added to attach to suspended particles and thus repel other particles. An example of such an anionic dispersant is polymethacrylate which has good dispersive characteristics.

There are a wide variety of relatively low molecular weight dispersants ("low" meaning small for a polymer) available today for specific fouling problems. This approach to deposit control will greatly improve the efficiency of heat exchange equipment and safeguard against under-deposit corrosion.

(2) Surfactants Surfactants, or wetting agents, can aid dispersants in maintaining suspended matter in suspension, and further allow the dispersants to remove old deposits. Generally, they are used for the control of oil and suspended matter, and as an aid to control microbiological activity (e. g. algae, slime, bacteria etc.) in open recirculating systems. They may also be used to remove oil from new and existing closed recirculating systems.

These wetting agents are polymers that are electrically neutral but have some sites along the molecule that are negatively charged and other sites that are positively charged. Some areas of the polymer molecule (termed hydrophilic) then become attracted to water while other areas (termed hydrophobic) repel water, but attract oil and suspended matte. The hydrophilic sites on the polymer molecule (i. e. negatively charged sites) keep other negatively charged particles in the water from combining to form large dense particles that could settle out of the system water.

These types of products also reduce the surface tension of water, allowing the water better to penetrate microbiological deposits, especially slime deposits.

(3) Flocculants While dispersants are usually relatively low molecular weight polymers, flocculants are high molecular weight polymers which may be used in the control of fouling deposits.

Care must be taken when using flocculants as low water velocities in any part of the system, including heat exchange equipment will cause sedimentation of the floc.

1.4 Waste water treatment processes

Raw or untreated sewage is mostly pure water. In fact, sanitary waste water comprises about 99.9% water and only about 0.1% impurities. In other word, if 1 L sample of waste water is to evaporate, only about 1 g, or 1, 000 mg, of solids will remain behind.

This chapter describes the most common types of waste water treatment systems. These treatment methods are grouped into three general categories: *primary treatment, secondary or biological treatment, and tertiary or advanced treatment.*

A flow chart showing the typical sequence of waste water treatment processes is given in Figure 1-10.

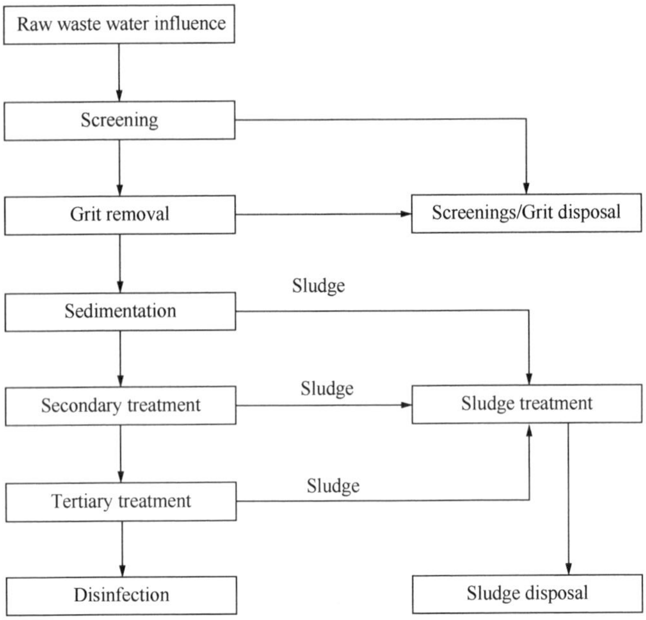

Figure 1-10 Schematic overview of a waste water treatment system

Screening, grit removal, and sedimentation (settling) are primary treatment processes. Secondary treatment usually involves biological processes and additional settling. Not all sewage treatment plants require tertiary (advanced) treatment.

1.4.1 Preliminary and primary treatment

Untreated or raw waste water usually flows by gravity from an interceptor or trunk sewer into the head works of a treatment facility; sometimes waste water may be pumped to the treatment plant in a force main. The head works of a treatment plant include a flow measurement device such as a parshall flume (see Figure 1 – 11) and mechanical systems that provide preliminary treatment. Preliminary treatment systems typically include screens, comminutors, and grit chambers.

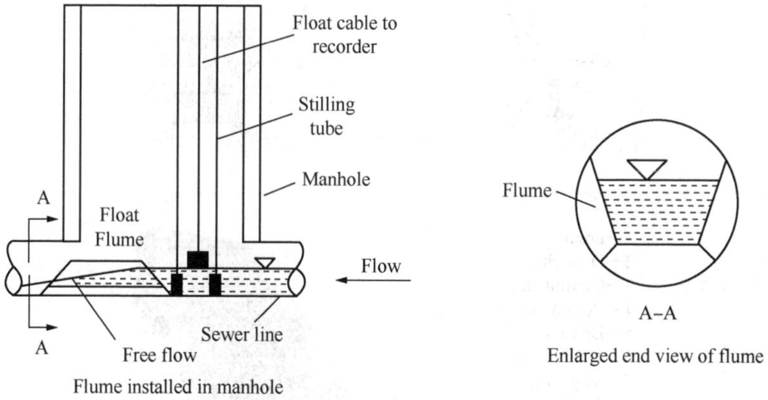

Figure 1–11 A typical palmer–bowlus flume installation

The first treatment process for raw waste water is coarse screening. *Bar screens* (*or racks*), as they are called, are made of long, narrow metal bars spaced about 25 mm (1 in.) apart. They retain floating debris, such as wood, rags, or other bulky objects, that could clog pipes or damage mechanical equipment in the rest of the plant. In most large sewage treatment plants, the bar screens are cleaned automatically by a mechanical device, as shown in Figure 1 – 12. The collected debris or screenings are promptly disposed of, usually by burial on the plant grounds.

In some treatment plants, a mechanical cutting or shredding device, called a *comminutor*, is installed just after the coarse screens. A typical comminutor, consists of a slotted cylindrical screen with a moving cutter blade. The comminutor shreds and chops solids or rags that passed through the bar screen. The shredded material is removed from the waste water by sedimentation or flotation later in the treatment plant. In small sewage treatment plants, a manually cleaned bar screen is generally installed in a channel next to the comminutor to serve as an emergency bypass when the comminutor needs repair.

(1) Grit removal A portion of the suspended solids in raw sewage consists of gritty material, such as sand, coffee grounds, eggshells, and other relatively inert material. In cities with combined sewer systems, sand and silt may be carried in the sewage.

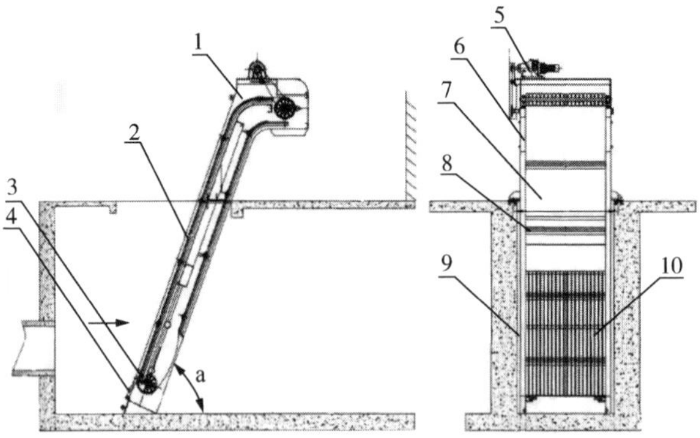

1—Frame
2—Drag chain
3—Guiding device
4—Bottom movable fence
5—Drive device
6—Manhole cover
7—Slag baffle
8—Tooth harrow
9—Side baffle
10—Fence

Figure 1-12 Typical bar screen installations, which serve to remove sticks, rags, and other debris from the flow of sewage.

Suspended grit can cause excessive wear and tear on pumps and other equipment in the plant. Most of it is nonbiodegradable and will accumulate in treatment tanks. For these reasons, a grit removal process is usually used after screening and/or comminuting.

In the sewers, the flow velocity is generally not less than 0.6 m/s (2 ft/s), the self-cleansing velocity. Most gritty material is dense enough to settle out of the flow by gravity if the velocity is reduced to about 0.3 m/s (1 ft/s). Although the grit will settle out at this reduced velocity, the lighter suspended organic solids will still be carried through to the next treatment unit.

The reduction in velocity and the collection of the grit is usually accomplished in long, narrow tanks called grit chambers. Even though the sewage flow rate varies throughout the day, the flow velocity can be kept almost constant at 0.3 m/s by the specially shaped outlet weir. Mechanical collectors made of buckets on a continuous chain serve to remove the grit from the tank; the grit is promptly disposed of by burial on the plant grounds. An aerated grit chamber is shown in the following figure. The rising air bubbles help to keep the organic solids in suspension while the grit settles to the bottom(Figure 1-13).

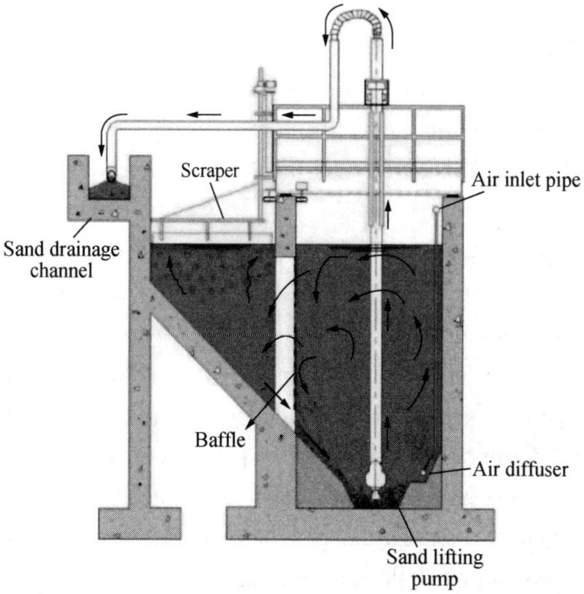

Figure 1-13 A mechanically cleaned grit chamber

(2) Primary sedimentation (settling) After preliminary treatment by screening, comminuting, and grit removal, the waste water still contains suspended organic solids that can be removed by plain sedimentation. The basic principles of sedimentation, or plain gravity settling, are discussed as follows:

The impurities in water may be either dissolved or suspended. The easiest way to remove the suspended material is to let the force of gravity do the work. Under quiescent conditions, when flow velocities and turbulence are minimal, particles that are denser than water will be able to settle to the bottom of a tank. This process is called *sedimentation*, and the layer of accumulated solids at the bottom of the tank is called *sludge*. The tank may be called a *sedimentation tank*, a *settling tank*, or a *clarifier*.

The speed at which suspended particles settle toward the bottom of a tank depends on their size as well as on their density. The larger and heavier particles will naturally settle faster than smaller or lighter particles. The forces opposing the downward force of gravity include buoyancy and friction(drag). The temperature and viscosity of the water are additional factors that affect the particle-settling rate.

The nature of the sedimentation process also varies with the concentration of suspended particles and their tendency to interact with one another. In a dilute suspension, where the particles are free to settle without interference, the process is called free settling or discrete settling. As the concentration increases, the particles tend to interact and interfere with the free movement of one another; this is sometimes called *hindered settling*. In a sedimentation tank there may be up to four different zones or types of settling that occur at different depths, and exact mathematical analysis of the process can be quite complicated. This section discusses come common factors related to discrete particle settling.

For the most part, the same principles apply to sewage treatment. There are, however, differences in the recommended design values for detention time, overflow rate, and tank configuration.

Settling tanks that receive sewage after grit removal are called *primary clarifiers*. They usually provide between 1 and 2 h of detention time; side water depth (SWD) is generally between 2.5 and 5 m (8 and 16 ft). The tanks may be circular or rectangular shape. In addition to mechanical sludge collectors that continually scrape the settled solids along the bottom to a sludge hopper for removal, a surface skimming device is used to remove grease and other floating materials from the liquid surface. Two or more clarifiers should be provided at all but the smallest treatment plants.

The combination of preliminary screening and gravity settling is called *primary treatment*. Chemicals may sometimes be added to the primary clarifiers to promote the removal of very small (or colloidal) particles: Primary treatment usually can remove up to 60% of the suspended solids and about 35% of the BOD from waste water, but this relatively low level of treatment is no longer adequate in the United States. In almost all cases, primary treatment must be followed by secondary treatment processes; tertiary treatment may also be required to protect sensitive bodies of water that receive the treated effluent.

Example

A primary clarifier has an average influent TSS (total suspended solids) concentration of 250 mg/L. If its TSS removal efficiency is expected to be 60%, what is the expected average effluent TSS concentration?

$$\text{efficiency} = \frac{P_{\text{IN}} - P_{\text{OUT}}}{P_{\text{IN}}} \times 100$$

P_{IN}: concentration of pollutant flowing into the treatment system

P_{OUT}: concentration of pollutant flowing out of the system

$$60 = \frac{250 - P_{\text{OUT}}}{250} \times 100$$

$$P_{OUT} = 250 - \frac{250 \times 60}{100} = 100 \text{ mg/L}$$

The removal efficiencies of primary settling tanks can be increased by the addition of chemical coagulants, although this is not common practice in most modern waste water treatment plants. To achieve BOD and TSS removal efficiencies of at least 85%, as required by the *Clean Water Act*, primary treatment must be followed by an additional treatment process. Generally, this next step is characterized as *secondary treatment*.

1.4.2 Secondary (biological) treatment

Primary treatment processes remove only those pollutants that will either float or settle out by gravity, but about half of the raw pollutant load still is remained in the primary effluent. The purpose of *secondary treatment* is to remove the suspended solids that did not settle out in the primary tanks and the dissolved BOD that is unaffected by physical treatment. Secondary treatment is generally considered to mean 85% BOD and TSS removal efficiency and represents the minimum degree of treatment required in most cases. In the United States, secondary treatment processes are almost always *biological* systems.

Biological treatment of sewage involves the use of microorganisms. The microbes, including bacteria and protozoa, consume the organic pollutants as food. They metabolize the biodegradable organics, converting them into carbon dioxide, water, and energy for their growth and reproduction. This natural aerobic process requires oxygen.

A biological sewage treatment system must provide the microorganisms with a comfortable home. In effect, the treatment plant allows the microbes to stabilize the organic pollutants in a controlled, artificial environment of steel and concrete, rather than in a stream or lake. This helps to protect the dissolved oxygen balance of the natural aquatic environment.

To keep the microbes happy and productive in their task of waste water treatment, they must be provided with enough oxygen, adequate contact with the organic material in the sewage, suitable temperatures, and other favorable conditions. The design and operation of a secondary treatment plant is accomplished with these factors in mind.

Two of the most common biological treatment systems are the *trickling filter* and the *activated sludge* process. The trickling filter is a type of *fixed-growth* system: The microbes remain fixed or attached to a surface while the waste water flows over that surface to provide contact with the organics. Activated sludge is characterized as a *suspended-growth* system, because the microbes are thoroughly mixed and suspended in the waste water rather than attached to a particular surface.

(1) Trickling filters　A trickling filter consists basically of a layer or bed of crushed rock about 2 m (6 ft) deep. It is usually circular in shape and may be built as large as 60 m (200 ft) in diameter. Trickling filters are always preceded by primary treatment to remove coarse and settleable solids. The primary effluent is sprayed over the surface of the crushed stone bed and trickles downward through the bed to an *underdrain system*.

A rotary distributor arm with nozzles located along its length is usually used to spray the sewage, although sometimes fixed nozzles are used. The rotary distributor arm is mounted on a center column in the trickling filter; it is driven around by the reaction force or jet action of the waste water that flows through the nozzles.

The underdrain system serves to collect and carry away the waste water from the bottom of the bed and to permit air circulation upward through the stones. Once topography permits, the sewage flows from the primary tank to the trickling filter by the force of gravity, rather than by pumping. A cutaway view of a typical trickling filter unit is shown in Figure 1-14.

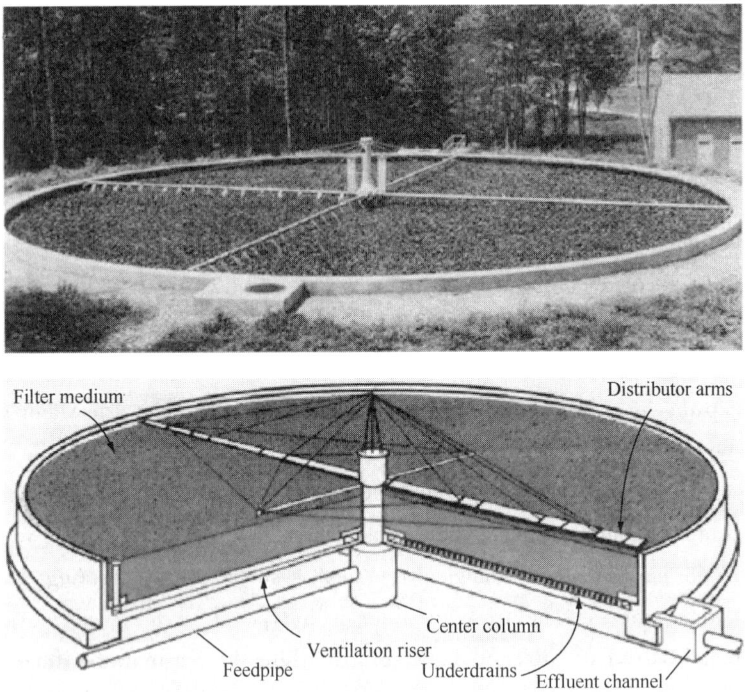

Figure 1-14　Cutaway view of a trickling filter

As the primary effluent trickles downward through the bed of stones, a biological slime of microbes develops on the surfaces of the rocks. The continuing flow of the waste water over these fixed biological growths provides the needed contact between the microbes and the organics. The microbes in the thin slime layer absorb the dissolved organics, thus removing oxygen - demanding substances from the waste water. Air circulating through the void spaces in the bed of stones provides the needed oxygen for stabilization of the organics by the microbes.

Note, however, that the trickling filter is not really a filter at all, in the true sense of the word. The stones are usually about 75 mm (3 in.) in size, much too large to strain or filter out suspended solids. And, by definition, filters have no effect on dissolved solids. The stones in a trickling filter only serve to provide a large amount of surface area for the biological growths, and the large voids allow ample air circulation. Sometimes materials other than rock, such as modules of corrugated plastic or redwood slats, are used to provide the needed surface area and void spaces, but the basic purpose and operation remain the same.

As the microorganisms grow and multiply, the slime layer gets thicker. Eventually, it gets so thick that the flowing waste water washes it off the surfaces of the stones. This is called *sloughing* (pronounced "sluffing"). Since sloughing does occur periodically, there is a need to provide settling time for the trickling filter effluent to remove the sloughed biological solids. These solids consist basically of billions of microorganisms that have absorbed the dissolved organics into their bodies.

The trickling filter effluent is collected in the underdrain system and then conveyed to a sedimentation tank called a *secondary clarifier*. The secondary clarifier, or *final clarifier* as it is sometimes called, is similar in most respects to the primary clarifier, although there are differences in detention time, overflow rate, weir loading, and other details.

To maintain a relatively uniform flow rate through the trickling filter and to keep the distributor arm rotating even during periods of low sewage flow, some of the waste water may be recirculated. In other words, a portion of the effluent is pumped back to the trickling filter inlet so that it will pass through the bed of stones more than once.

Recirculation can also serve to improve the pollutant removal efficiency; it allows the microbes to remove organics that flowed by them during the previous pass through the bed. There are many recirculation patterns and configurations of trickling filter plants. One common pattern, called *direct recirculation*, is shown in a flow diagram in Figure 1-15.

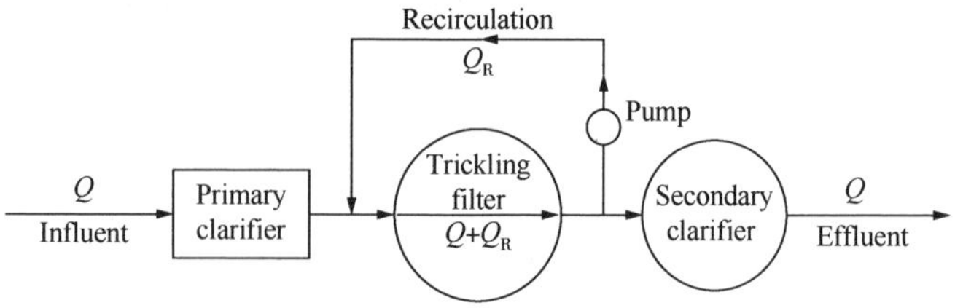

Figure 1-15 Direct recirculation

Schematic diagram showing recirculation of flow through a trickling filter. The rate of sewage flow applied to the filter is the sum of the influent flow rate and the recirculated flow rate.

① Recirculation. The amount of recirculation is variable. It is characterized by a *recirculation ratio*, which is the ratio of recycled flow to the raw waste water flow. In formula form, it is given by

$$R = \frac{Q_R}{Q}$$

where R = recirculation ratio;

Q_R = recirculated flow rate;

Q = raw sewage flow rate.

The recirculation ratio, R, is generally in the range of 0.0 to 3.0.

② Hydraulic Load. The rate at which the waste water flow is applied to the trickling filter surface is called the *hydraulic load*. The hydraulic load includes the recirculated flow Q_R; the total flow through the trickling filter is equal to $Q + Q_R$. In formula form, it is given by

$$\text{hydraulic load} = \frac{Q+Q_R}{A_S}$$

where Q = raw sewage flow rate;

Q_R = recirculated flow rate;

A_S = trickling filter surface area (plan view).

Hydraulic load may be expressed in terms of cubic meters per day per square meter of surface area, or $m^3/(m^2 \cdot d)$. It also may be expressed in terms of million gallons per acre of surface area per day, or mil gal/(ac · d). A typical value for a conventional trickling filter is 20 $m^3/(m^2 \cdot d)$ [19 mil gal/(ac · d)].

③ Organic (BOD) load. The rate at which organic material is applied to the trickling filter is called the *organic* or *BOD load*. It does not include the BOD added by recirculation. Organic load is expressed in terms of kilograms of BOD per cubic meter of bed volume per day, or kg/(m^3 · d). It is also expressed in terms of pounds of BOD per thousand cubic feet of bed volume per day, or 1b/(1000ft^3 · d). A typical value for organic load on a trickling filter is 0.5 kg/(m^3 · d), or 30 1b/(1000ft^3 · d.) In formula form, the organic load may be expressed as

$$\text{organic load} = \frac{Q \times \text{BOD}}{V} \quad (\text{SI metric units})$$

$$\text{organic load} = 8340 \times \frac{Q \times \text{BOD}}{V} \quad (\text{U.S. Customary units})$$

where Q = raw waste water flow, ML/d (mgd);

BOD = BOD_5 in the primary effluent, mg/L (ppm);

V = volume of trickling filter bed, m^3(ft^3).

Example

A 2-m-deep trickling filter with a diameter of 18 m is operated with a recirculation ratio of 1.5. The raw waste water flow rate is 2.5 ML/d, and the 5-day BOD of the raw sewage is 210 mg/L. It is assuming that the primary tank BOD removal efficiency is 30%, compute the hydraulic load and the organic load on the trickling filter.

Solution

First, compute the surface area of the trickling filter,

$$A_S = \frac{\pi \times D^2}{4} = \frac{\pi \times 18^2}{4} = 254.5 \text{ m}^2$$

Since volume = area× depth,

$$V = 254.5 \text{ m}^2 \times 2 \text{ m} = 509 \text{ m}^3$$

From Equation $R = \frac{Q_R}{Q}$, write $Q_R = R \times Q$ and $Q_R = 1.5 \times 2.5 = 3.75$ ML/d

Also, $Q + Q_R = 2.5 + 3.75 = 6.25$ ML/d = 6250 m^3/d

From Equation hydraulic load $= \frac{Q + Q_R}{A_S}$

$$\text{hydraulic load} = \frac{6250 \text{ m}^3/\text{d}}{254.5 \text{ m}^2} \approx 25 \text{ m}^3/(\text{m}^2 \cdot \text{d})$$

The primary effluent BOD can be computed with Equation efficiency $= \frac{P_{IN} - P_{OUT}}{P_{IN}} \times 100$

$$30 = \frac{210 - P_{OUT}}{210} \times 100$$

and $P_{OUT}=210-\frac{210\times30}{100}=147$ ppm

From Equation organic load $=\frac{Q\times BOD}{V}$ (SI metric units)

organic load $=\frac{2.5\times147}{509}=0.72$ kg/(m^3 · d)

④ Efficiency. BOD removal efficiency of a trickling filter unit depends primarily on the organic load, the recirculation ratio, and the temperature of the waste water. Generally, the efficiency increases with decreasing organic load, increasing recirculation, and increasing temperature. For example, with no recirculation ($R=0$) and a temperature of 20℃, a typical trickling filter will have an efficiency of about 60% when the organic load is about 2 kg/(m^3 · d). But if the organic load is 0.5 kg/(m^3 · d), at the same conditions of recirculation and temperature, the efficiency will be 75%.

Furthermore, at the 0.5 kg/(m^3 · d) loading, a recirculation of $R=1$ instead of $R=0$ could raise the efficiency to about 80%. If the temperature increased to 22℃, the efficiency would be raised to about 85%. Because of the marked effect of temperature on treatment efficiency, trickling filters in northern climates are often enclosed under fiberglass domes to provide protection against wind and snow and to reduce the rate of heat loss from the waste water.

Example

The BOD removal efficiency of a trickling filter system is 79%, and the efficiency of the primary treatment that precedes it is 35%. If the raw BOD is 200 mg/L, what is the effluent BOD? Is the treatment plant providing an efficiency that meets the requirement for secondary treatment?

Solution

Although the biological treatment and secondary clarification provide only 79% BOD removal, it should be remembered that 35% of the raw BOD was recently removed by primary settling. Thus, 65% or $0.65\times200=130$ mg/L remains as BOD in the primary effluent.

But 79% of the BOD entering the trickling filter is removed, leaving 21% or $0.21\times130=27$ mg/L in the secondary effluent. The overall plant efficiency is therefore computed as

$$\text{efficiency}=\frac{200-27}{200}\times100\approx87\%$$

This is greater than 85%, so the treatment plant is providing secondary treatment.

(2) Activated sludge treatment The basic components of an activated sludge sewage treatment system include an aeration tank and a secondary settling basin or clarifier. Primary effluent is mixed with settled solids that are recycled from the secondary clarifier and then introduced into the aeration tank. Compressed air is injected continuously into the mixture through porous diffusers located at the bottom of the tank along one side. This is illustrated in Figure 1-16.

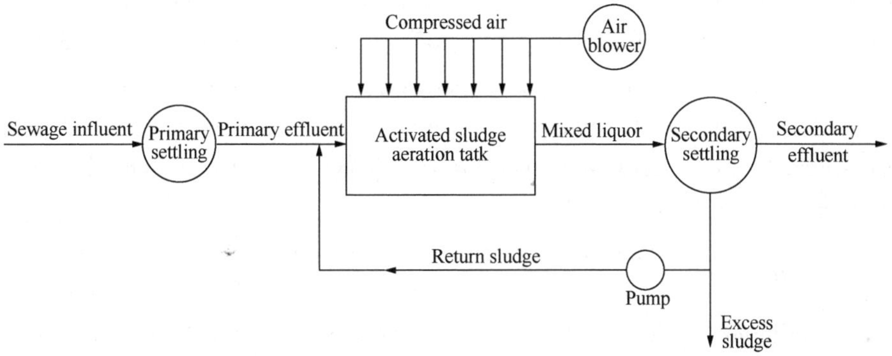

Figure 1-16 A flow diagram for a conventional activated sludge plant.

In the aeration tank, microorganisms consume the dissolved organic pollutants as food. The microbes absorb and aerobically decompose the organics, using oxygen provided in the compressed air, water, carbon dioxide, and other stable compounds are formed. In addition to providing oxygen, the compressed air thoroughly mixes the microbes and waste water together as it rapidly bubbles up to the surface from the diffusers. Sometimes, mechanical propeller like mixers, located at the liquid surface, are used instead of compressed air and diffusers. The churning action of the propeller blades mixes air with the waste water and keeps the contents of the tank in a uniform suspension.

The aerobic microorganisms in the tank grow and multiply, forming an active suspension of biological solids called *activated sludge*. The combination of the activated sludge and waste water in the aeration tank is called *the mixed liquor*. In the basic or conventional activated sludge treatment system, a tank detention time of about 6 h is required for thorough stabilization of most of the organics in the mixed liquor.

After about 6 h of aeration, the mixed liquor flows to the secondary or final clarifier, in which the activated sludge solids settle out by gravity. The clarified water near the surface, called the *supernatant*, is discharged over an effluent weir, the settled sludge is pumped out from a sludge hopper at the bottom of the tank. Recycling a portion of the sludge back to the inlet of the aeration tank is an essential characteristic of this treatment process. The settled sludge is in an *active state*. In other words, the microbes are well acclimated to the waste water and, given the opportunity, will readily absorb and decompose more organics by their metabolism.

By pumping about 30% of the waste water flow from the bottom of the clarifier back to the head of the aeration tank, the activated sludge process can be maintained continuously. When mixed with the primary effluent, the hungry microbes quickly begin to absorb and metabolize the fresh food in the form of BOD-causing organics. Since the microbes multiply and increase greatly in numbers, it is not possible to recycle or return all the sludge to the aeration tank. The excess sludge, called *waste activated sludge*, must eventually be treated and disposed of (along with sludge from the primary tanks). Sludge management is a major aspect of waste water treatment.

① *F/M* ratio. An important factor used in the design and operation of activated sludge systems is known as the *food-to-microorganism (F/M) ratio*. The food is measured in terms of kilograms (pounds) of BOD added to the tank per day. Since the suspended solids in the mixed liquor consist mostly of living microorganisms, the suspended solids concentration is used as a measure of the content of microorganisms in the tank. This concentration is called the *mixed liquor suspended solids*, or MLSS.

The *F/M* ratio is an indicator of the organic load onthe system with respect to the content of biological solids in the tank. For conventional aeration tanks, the ratio is in the range of 0.2 to 0.5. It can be computed from the following formula:

$$F/M=\frac{Q\times \mathrm{BOD}}{\mathrm{MLSS}\times V}$$

where *F/M*—food-to-microorganism ratio, in units of kilograms of BOD per kilogram of MLSS per day;

Q—raw sewage flow rate, ML/d (mgd);

BOD—applied 5-day BOD, mg/L (ppm);

MLSS—mixed liquor suspended solids, mg/L;

V—volume of aeration tank, ML (million gal).

Example

An activated sludge tank is 30 m long and 10 m wide and has an SWD of 4 m. The waste water flow is 4.0 ML/day and the raw 5-day BOD is 200 mg/L. The MLSS concentration is 2,000 mg/L. Compute the food-to-microorganism ratio for the system.

Solution

A conventional activated sludge aeration tank is preceded by primary treatment. It is assuming that 35% of the raw BOD is removed in the primary clarifier, 65% of the BOD will be applied to the aeration tank:

$$0.65 \times 200 = 130 \text{ mg/L}$$

The tank volume is the product of length, width, and depth, or

$$V = 30 \text{ m} \times 10 \text{ m} \times 4 \text{ m} = 1,200 \text{ m}^3 \text{ or } 1.2 \text{ ML}$$

Applying Equation $F/M = \frac{Q \times \text{BOD}}{\text{MLSS} \times V}$

$$F/M = \frac{4.0 \times 130}{2,000 \times 1.2} = 0.22$$

Example

A conventional aeration tank is to treat a flow of 800,000 gpd of primary effluent with a BOD of 125 ppm. The MLSS concentration is to be maintained at 1,800 ppm, and a food-to-microorganism ratio of 0.4 is specified. Compute the required volume of the aeration tank. If the side water depth is to be 15 ft and the tank length is to be three times its width, how long should the tank be?

Solution

Rearranging the terms of Equation $F/M = \frac{Q \times \text{BOD}}{\text{MLSS} \times V}$ gives

$$V = \frac{Q \times \text{BOD}}{F/M \times \text{MLSS}} = \frac{0.8 \times 125}{0.4 \times 1800} = 0.139 \text{ million gal} \approx 140,000 \text{ gal}$$

and $140,000\text{gal} \times \frac{1 \text{ ft}^3}{7.5 \text{ gal}} = 18,500 \text{ ft}^3$

Since volume = length×width×depth, or $V = L \times W \times \text{SWD}$, and since the length is to be three times the width, or $L = 3W$, volume is expressed as

$$V = 3W \times W \times \text{SWD} = 3 \times W^2 \times \text{SWD}$$

Solving for the tank width yields

$$W = \left(\frac{18,500 \text{ ft}^3}{3 \times 15 \text{ ft}}\right)^{\frac{1}{2}} \approx 20 \text{ ft}$$

and solving for lengths yields $L = 3 \times 20 = 60$ ft

② Sludge settling. In the activated sludge process, the organic pollutants are absorbed by the billions of microorganisms in an aeration tank. These microorganisms essentially are the *activated sludge*. But without proper clarification or separation of the sludge from the liquid portion of the mixed liquor, the treatment process will not be effective at all. For this reason, gravity settling in the secondary clarifier is a most important part of the activated sludge treatment system. If the sludge does not settle fast enough, some of it will be carried over the effluent weirs of the clarifier and cause pollution of the receiving body of water.

Under certain conditions in an activated sludge sewage treatment plant, filamentous or stringy bacteria, usually of the species *Sphaerotilus natans*, grow prolifically in the aeration tank, making the sludge very fluffy and light. Sludge with excessive growth of these filamentous organisms settles very slowly, and a clear supernatant is not formed in the secondary clarifier. Much of the sludge flows out with the effluent. This condition is called *sludge bulking*. Bulking of activated sludge may be controlled or limited by appropriate adjustments in the MLSS concentration and *F/M* ratio; this is accomplished by regulating the rate of sludge return from the clarifier. The amount of aeration may also be a factor; sludge bulking is sometimes associated with too much aeration. Occasionally, adjustments in the mixed liquor pH are made to solve the problem.

A number called the *sludge volume index* (*SVI*) is used to evaluate the settling property of the activated sludge. It is equal to the volume occupied by 1 g of settled sludge and is expressed in units of milliliters per gram (mL/g) (Figure 1-17).

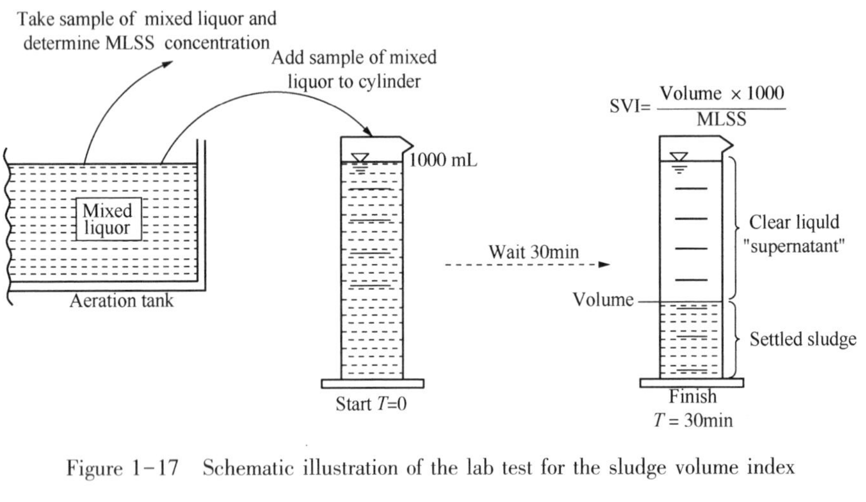

Figure 1-17 Schematic illustration of the lab test for the sludge volume index

The determination of SVI involves taking a sample of mixed liquor from the aeration tank and allowing it to settle for 30 min in a 1-L graduated glass cylinder. This is illustrated in the following figure. The volume of settled sludge is read from the markings on the cylinder. The MLSS concentration in the mixed liquor is also measured.

The following formula is used to compute SVI:

$$SVI=\frac{V\times 1,000}{MLSS}$$

where SVI—sludge volume index, mL/g;

V—volume of settled sludge, mL/L;

MLSS—mixed liquor suspended solids, mg/L.

Example

An aeration tank has an MLSS concentration of 2, 000 mg/L. After settling for 30 min in a 1 – L graduated cylinder, the sludge volume is measured to be 150 mL. Compute the SVI of the sludge.

Solution

Applying Equation $SVI=\frac{V\times 1,000}{MLSS}$ yields

$$SVI=\frac{150\ mL/L\times 1,000}{2,000\ mg/L}=75\ mL/g$$

A normal sludge with good settling characteristics generally has an SVI of less than 100. As the SVI increases above 100, sludgesettleability decreases and some solids get carried over the effluent weir of the clarifier. Severely bulking sludges have SVI values of over 200. A very high SVI is an indication to a treatment plant operator that the sludge return rate should be increased, the aeration rate should be decreased, or some other process adjustment should be made. Sometimes chlorine is added to the aeration tank to destroy the filamentous organisms, but this is a last resort to control the problem.

In general, a well – operated activated sludge treatment system can remove about 90% of the raw sewage BOD and TSS; sometimes removal efficiency may be as high as 95%. In contrast to the simpler trickling filter sewage treatment system, however, an activated sludge plant requires careful operational control. Energy requirements in an activated sludge plant are also high because of the power consumed for aeration.

1.4.3 Tertiary (advanced) treatment

Secondary treatment can remove between 85% and 95% of the BOD and TSS in raw sanitary sewage. Generally, this leaves 30 mg/L or less of BOD and TSS in the secondary effluent. But sometimes this level of sewage treatment is not sufficient to protect the aquatic environment. For example, periodic low flow rates in a trout stream may not provide the amount of dilution of the effluent that is needed to maintain the necessary DO levels for trout survival.

Another limitation of secondary treatment is that it does not significantly reduce the effluent concentrations of nitrogen and phosphorus in the sewage. Nitrogen and phosphorus are important plant nutrients. If they are discharged into a lake, algal blooms and accelerated lake aging or cultural eutrophication may be the result. Also, the nitrogen in the sewage effluent may be present mostly in the form of ammonia compounds. These compounds are toxic to fish if the concentrations are high enough. Yet another problem with the ammonia is that it exerts a *nitrogenous oxygen demand* in the receiving water as it is converted to nitrates.

When pollutant removal greater than that provided by secondary treatment is required, either to further reduce the BOD or TSS concentrations in the effluent or to remove plant nutrients, additional or *advanced treatment* steps are required. This is also called tertiary treatment, because many of the additional processes follow the primary and secondary processes in sequence.

In high economic countries, the discharge standard is very strict due to the closed bays or the lakes as the water resource. For example, the discharge standard in Tokyo Bay is as followed: TN = 10 mg/L, TP = 0.5 mg/L, COD = 12 mg/L. The discharge standard in Lake Biwa is that: BOD = 5 mg/L, TN = 10 mg/L, TP = 0.5 mg/L, SS = 6 mg/L. For the standard, only the tertiary treatment can meet the demand. It is said the treatment rate of the waste water in China is up to 85%, but the total drainage load of COD is similar as that in 1997. The main reason is that the tertiary treatment can remove 70% of the N, P and COD load.

Tertiary treatment of sewage can remove more than 99% of the pollutants from raw sewage and can produce an effluent of almost drinking water quality (Table 1-16). But the cost of tertiary treatment, for operation and maintenance as well as for construction, is very high, sometimes doubling the cost of secondary treatment. The benefit-to-cost ratio is not always big enough to justify the additional expense. Nevertheless, application of some form of tertiary treatment is not uncommon. The removal of main pollutants and the clarified technologies in tertiary treatment.

Some of the more common tertiary processes are discussed in the following sections.

Table 1-16 The removal of main pollutants and the clarified technologies in tertiary treatment

Pollutants	Advised technologies
NH_4^+-N, NO_2^--N, NO_3^--N, TN	Preanaerobic tank - anoxic - oxic activated sludge, Immobilization anoxic-oxic activated sludge.
PO_4^{3-}-P, TP	Anaerobic-oxic activated sludge, activated sludge with coagulant, coagulation, Crystal dephosphorization.
NH_4^+-N, NO_5^--N, NO_2^--N, TN, PO_4^{3-}-P, TP	Anaerobic-anoxic-oxic activated sludge (A/A/O) Anoxic-oxic activated sludge with coagulant (A/O)
Soluble BOD (SS)	Filtration with sand, Coagulation and flocculation and sedimentation, Ultrafiltration, Microfiltration
Non-biodegradable, soluble BOD, COD	Ozone, Activated carbon, Biofilm, Reverse osmosis

(1) Effluent polishing The removal of additional BOD and TSS from secondary effluents is sometimes referred to as *effluent polishing*. It is most often accomplished using a granular-media filter, much like the filters used to purify drinking water. Since the suspended solids consist mostly of organic compounds, filtration removes BOD as well as TSS.

Generally, mixed-media filters are used to achieve in-depth filtration of the effluent. Because of the organic and biodegradable nature of the suspended solids in the secondary effluent, tertiary filters must be backwashed frequently. Otherwise, decomposition would cause septic or anaerobic conditions to develop in the filter bed. In addition to the conventional backwash cycle, an auxiliary surface air-wash is used to thoroughly scour and clean the filter bed. Filtration may be done by gravity in an open tank or by pressure in closed pressure vessels.

A schematic diagram of an automatic-backwash tertiary filter is shown in Figure 1-18. The filtered water may be stored in an adjacent tank and used for backwash water when the head loss through the filter reaches a predetermined level. Some filter manufacturers mount the backwash storage tank directly above the filter, forming a single self-contained unit.

Another process, called microstraining, also finds application as a tertiary step in waste water treatment for suspended solids reduction. The microstrainers, also called *microscreens*, are composed of specially woven steel wire cloth mounted around the perimeter of a large revolving drum. The steel wire cloth acts as a fine screen, with

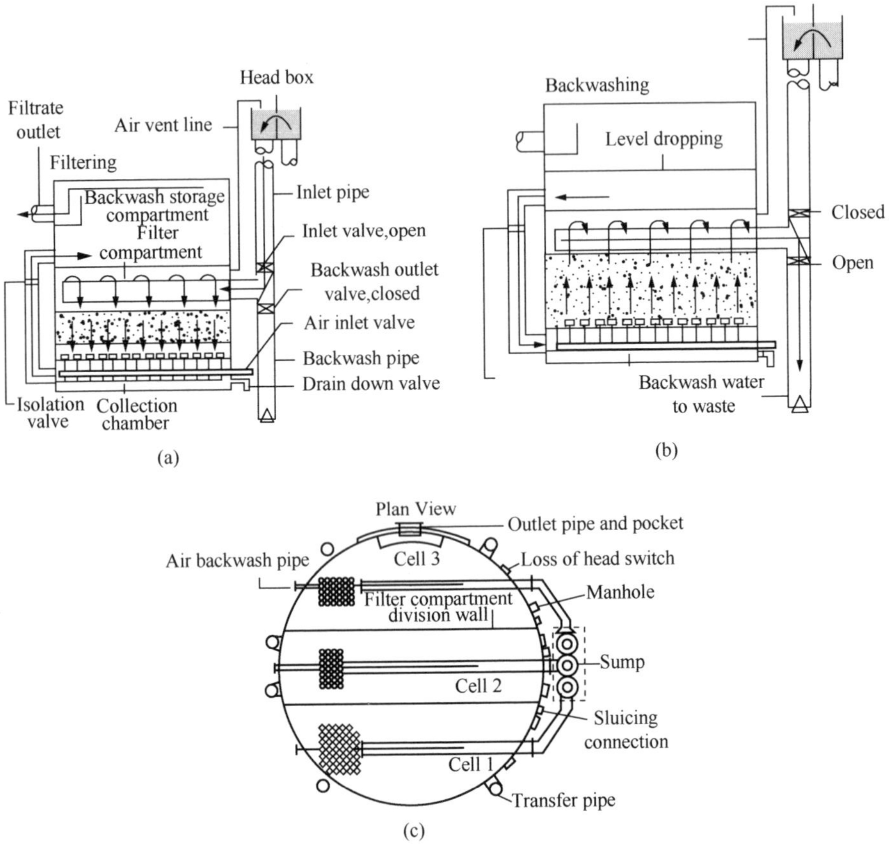

Figure 1-18 A schematic diagram of an automatic-backwash tertiary filter

openings as small as 20 micrometers.

The rotating drum is partially submerged in the secondary effluent, which must flow into the drum and then outward through the microscreen. As the drum rotates, captured solids are carried to the top, where a high-velocity water spray flushes them into a hopper mounted on the hollow axle of the drum.

(2) Phosphorus removal Phosphorus is one of the plant nutrients that contributes to the eutrophication of lakes. Raw sewage contains about 10 mg/L of phosphorus, from household detergents as well as from sanitary wastes. The phosphorus in waste water is primarily in the form of organic phosphorus and as phosphate, PO_4^{3-}, compounds. Only about 30% of this phosphorus is removed by the bacteria in a conventional secondary sewage treatment plant, leaving about 7 mg/L of phosphorus in the effluent.

When stream or effluent standards require lower phosphorus concentrations, a tertiary treatment process must be added to the treatment plant. This usually involves*chemical precipitation* of the phosphate ions and coagulation. The organic phosphorus compounds are entrapped in the coagulant flocs that are formed and settle out in a clarifier.

One chemical frequently used in this process is aluminum sulfate, Al_2SO_4. This is called *alum*, the same coagulant chemical used to purify drinking water. The aluminum ions in the alum react with the phosphate ions in the sewage to form the insoluble precipitate called aluminum phosphate. Other coagulant chemicals that may be used to precipitate the phosphorus include ferric chloride, $FeCl_3$, and lime, CaO.

Adding the coagulant downstream of the secondary processes provides the greatest overall reliability for phosphorus reduction. It not only removes about 90% of the phosphorus, but it removes additional TSS and serves to polish the effluent as well. But when applied in this manner, as a third or tertiary treatment step, additional flocculation and settling tanks must be built. In some cases, even filters may have to be added to remove thenonsettleable floc.

To avoid the need for construction of additional tanks and filters, in most plants requiring phosphorus removal the coagulant is added to the waste water at some point in the conventional process. For example, alum may be added just before the primary settling tanks. The resulting combination of primary and chemical sludge is removed from the primary clarifiers.

In activated sludge plants, the coagulant may be added directly into the aeration tanks. In this case, the precipitation and flocculation reactions occur along with the biochemical reactions. Sometimes, the coagulant may be added to the waste water just before the secondary or final clarifiers. Regardless of the point in the process at which coagulant is added, the total volume and weight of sludge requiring disposal increase significantly. The options for phosphorus removal by chemical precipitation are illustrated in Figure 1-19.

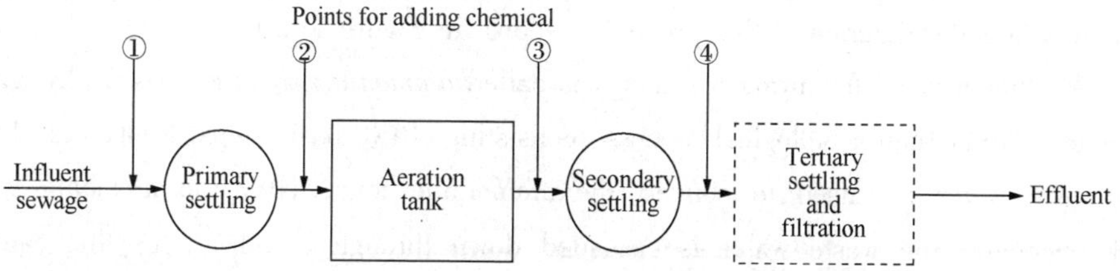

Figure 1-19 The options for phosphorus removal by chemical precipitation

(3) Nitrogen removal Nitrogen can exist in waste water in the form of organic nitrogen, ammonia, or nitrate compounds. The effluent from a conventional activated sludge plant contains mostly the ammonia nitrogen (ammonium ion) form, NH_4^+. Effluents from a trickling filter or rotating biodisc may contain more of the nitrate ion form, NO_3^-. This is because the nitrifying bacteria, those microbes that convert ammonia to nitrate, have a chance to grow and multiply on some of the surfaces in the trickling filter or biodisc units. They do not survive in a mixed-growth aeration tank, where they are crowded out by faster growing bacteria that consume carbonaceous organics.

Nitrogen in the form of ammonia can be toxic to fish, and it exerts an oxygen demand in receiving waters as it is converted to nitrate. Nitrate nitrogen is one of the major nutrients that causes algal blooms and eutrophication. For these reasons, it is sometimes necessary to remove the nitrogen from the sewage effluent before discharge. This is particularly important if it is discharged directly into a lake.

One of the methods used to remove nitrogen is called *biological nitrification – denitrification*. It consists of two basic steps. First, the *secondary* effluent is introduced into another aeration tank, trickling filter, or biodisc. Since most of the carbonaceous BOD has already been removed, the microorganisms that will now thrive in this tertiary step are the nitrifying bacteria, *Nitrosomonas* and *Nitrobacter*. In this first step, called *nitrification*, the ammonia nitrogen is converted to nitrate nitrogen, producing a *nitrified effluent*. At this point, the nitrogen has not actually been removed, but only converted to a form that is not toxic to fish and that does not cause an additional oxygen demand.

A second biological treatment step is necessary to remove the nitrogen from the waste water. This is called *denitrification*. It is an aerobic process in which the organic chemical methanol is added to the nitrified effluent to serve as a source of carbon. The denitrifying bacteria *Pseudomonas* and other groups use the carbon from the methanol and the oxygen from the nitrates in their metabolic processes. One product of this biochemical reaction is molecular nitrogen, N_2, which escapes into the atmosphere as a gas. A schematic diagram of this process is shown in Figure 1-20.

Another method for nitrogen removal is called *ammonia stripping*. It is a physical-chemical rather than a biological process, consisting of two basic steps. First, the pH of the waste water is raised to convert the ammonium ions, NH_4^+, to ammonia gas, NH_3. Second, the waste water is cascaded down through a tall tower; this causes turbulence and contact with air, allowing the ammonia to escape as a gas. Large volumes of air are circulated through the tower to carry the gas out of the system. The combination

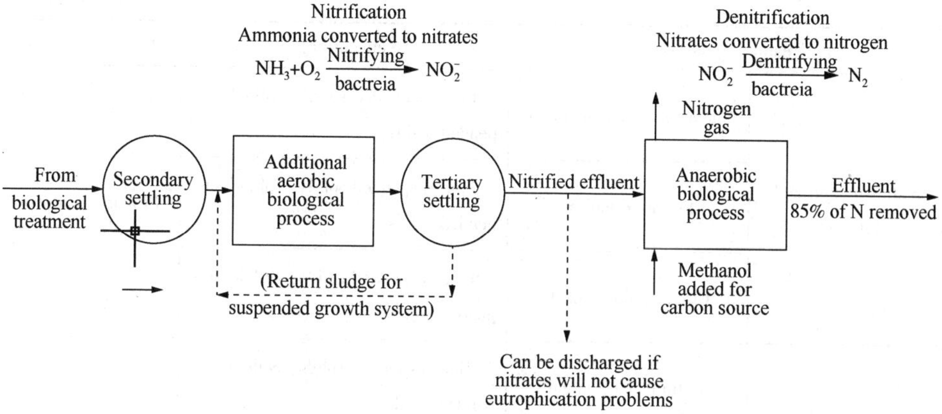

Figure 1-20 A schematic diagram of this process

of ammonia stripping with phosphorus removal using lime as a coagulant is beneficial since the lime can also serve to raise the pH of the waste water. Ammonia stripping is less expensive than biological nitrification-denitrification, but it does not work very efficiently under cold weather conditions (Table 1-17).

Table 1-17 Treatment processes, treatment object, treatment degree

Treatment processes		Treatment object	Treatment degree
Filter	Screen	Floating object, thick solids, fiber	Pretreatment
	Sieve		
	Filter		
Regulation, homogenization	Regulation tank	Waste water quality change, waste water quantity change	Pretreatment
	Homogenization tank		
	Neutralization tank	Regulating pH	
Sedimentation and filtration	Grit chamber	Thick inorganic solids, such as grit, residuals	Pretreatment
	Primary sedimentation tank	Suspended solids	Primary treatment
	Floatation tank	Separating oil or suspended solids	
	Oil separator	Separating oil	
	Clarification tank	Fine suspended solids	
	Coagulation and clarification tank	Fine suspended solids, colloid suspended solids	

continued

Treatment processes		Treatment object	Treatment degree
Sedimentation and filtration	Chemical sedimentation *	Fine suspended solids, colloid suspended solids	Primary treatment
	Filtration tank *	Fine suspended solids, colloid suspended solids	
	Ultrafiltration tank *	Fine suspended solids, colloid suspended solids	
Aeration	Natural flocculation	Fine suspended solids, colloid suspended solids	Primary treatment
	Biological flocculation	Fine suspended solids, colloid suspended solids	
	Oxidization	Reduction matters	
	Increasing oxygen	Corrupt things	
Biological technologies	Activated sludge (oxic and anoxic)	Colloid organic matters, dissolved organic matters	Secondary treatment
	Biological membrane (oxic and anoxic)	Colloid organic matters, dissolved organic matters	
	Secondary sedimentation tank	Suspended solids	
	Dissolved air flotation	Colloid matters and suspended solids	
Chemical technologies	Chemical oxidization	Colloid matters and suspended solids, disinfection	Secondary treatment
	Land treatment system	Colloid, organic matters, dissolved organic matters, suspended organic matters	
	Stable pond	Colloid, organic matters, dissolved organic matters, suspended organic matters	

continued

Treatment processes		Treatment object	Treatment degree
Physical technologies	Evaporation	Volatile organic compound	Secondary treatment
	Solvent extraction	Dissolved organic matter and heavy metals	
	Blow off	Volatile organic compound	
Biological technologies	Oxic nitrification	Nitrogen compound and part of the organic matter	Tertiary treatment
	Anaerobic denitrification	Nitrogen compound and part of the organic matter	
	Anaerobic−anoxic−oxic	Nitrogen and phosphorus compound and part of the organic matter	
Chemical, physical technologies	Coagulation and sedimentation	Dispersed, small particles and phosphorus	Tertiary treatment
	Ammonia blow off	Ammonia	
	Filtration	Fine particles	
	Activated carbon	Insoluble organic matters and color, taste, odour	
	Ion exchange	Ions, ammonia, phosphorus, heavy metals	
	Reverse osmosis	Soluble salts and organic matters	
	Electroosmosis	Inorganic salts	
	Ozone	Organic matters, color, virus and-pathogene	
Land treatment		Suspended solids, ammonia, phosphorus, virus, pathogene and organic matters	Tertiary treatment
Stable ponds		Suspended solids, ammonia, phosphorus, virus, pathogene and organic matters	Tertiary treatment

* It is referred that there is the biological treatment before the unit.

Questions

1. What are the typical waste water and what are their characteristics?

2. What are the indicators for evaluating water quality and what are the contents of water quality standards in China? What are the differences between China and other countries?

3. What are the characteristics of urban domestic waste water, industrial waste water, rural waste water? What are the main treatment methods for these three types of waste water.

4. What are the main hazards for the domestic waste water?

Chapter 2 Treatment theory of the activated sludge

Emphasis: Characteristics of activated sludge. The basic processes of activated sludge. The decomposition process of activated sludge. Aeration designment and management.

Nut: The designment of activated sludge. The controlling parameters of activated sludge.

2.1 The traditional activated sludge

The activated sludge process is now the most widely used biological process for treatment of municipal and industrial waste waters. Activated sludge process was developed in England in 1914 and was so named because it involved the production of an activated mass of microorganisms capable of aerobic stabilization of the organic content of waste water.

2.1.1 The basic conception and the process

The basic activated sludge process is illustrated by Figure 2-1. The mixture of influent waste water commonly coming from primary clarifier and returning activated sludge is known as mixed liquor. In activated sludge process, the mixed liquor flows to aeration tank in which the microorganisms responsible for treatment are kept in suspension and aerated. In the aeration tank, contact time is provided for mixing and aerating influent waste water. Aeration serves at least three important functions: ① mixing the mixed liquor, ② keeping the activated sludge in suspension, and ③ supplying the oxygen to the biochemical reactions necessary for the stabilization of the waste water. Once the mixed liquor has received sufficient treatment, excess mixed liquor is discharged into secondary clarifier. The function of the secondary clarifier is to separate the activated sludge solids from the mixed liquor. These solids represent the colloidal and dissolved solids that were originally present in the waste water. In the aeration unit they were incorporated into the activated sludge floc, which are settleable solids. The separation of these solids, a critical step in the activated sludge process, is accomplished in the secondary or final settling tanks. A part of this settled biomass, described as activated sludge, is returned to the aeration tank which is important to maintain a sufficient concentration of activated sludge in the aeration tank for the desired degree of treatment. The excess sludge must be removed before it loses its activity because of the death of the aerobic organisms resulting from lack of oxygen at the bottom of the tank. The most common practice is to waste from the sludge return line. Sometimes waste sludge is removed by withdrawing mixed liquor directly from the aeration tank. The wasted mixed liquor can then be discharged to a thickening tank or to the primary tanks where the sludge settles and mixes with the raw primary sludge. The waste sludge is further thickened by final sedimentation, centrifuging, or flotation thickening and then treated by biological or chemical means.

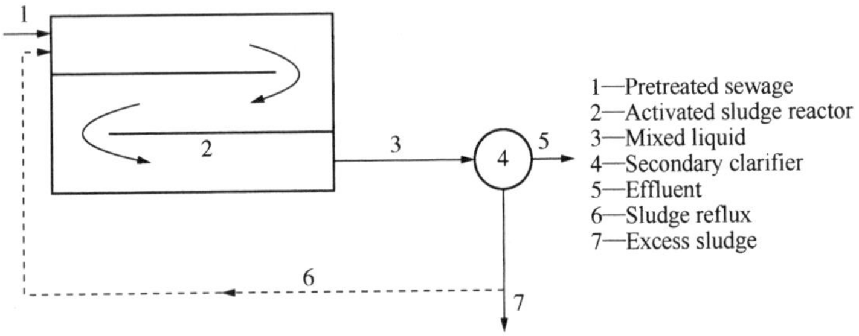

Figure 2-1 Basic activated sludge process

2.1.2 The forms and component of the activated sludge

The activated sludge is the main part in the activated sludge system. In fact, the activated sludge is not sludge, while it is the microorganism group with high vitality. The organic matter can be transformed into inorganic matter by the activated sludge under the metabolism of the microorganism. This is called activated sludge.

The appearance of the activated sludge is the particle with villus, and it is called biological flocculant. Under the standing, the small particles start to form the big particles to settle due to the condensation of the activated sludge. In general, the color of the activated sludge is yellow or brown. In addition, the water content of the activated sludge is high, up to more than 99%. The density of the activated sludge is between

1.002 and 1.006. The specific surface area of the activated sludge is between 20 ~ 100 cm^2/mL.

For the activated sludge, the solid is only less than 1%, while the solid consists the organic matter and inorganic matter. The component of the activated sludge is variable depending on the raw water quality. For example, the organic matter is about 75% ~ 85% in the activated sludge in the domestic waste water. The organic matter is mainly the microorganisms in the activated sludge. In addition, there are some organic matters in the influent, that is the nonbiodegradable organic matters. After the self-oxidation of the microorganisms, the residual is nonbiodegradable. As for the inorganic matters in the activated sludge, they mainly come from the influent.

The component of the activated sludge are four parts. ① the microorganism group with high vitality (M_a). ② the residual of the self-oxidation of the microorganisms (M_e). ③ the organic matter adsorbed by the microorganisms due to the influent (M_i). ④ the inorganic matter due to the influent (M_{ii}).

2.1.3 The microorganism in the activated sludge and its function

The bacteria are the main part for the microorganism group in the activated sludge, and the ratio to the total weight of the microorganism is up to 90% ~ 95%. In some industrial waste water, the ratio is up to 100%. Except of the bacteria, there is the fungus, protozoa and metazoan. Therefore, the component of the activated sludge is organic matter, bacteria, protozoa and metazoan. They build a small and stable ecology and food chain (Figure 2-2).

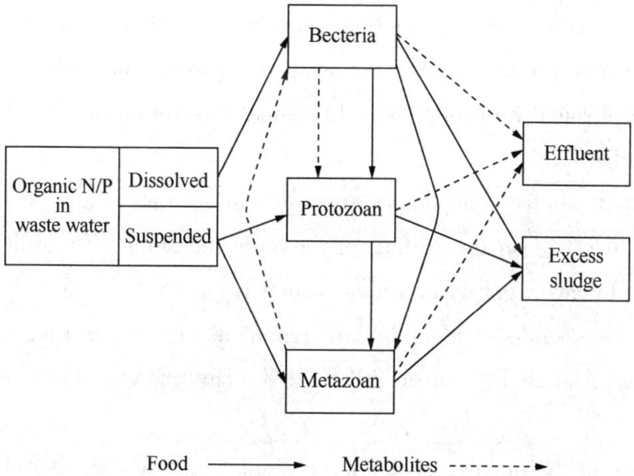

Figure 2-2 The food chain of the microorganism group in the activated sludge

Heterotropic microorganisms are the main part in the activated sludge. The bacteria quantity is about $10^7 \sim 10^8$ number/mL in the activated sludge. The dominant bacteria are *Alcaligenes*, *Bacillus*, *Flavobacterium*, *Zoogloea*, *Pseudomonas*, *Escherichia Coli*. In addition, the bacteria maybe *Achromobacter*, *Microbaccus*, *Nocardia*, *Sarcina*, et al. The dominant bacteria depend on the organic matter in the raw waste water. For example, the dominant bacteria are *Alcaligenes* in the waste water with high content of protein, while the *Pseudomonas* will be the dominant bacteria in the waste water with high content of sugar and hydrocarbon.

The *Flavobacterium*, *Zoogloea*, *Pseudomonas* can form the flocculant granule. The bacteria are the main part of the activated sludge. This is called zoogloea group. They have strong adsorption ability and oxidization on the organic matter. The zoogloea group can avoid of the phagocytosis by the microfauna. In addition, the zoogloea group can decrease of the effect of the toxic matter. The settling is very good for the zoogloea group and it is easily separated in the second sedimentation tank.

2.1.4 The clarification process by the activated sludge

The removal of organic matter from the waste water is the metabolism by the microorganism as the nutrient for the activated sludge. The results of the process are the clarification of the waste water, obtaining energy by the microorganism and compounding the new cell, the growth of the activated sludge.

The clarification process is divided into 3 steps.

① The initial adsorption. Lots of organic matter can be removed by the activated sludge after 5 ~ 10 min of the contacting between the activated sludge and the waste water in the activated sludge system, showing the removal of BOD is very high. It is attributed to the physical adsorption and biological adsorption on the organic matter by the activated sludge.

The activated sludge has huge specific surface area and there are lots of microorganisms adsorbed on the surface of the activated sludge. In addition, the outside of the activated sludge is polysaccharides. When the activated sludge contacts with the waste water, the suspended and colloidal organic matter is adsorbed by the activated sludge. The removal of BOD is up to 20% ~ 70%. The phenomenon is called the initial adsorption.

The removal on the organic matter is very quick, while the adsorption rate and the adsorption degree depend on the microorganism activity, the component and physical forms of the organic matter. The microorganism activity is related closely with the adsorption ability and the condensation ability. The component and physical forms of the organic matter are related closely with the adsorption difficulty level of the organic matter. The huge specific surface area presents the strong activity of the activated sludge. In addition, the proliferating phase plays the important role in the initial adsorption. In general, the microorganism in the endogenous respiration phase has the strong activity and the strong adsorption on the organic matter.

After the several hours' aeration, the adsorbed organic matter can come inside of the microorganism and then be degraded. Therefore, the initial adsorption on the organic matter is limit. The best method to increase the initial adsorption is to provide adequate aeration for the activated sludge. The adsorbed organic matter on the surface of the activated sludge is degraded to fast the microorganism coming into the endogenous respiration phase. However, once the aeration is excessive, the activity of the activated sludge decreases, resulting in the initial adsorption low.

② The metabolism of activated sludge. At first, the organic matter is adsorbed onto the surface of the activated sludge with lots of microorganisms. Under the permease, the organic matter passes the cell wall to come inside of the microorganism. For the micromolecule organic matters, they can come into the microorganism directly through the cell wall. For the macromolecule organic matters, they have to be hydrolyzed to the micromolecule under the extracellular enzyme and hydrolytic enzyme and come into the microorganism directly through the cell wall.

Under many kinds of enzymes (such as dehydrogenase and oxidase) in the cell, the metabolism on the organic matter happens in the microorganism. The metabolism consists two kinds. The one is catabolism. That is the oxygenolysis of the part of the organic matter. The organic matter is transformed into stable inorganic matters, such as CO_2 and H_2O. During the process, the new cell gets the energy for the synthesis. It is described by the equation.

$$C_xH_yO_z+\left(x+\frac{y}{4}-\frac{z}{2}\right)O_2 \longrightarrow xCO_2+\frac{y}{2}H_2O+\Delta H$$

The $C_xH_yO_z$ presents the organic matter.

The other part of organic matter is used for the synthesis of the new cell. That is anabolism. The energy mainly comes from the oxygenolysis. It is described by the equation.

$$nC_xH_yO_z+nNH_3+n\left(x+\frac{y}{4}-\frac{z}{2}-5\right)O_2 \longrightarrow$$

$$(C_5H_7NO_2)_n+n(x-5)CO_2+\frac{n}{2}(y-4)H_2O-\Delta H$$

When the organic matter is enough in the system, and lots of new cells increase, and the endogenous respiration phase is not obvious. When the organic matter is used up, the energy mainly comes from the endogenous respiration phase. The oxygenolysis and the synthesis and their products can be described by the equation.

$$(C_5H_7NO_2)_n+5nO_2 \longrightarrow 5nCO_2+2nH_2O+nNH_3+\Delta H$$

This is described in Figure 2-3.

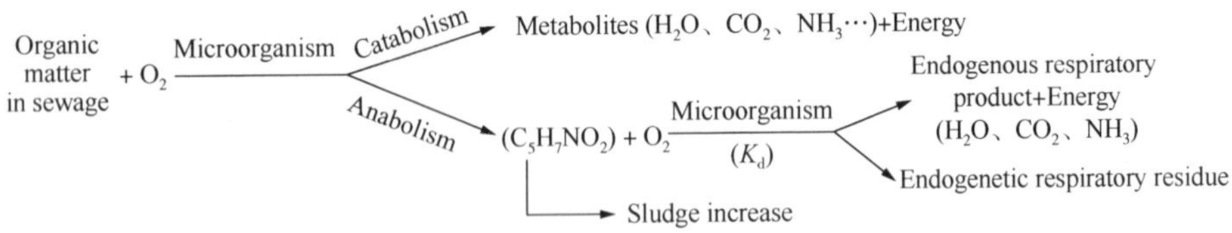

Figure 2-3 The degradation and the anabolism by the microorganisms and their products

③ Separation of the activated sludge. The last process for the activated sludge is the solid-liquid separation, and the process is realized in the second sedimentation tank.

After the organic matters are transformed into the stable inorganic matters under the metabolism of activated sludge, the effluent is drained into the natural water body. It is necessary to separate the activated sludge from the waste water. The step is very important, since the separation affects the normal running of the activated sludge system. Once the separation of solid and liquid is not good, the activated sludge comes into the natural water body and destroys the aquatic system. In addition, the residual sludge should be controlled, otherwise, it causes the second pollution on the environment.

2.1.5 The effect of the environmental factors on the activated sludge

Under the suitable environmental conditions, the bacteria in the activated sludge can survive and degrade the organic matter. The technology of the activated sludge is based on the theory, but there are some factors affecting the activity of the bacteria.

(1) Nutrient The bacteria take in continuously the nutrient (such as C, N, P, inorganic salts and other elements) during their life. C is the important matter for the synthesis of new cells. The demand on the C is very big for the bacteria in the activated sludge. Generally, the quantity of C in the waste water is presented by the BOD_5. N is another important matter for the bacteria, especially for the synthesis of protein and

nucleate. N mainly comes from the inorganic N compound, such as N_2, NH_3, NO_3^-. It comes also from the organic compound, such as protein, peptone and amino acid. P is the important matter for the synthesis of nucleoprotein, lecithin and other compound with P. P plays the important role in the metabolism and the transformation of different matters. In addition, the bigger demand of inorganic element for the bacteria is P.

The potassium, sodium, sulphur, calcium, iron, magnesiumis necessary for the bacteria, but the demand quantity is very small. Most waste water can meet the demand.

The balance of the nutrient in the waste water is described by the ratio of BOD_5 : N : P. For domestic waste water, the ratio can be considered by the 100 : 5 : 1. The quantity is related with the sludge load and sludge age.

In general, the nutrient for the bacteria in the activated sludge is enough in the domestic waste water and municipal waste water, but for the industrial waste water, it may be not. For the industrial waste water with low content of C, the C source should be replenished to meet the demand of the bacteria. For example, the domestic waste water, washing water of rice or starch. For the industrial waste water with low content of N, the N source should be replenished to meet the demand of the bacteria. For example, the urea, ammonium sulfate. For the industrial waste water with low content of P, the P source should be replenished to meet the demand of the bacteria. For example, the potassium phosphate, sodium phosphate, superphosphate or phosphoric acid.

(2) Dissolved oxygen The content of the DO keeps above 0.3 mg/L for the free bacteria in the mixed liquid. But the bacteria group in the activated sludge lives in community, the DO must penetrate the inside of the activated sludge flocculant. According to the running experience in many plants, the content of DO should be kept above 2 mg/L (the exit in the aeration tank) in the aeration tank to make the bacteria for physiological activity. For the entrance of the aeration tank, the organic matter content is high and the oxygen uptake rate is high. Therefore, the content of DO is difficult to be above 2 mg/L. If the DO is less than 1 mg/L, it is not suitable for the running of the plant. However, the DO with too high content in the aeration tank is not beneficial since it is not economic.

(3) pH The function of pH on the bacteria is that the pH can change the electric charge on the cytomembrane, affecting the adsorption on the nutrient by the bacteria. The other function is that the pH can change the activity of the enzyme during the metabolism for the bacteria in the activated sludge. Another function is that the pH

can change the bioavailability and the toxicity of the matters in the waste water. High pH can cause the hydrolysis of the protein and nucleate on the surface of the bacteria.

The pH is different for the different bacteria. The practice experience show that the suitable pH is 6.5 ~ 8.5. After the bacteria domestication, the range of the pH for the bacteria increases. If the pH is too high or too low, it is necessary to place a regulating tank and then the waste water come into the aeration tank.

(4) Temperature　The suitable temperature can promote and strengthen the physiological activity of bacteria, while the non-suitable temperature can decrease or destroy the physiological activity of bacteria. The non-suitable temperature can cause thebiomorphology and physiological property of the bacteria change, even die.

The physiological activity of bacteria is strong and the breeding of the bacteria is fast and the generation time is quick under the suitable temperature (Table 2-1).

Table 2-1　The generation time for coliform at different temperature

Temperature/℃	Generation time/min	Temperature/℃	Generation time/min
20	60	40	19
25	40	45	32
30	29	50	No breeding
37	17		

According to the above table, the suitable temperature is 30 ~ 40℃, and the generation time is short, 17 ~ 19 min.

Most bacteria is attributed to the mesophile, and the suitable temperature is 15 ~ 30℃. Therefore, the highest temperature and the lowest temperature is 35℃ and 40℃, respectively. For those plants with the activated sludge in cold region, the small can be placed in the room, while the middle and big which were placed outside have to be considered the insulation measure. In addition, increasing the content of the activated sludge in the mixed liquid or decreasing the BOD load or lengthen the aeration time, these measures are favor of the bad effects due to the low temperature.

If the influent temperature is too high, the cooling measures should be taken.

(5) Toxic matters　The toxic matters for the bacteria in the activated sludge include inorganic matters, such as, heavy metals, cyanide, H_2S and organic matters, such as phenol, aldehyde, alcohol and dyes.

The heavy metals (Pr, Cr, Cd, Fe, Cu, Zn et al) can react with the protein in the cells, resulting in the protein denaturation or sedimentation. The heavy metals (Hg,

Ag, As) can bond with the —SH group of the protein of the enzyme, resulting in the decrease of the mentalism of the microorganism in the activated sludge.

The phenol can damage the cytomembrane and fast the solidification of the protein. In addition, the phenol can destroy the enzyme, such as deaminase and oxidase, and damage the normal metabolism in the bacteria. The other derivatives of the phenol have the function as the disinfectant, such as para-cresol, ortho-cresol, Offset cresol, propenyl phenol, butylphenol. The methanol bonds with amidogen of the protein, resulting in the protein denaturation.

However, in the practice, the amount of the toxic matter is not neglected. That is, the amount of some toxic matter is specific content and then the damage and inhibition on the bacteria appears. Therefore, the content is limit allowable intensity of toxic substances. If the content of the toxic matter is lower than the limit allowable intensity of toxic substances, there is no damage on the bacteria by the toxic matter. But there are no uniform standards for the content. After the bacteria domestication, the content maybe higher than that no bacteria domestication.

The damage of the toxic matters is related with the pH, temperature, dissolved oxygen, the other toxic matter, the quantity of the bacteria, bacteria domestication. In a word, the mechanism of the damage of the toxic matters is very complex, depending on many factors(Table 2-2).

Table 2-2 The threshold value on the bacteria by the different chemical matters

mg/L

Chemical matter	Maximum concentration	Chemical matter	Maximum concentration
pH (acid)	5.0	methyl alcohol	200
pH (alkali)	8.0	ethyl alcohol	15, 000
sodium chloride	8, 000~9, 000	pentanol	3
sodium sulfate (Na_2SO_4)	3, 000	ethanediol	1, 000
sodium sulfate (Na_2SO_3)	300	propanediol	1, 000
magnesium sulphate	10, 000	monochloroacetic acid	100
sulfide (S^{2-})	5~25	dichloroacetic acid	100
sulfide (H_2S)	20	butyric acid	500
thiocyanide	36	citric acid	2, 500
ammonium thiocyanate	500	oxalic acid	1, 000

continued

Chemical matter	Maximum concentration	Chemical matter	Maximum concentration
editpotassium cyanide	1~8	lauric acid	340
chlorine	0	*m*-toluic acid	120
magnesium chloride	16, 000	crylic acid	100
iron compound (Fe)	5~100	sodium benzoate	250
copper compound (Cu)	0.5~1.0	ammonium acetate	500
silver compound (Ag)	0.25	unsym-dimethylhydrazine	1.0
zinc compound (Zn)	5~13	gasoline	100
lead compound (Pr)	1.0	lamp kerosene	500
methanal	1, 000	vinyl chloride	5
acetaldehyde	1, 000	dichloromethane	1, 000
crotonaldehyde	250	trichloromethane	120
vinylidene chloride	1, 000	vinyl acetate	250
methyl acrylate	100	methyl methacrylate	100
alkyl sulfur compounds	50~100	trichlorphon	100
water-soluble petroleum sulfonic acid	50	chromium compound (Cr)	2~5
chromium compound (Cr^{3+})	2.7	chromium compound (Cr^{6+})	0.5
stibium compound (Sb)	0.2	cadmium compound (Cd)	1~5
mercury compound (Hg)	0.5	vanadium compound (V)	5
arsenic compound (As^{3+})	0.7~2.0	lanthanum nitrate	1.0
aniline	100~250	acetonitrile	600
melamine	50	methacrylamide	25
2,4,6-trinitrotoluene	12	dimethylamine	200
diethylamine	100	triethylamine	85
oil	100	benzene	100
chlorobenzene	10	hydroquinone	15
resorcin	450	catechol	100
p-cresol	243	phenol	250~1, 000
phloroglucinol	100	pyrogallol	100

continued

Chemical matter	Maximum concentration	Chemical matter	Maximum concentration
hydroquinone	600	acetone	800
glycerine	500	diglycol	300
pyridine	400	salicylic acid	500
stearic acid	300	styrene	65
triphenyl phosphate	10	triethanolamine	890
ammonium acetate	500	tetrachloromethane	50
ethyl acetate	500	ethylene chlorohydrin	350
non-ionic detergent	9~100	open powder	100

(6) Organic load (*F/M*) The definiti on is that per unit activated sludge in the aeration tank in unit time can receive the amount of the organic matter. It is described by N_s.

F/M is an important factor affecting the growth of the activated sludge and organic matter degradation. Increasing *F/M*, it is faster the growth of the activated sludge and organic matter degradation. It is favor of the decrease of the aeration tank. The high *F/M* is not favor of the requirement of water quality. The low *F/M* is that the organic matter degradation rate is low and the aeration tank is big, resulting in the increase of the invest of the construction.

In practice, the *F/M* can be controlled at the range from 0.15 ~ 0.40 kgBOD/(kgMLSS · d). However, the *F/M* maybe different for the removal of organic matter, nitrification and removal of N and P.

2.2 Some performance indexes of the activated sludge

2.2.1 Some performance indexes of the activated sludge

(1) The indexes for the microorganisms in the activated sludge The microorganism is the key for the activated sludge. The quantity of the microorganism is necessary for the normal running of the activated sludge. In the process, the microorganism is adsorbed on the surface of the activated sludge. That is, the key of the activated sludge is the microorganism. Therefore, the content of the activated sludge can present the quantity of the microorganism.

The content of the activated sludge is stable by the growth and breeding of the microorganism in the activated sludge and the backflow of the activated sludge from the second sedimentation tank. There are two indexes to describe the content of the activated sludge.

① Mixed liquor suspended solids (MLSS). The index shows the total mass of the activated sludge per volume mixed liquor in the aeration tank.

$$\text{MLSS} = M_a + M_e + M_i + M_{ii}$$

MLSS—mixed liquor suspended solids, unit, mg/L, g/L, g/m^3, kg/m^3;

M_a—the microorganism group with high vitality;

M_e—the residual of the self-oxidation of the microorganisms;

M_i—the organic matter adsorbed by the microorganisms due to the influent;

M_{ii}—the inorganic matter due to the influent.

Although the method is simple and easy, the use is widely. But the M_e and M_i are not active matter in the index, the M_{ii} is the inorganic matter. Therefore, only M_a is related with the bacteria and the exact quantity of the activated sludge is not presented by the index. In a word, the index describes the relative value of the activated sludge. But the index is enough to be the designment and running parameter.

② Mixed liquor volatile suspended solids (MLVSS). The index shows the mass of the organic matter in the activated sludge per volume mixed liquor in the aeration tank.

$$\text{MLVSS} = M_a + M_e + M_i$$

MLSS—mixed liquor suspended solids, unit, mg/L, g/L, g/m^3, kg/m^3;

M_a—the microorganism group with high vitality;

M_e—the residual of the self-oxidation of the microorganisms;

M_i—the organic matter adsorbed by the microorganisms due to the influent.

Compared with MLSS, MLVSS is more accurate since there is no M_{ii} in the index. But the M_e and M_i are inert matter in the index. Therefore, MLVSS is still the relative value.

The difference between MLSS and MLVSS is as Figure 2-4.

The ratio of MLVSS to MLSS is described by f.

$$f = \frac{\text{MLVSS}}{\text{MLSS}}$$

Generally, f is stable. For the domestic waste water and the municipal waste water with main part of the domestic waste water, f is about 0.75.

In practice, the MLSS and MLVSS are the key index for the designment and

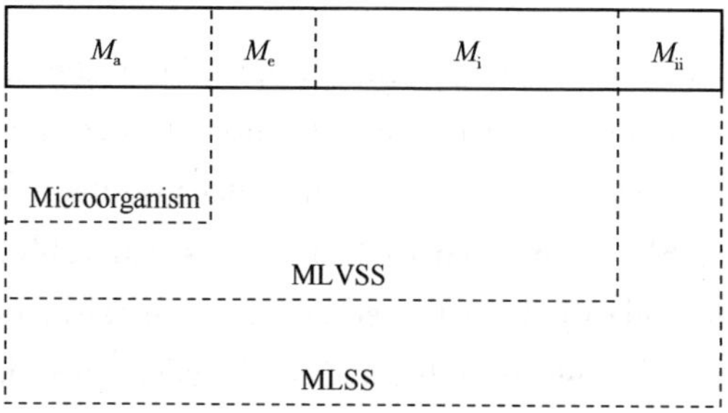

Figure 2-4 The component of the activated sludge

running of the waste water treatment plant. Generally, as the running parameter, MLSS is measured each day in waste water treatment plant, while MLVSS is one time one week.

If MLSS is too high, it is indicating that the content of the activated sludge in the aeration tank is big. The drainage of residual sludge should be strengthened. This affects the separation of solid and liquid in the second sedimentation tank, resulting in the quality of the effluent worsen. On the other hand, If MLSS is too low, it is indicating that the content of the activated sludge in the aeration tank is small. The activated sludge is not enough to degrade the organic load. Therefore, the effluent quality is worse.

In the running of the waste water treatment plant, for the traditional activated sludge, MLSS is about 3, 500mg/L, and it is favor of the running. For the AB process, MLSS is about 1, 500~2, 000 mg/L. for the SBR process, MLSS is about 1, 500~5, 000 mg/L. For the oxic ditch, MLSS is about 4, 000 mg/L. For the MBR process, MLSS is more than 10, 000 mg/L.

(2) The indexes for the sedimentation of the activated sludge The sedimentation of the activated sludge passes the flocculation settling, stratified settling, compression settling.

In general, the activated sludge finishes the flocculation settling and stratified settling process in 30 min, and then the compression settling starts. But the process of the compression settling is slow and it is long. Because the separation of solid and liquid in the second sedimentation tank depends on the settling property, the two indexes are invented to evaluate the settling property of the activated sludge according to the sedimentation of activated sludge for 30 min.

① Settling velocity (SV). The ratio of the volume of settling sludge after the sedimentation for 30 min in the cylinder to the total volume of the mixed liquid is called

SV. The unit is %.

Generally, the value of SV can present the quantity of the activated sludge in the aeration tank in the running. The value can be used to evaluate the drainage of the residual sludge and to forecast the sludge bulking. Because the method for SV is simple and easy, in practice, SV is measured each day. It is noticeably, one value of SV in one day is not enough to clarify the activated sludge. Generally, the trend of the SV in several days can present the content of the activated sludge. In addition, the SV and the other indexes can be used to present the content of the activated sludge.

② Sludge volume index (SVI). The definition of the SVI is the volume of per gram dry sludges after the mixed liquid from the aeration tank settles for 30 min. It is calculated by the equation.

$$SVI=\frac{SV}{MLSS}$$

In the equation, the unit of SVI is mL/g. The unit of SV is mL/L, and the unit of MLSS is g/L.

SVI can present the flocculation and settling property, and the value is between 70 mL/g and 100 mL/g for the domestic waste water and municipal waste water. If SVI is too small, it is indicating that the activated sludge particle is fine and the content of the inorganic matter is big, resulting in the low activity of the activated sludge. If SVI is too big, it is indicating that the settling property of the activated sludge is bad and the possibility of the sludge bulking increases.

The dissolved sludge volume index (DSVI) is proposed and it is advised to replace the SVI because of the diameter of the cylinder, the initial content of the sludge and the mixing.

(3) The indexes for activity of the activated sludge The specific oxygen uptake rate (SOUR) is used to evaluate the activity of the activated sludge. The index is presented by OUR. The definition of the index is the consumption of oxygen in unit time for unit weight of activated sludge. The unit of the index is $mgO_2/(gMLVSS \cdot h)$. OUR is related with the content of DO, the content of the substrate, the sludge age, the degree of the biodegradable organic matter.

In practice, OUR can be used to present the degradable rate of organic matter and the poisoning of the activated sludge. Generally, the value of OUR should be at the range from 8 to 20 $mgO_2/(gMLVSS \cdot h)$.

2.2.2 Some indexes of the activated sludge for designment and running

Activated sludge process is a complex engineering system. There are some indexes for the designment and running.

(1) Sludge load rate of BOD and sludge volume rate of BOD According to themultiplication rule of bacteria, the growing period of the bacteria depends on the ratio of F to M. On the other hand, the property for degradation of organic matter is different at different growing period of the bacteria. Therefore, regulating the ratio of F to M is favor of the perfect growing period of the bacteria and is beneficial the treatment on the pollutants in the waste water.

In engineering, the ratio of F to M can be described by the sludge load rate of BOD (N_s). The unit is kg BOD/(kg MLSS · d). The equation is as followed.

$$N_s = \frac{F}{M} = \frac{QS_0}{VX}$$

In the equation, Q—The amount of the waste water, m^3/d;

S_0—The content of the organic matter, mg/L;

V—The volume of the aeration tank, m^3;

X—The content of the activated sludge in the aeration tank, MLSS or MLVSS.

The definition of N_s is that per unit activated sludge in the aeration tank in unit time can receive the amount of the organic matter. Sometime, the S_0 is presented by COD. The X is presented by the MLSS or MLVSS.

There is another index to describe the sludge load. That is the sludge volume rate of BOD (N_v). The unit is kg BOD/(m^3 · d). The equation is as followed.

$$N_v = \frac{QS_0}{V}$$

In the equation, Q—The amount of the waste water, m^3/d;

S_0—The content of the organic matter, mg/L;

V—The volume of the aeration tank, m^3.

The definition of N_v is that per unit volume of the aeration tank in unit time can receive the amount of the organic matter.

The relationship between N_s and N_v is as followed.

$$N_v = N_s X$$

Sludge load rate of BOD and sludge volume rate of BOD have price for the engineering. In especial, the former comes from the ratio of F to M, so it has theory meaning.

Sludge load rate of BOD in the important factor to affect the organic matter degradation and the growth of activated sludge. In practice, higher N_s, faster the degradation of organic matter and the growth of activated sludge. It is indicating that the volume of the aeration tank is reduced. It is economically feasible, but the effluent quality maybe not good and even it does not meet the demand of the effluent standard. On the other hand, N_s is low, resulting in the degradation of organic matter and the growth of activated sludge down. It is indicating that the volume of the aeration tank increases and the fee on the building increases. The effluent quality is good.

(2) Sludge age The average time for the bacteria from the generation to drainage in the aeration tank is called sludge age. It is also referred to the time of all the bacteria in the aeration tank renewed. Therefore, the sludge age is also called sludge retention time (SRT). In the engineering, the SRT is the ratio of the total activated sludge in the aeration tank to the generation amount under the stable conditions each day. The equation is as followed.

$$V_c = \frac{VX}{\Delta X}$$

In the equation, V_c—SRT, the unit is d;

ΔX—the generation amount of activated sludge. The drainage of the activated sludge under the stable conditions, kg/d.

In activated sludge reactor, the bacteria generation is continuous. Therefore, the new cells are generated each time, at the same time, a part of bacteria are old and their activity is down. The amount of activated sludge that is same as the new generation of bacteria should be drained from the reactor, to keep the activity of the activated sludge and the constant bacteria amount.

According to the above, the amount of activated sludge that includes the residual sludge and those drained with the effluent is drained from the reactor. The equation is presented.

$$\Delta X = Q_w X_w + (Q - Q_w) X_e$$

In the equation, Q_w—The drainage of the amount as residual sludge, m^3/d;

X_w—The content of the residual sludge, kg/m^3;

X_e—The content of the suspended solids in the effluent, kg/m^3.

Therefore, another equation is as followed.

$$V_c = \frac{VX}{Q_w X_w + (Q - Q_w) X_e}$$

In practice, the value of X_e is too small and it is neglected. The above equation is as followed.

$$V_c = \frac{VX}{Q_w X_w}$$

The approximate value of X_w is calculated by the equation.

$$X_w = \frac{10^6}{\text{SVI}} \cdot r$$

In the equation, SVI—The index of sludge volume, mL/g;

r—Correction factor, it is related with the SRT in the aeration tank and the depth of the aeration tank. Generally, it is about 1.2.

Sludge age is very important for the designment and running of the waste water treatment plant. The index can present the growth conditions of the activated sludge. If the generation time of the bacteria is longer than sludge age, the bacteria is not dominant in the aeration tank. Take nitrobacterium under 20℃ as the example, the generation time is 3 d. When the sludge age is less than 3 d, the nitrobacterium is never the dominant. It is forecasting that there is no nitration in the aeration tank.

(3) The ratio of the sludge backflow The ratio of the amount of the sludge backflow from the second sedimentation tank to the aeration tank to the amount to the waste water is called the ratio of the sludge backflow (R). The equation is as followed.

$$R = \frac{Q_R}{Q}$$

Another equation according to the content of the sludge in the aeration tank and the ratio of the sludge backflow and the content of the backflow sludge is as followed.

$$R = \frac{X}{X_r - X}$$

In practice, the R is an important index for the running of the waste water treatment plant. For the different waste water quality and the different effluent quality and the different treatment processes, the R is different. For the denitrification and removal of P from the waste water, the R is different. In addition, the R is regulated according to the real practice.

(4) Hydraulic retention time The hydraulic retention time is the average time in the aeration time for the waste water. It is called aeration time (t). The unit is h. The

equation is as followed.

$$t=\frac{V}{Q}$$

2.2.3 Aeration

For the activated sludge, aeration is necessary. The function of the aeration is to promote the oxygen in the air coming into the mixed liquid in the aeration tank and make the mixed liquid suspension.

(1) The function of aeration The function of the aeration. The one is oxygenation. Providing the enough dissolved oxygen to the bacteria in the activated sludge to meet the demand of the metabolism. The other is disturbing and mixing. Making the mixed liquid suspension and increasing the probability of the oxygen with the waste water.

The main designing parameters are oxygen transfer rate, oxygen transfer efficiency, aeration amount and oxygen demand.

①Oxygen transfer rate. The oxygen transfer rate (R_0) is calculated by the following equation.

$$R_0=\frac{d_c}{d_t}=K_{La(20)}[C_{sm(20)}-C_L]=K_{La(20)}C_{sm(20)}$$

In the equation, R_0—The oxygen transfer rate, kg/h;

C_L—The content of oxygen in the water, mg/L. For the clean water removing oxygen, $C_L=0$;

$C_{sm(20)}$—The content of oxygen in the water under the water temperature of 20℃, mg/L;

$K_{La(20)}$—The oxygen demand at the water temperature of 20℃, mg/L.

The real oxygen transfer rate (R) is calculated by the following equation.

$$R=\alpha K_{La(20)}\times 1.024^{T-20}[\beta\rho C_{sm(T)}-C_L]$$

In the equation, R—The real oxygen transfer rate, kg/h;

α—The ratio of the oxygen transfer rate in the waste water to the oxygen transfer rate in the clean water. In general, the value is 0.5~0.95;

β—The ratio of the saturated oxygen in the waste water to the saturated oxygen in the clean water. The value is 0.90~0.97;

ρ—The real air pressure, Pa;

T—Water temperature, ℃;

C_L—The content of oxygen in the mixed liquid, mg/L. In general, 2 mg/L.

Therefore,

Generally,

$$\frac{R_0}{R} = 1.33 \sim 1.61$$

$$R_0 = \frac{RC_{sm(20)}}{\alpha \times 1.024^{T-20}[\beta \rho C_{sm(T)} - C_L]}$$

② Oxygen transfer efficiency.

$$E_A = \frac{R_0}{O_c} \times 100\%$$

$$O_c = G_s \times 21\% \times 1.43 = 0.3\ G_s$$

In the equation. E_A—Oxygen transfer efficiency, %;

O_c—Aeration amount, kg/h;

21%—The percentage of the oxygen to the air;

1.43—The density of the oxygen, kg/m^3;

G_s—Aeration amount.

③ Aeration amount.

$$G_s = \frac{R_0}{0.3\ E_A} \times 100\%$$

④ Oxygen demand. The aeration amount is equal to the oxygen demand to keep the balance between the aeration rate and the consumption rate of the oxygen by the activated sludge.

(2) The aeration technology. In practice, there are three technologies used for the aeration. The first is blow aeration. The pressure air is sent into the pipes system at the bottom of the aeration tank and then the pressure air releases from the air diffusing equipment. The small are bubbles penetrate the waste water and the oxygen in the air is transformed into the mixed liquid to meet the demand of the bacteria in the activated sludge. Another function of the air bubble is to make the mixed liquid disturbing and mixing. For the aeration, the aeration blower is the key. The second is the mechanical aeration. The aeration is practiced by the impeller up or down the water table. The water is disturbed by the impeller and to renew the water table, resulting in the air

penetration. The third is the combination of blow aeration and mechanical aeration.

(3) The demand on the oxygen. For the designment and running of the waste water treatment plant, the aeration amount is very important. The demand rate on the oxygen by the bacteria in the aeration tank goes to the supply rate on the oxygen by the aeration, the demand on the oxygen in the aeration tank should be equate the oxygen provided by the aeration.

In the aeration tank, the need on the oxygen consists two parts. The first is the oxygenolysis on the organic matter by the activated sludge. The second is the self-metabolism of the activated sludge. Therefore, the oxygen amount is calculated by the equation.

$$\Delta O_2 = aQS_r + bVX$$

In the equation, ΔO_2—The demand on the oxygen in the mixed liquid, kgO_2/d;

a—The demand rate on the oxygen in the mixed liquid. The kg O_2 for the metabolism of per kgBOD by the bacteria in the activated sludge. In general, the value is 0.42~0.53;

b—The demand rate on the oxygen used for the self-metabolism of the activated sludge. The self-metabolism rate of the activated sludge, and the unit is d^{-1}. In general, the value is 0.10~0.20.

The above equation can be rewritten the following equations.

$$\frac{\Delta O_2}{XV} = a\frac{QS_r}{XV} + b$$

Or

$$\frac{\Delta O_2}{QS_r} = a + \frac{XV}{QS_r}b$$

In the equation, $\frac{\Delta O_2}{XV}$—The demand on the oxygen for unit weight activated sludge. The unit is $kgO_2/(kgMLVSS \cdot d)$;

$\frac{\Delta O_2}{QS_r}$—The demand on the oxygen for unit organic matter removal. The unit is $kgO_2/(kgBOD \cdot d)$.

In practice, the demand on the oxygen for unit organic matter removal (per kg BOD) is lower under the higher specific degradation rate of BOD and the shorter sludge age. The main reason is that a part of the organic matter is adsorbed onto the surface of the activated sludge and drained outside the aeration tank with the residual sludge. In addition, under the higher specific degradation rate of BOD, the self-metabolism is lower, and the demand on the oxygen is lower. On the other hand, under the lower specific degradation rate of BOD and longer sludge age, the degradation on the organic matter by the bacteria is complete and the self-metabolism is strong. Therefore, the demand on the oxygen for unit organic matter removal (per kgBOD) is higher.

2.3 The development and application of the activated sludge

Since the activated sludge process is used widely, lots of modifications of the treatment objectives.

2.3.1 Complete-mix activated sludge process

A completely mixed system can allow a more uniform aeration of the waste water in the aeration tank (Figure 2-5).

After the waste water and the backflow sludge come into the aeration tank, they are mixed immediately with the mixed liquid in the aeration tank. Therefore, the content of the pollutants and DO is the uniform in the tank.

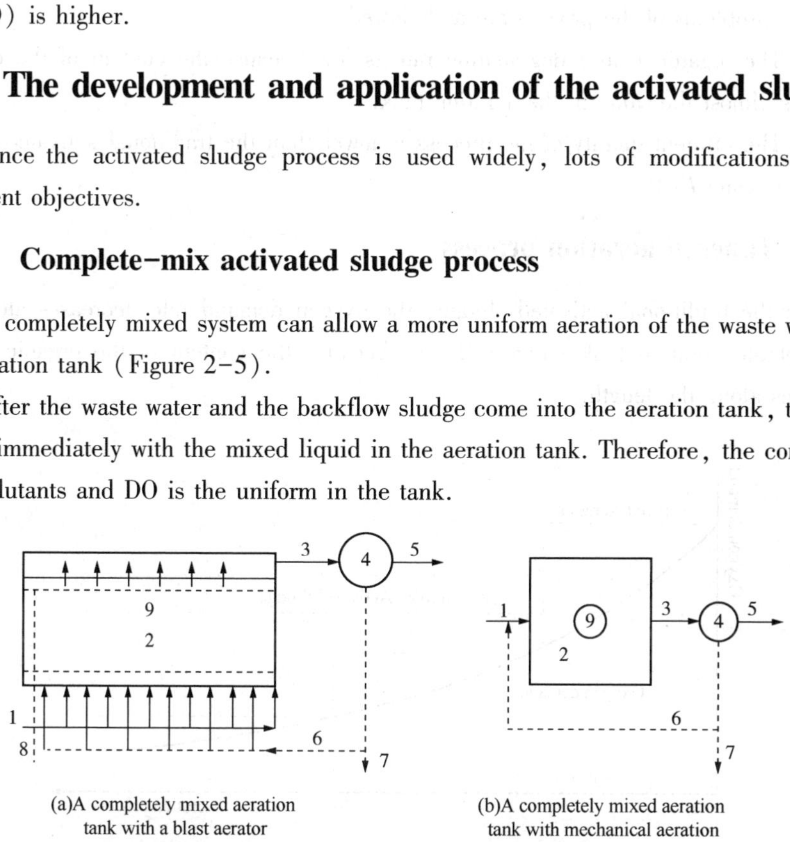

1—Pretreated sewage
2—A completely mixed aeration tank
3—Mixed liquid
4—Secondary clarifier
5—Effluent
6—Reflux sludge system
7—Excess sludge
8—Gas supply system
9—Aeration system and air diffuser device

Figure 2-5 Completely mixed activated sludge process

The mechanism of the process is as followed. When the waste water come into the aeration tank, it is immediately diluted and homogenized with the mixed liquid. It is showing that the effect of the water quality and quantity on the activated sludge is very little. Therefore, the adaptive capacity of process is very strong and it is suitable for treatment of industrial waste water, especially the higher content of organic matter. The waste water is distributed uniformly in the aeration tank so that the component and the amount of the bacteria is almost the same in the aeration tank. It is indicating that the degradation on the organic matter is the same. In practice, the regulation of F/M is enough to control the aeration tank in good status.

The problems of the process are as followed.

① The organic matter degradation rate is low because the content of the organic matter is almost the same in the aeration tank.

② The effluent quality of the process is lower than the traditional activated sludge under the same F/M.

2.3.2 Tapered aeration process

For the traditional activated sludge, the oxygen demand rate decreases along the length of the aeration tank (Figure 2-6) because the content of the organic matter decreases along the length.

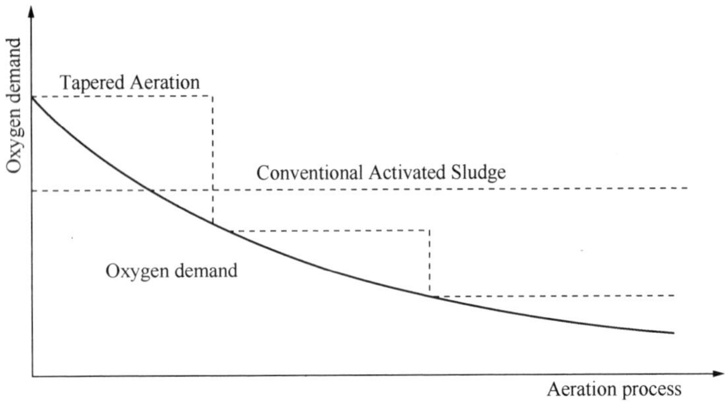

Figure 2-6 The comparison between the conventional activated sludge and the tapered aeration

In this modified system, influent waste water is introduced at several points along the aeration tank length. This leads to a relatively homogenous load distribution along the length of the aeration tank resulting in a more efficient use of dissolved oxygen. In practice, the aeration tank is divided into three parts. The aeration amount in the three parts to the total aeration amount is 50%, 35% and 15%, respectively.

2.3.3 Contact-stabilization activated sludge

The key of the process is that the degradation of the organic matter by the activated sludge is divided into two parts - adsorption and stabilization and is finished in the different reactors (Figure 2-7).

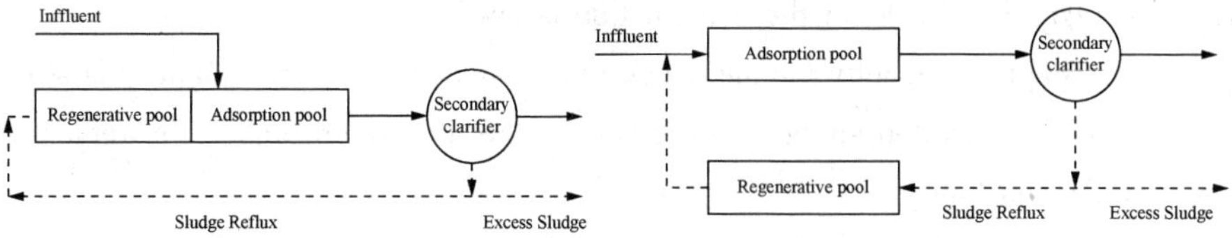

Figure 2-7 Contact-stabilization activated sludge process.

The former is combined system and the latter is separated system.

The process originated from the experiment of Smith. Smith aerated the waste water with soluble organic matter and insoluble organic matter and the activated sludge. He found the relationship of BOD_5 and aeration time (Figure 2-8).

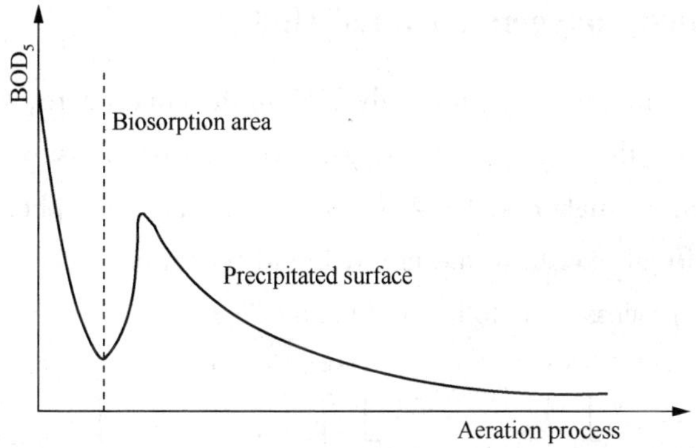

Figure 2-8 The change of BOD_5 after aeration for the waste water and activated sludge

According to the experiment phenomenon, the contact - stabilization activated sludge process is built (Figure 2-7). the key of the process is as followed. The waste water and the activated sludge with high activity after the stabilization come into the adsorption tank. The influent contacts with the activated sludge for a short period of time (30 to 60 min). After the adsorption, most suspended, colloid and part of the soluble organic matter is disappeared. Therefore, the content of organic matter decreases obviously. And then the mixed liquid goes to the secondary sedimentation tank for the separation of solid and liquid. The clarified water is drained and the sludge goes to the stabilization tank. After the degradation and anabolism of the organic matter, the activated sludge comes into the endogenous respiration phase. The aim is to restore the activity of the activated sludge.

The characteristics of the process are as followed.

① The contacting time is short, so the volume of the adsorption tank is small. The sum volume of the adsorption tank and the stabilization tank is less than the traditional activated sludge. The invest of the construction is low.

② The adaptive capacity for the shock of the water quality and quantity is strong.

③ Once the activated sludge is destroyed in the adsorption tank, it is replenished by the stabilization tank.

The problems of the process are as followed.

① The treatment effect is less than the traditional activated sludge.

② It is not suitable for the waste water with high content of organic matter.

③ The residual sludge is big.

④ No nitrification.

2.3.4 High-purity oxygen activated sludge

It is well known that the oxygen is only 21% in the atmosphere, while the oxygen is up to 90% ~ 95% for the high-purity oxygen. The partial pressure of oxygen for the high-purity oxygen is higher 4.4 ~ 4.7 times than the atmosphere. Therefore, the transmission capacity of oxygen to the mixed liquid increases obviously using the high-purity oxygen. The process is as followed (Figure 2-9).

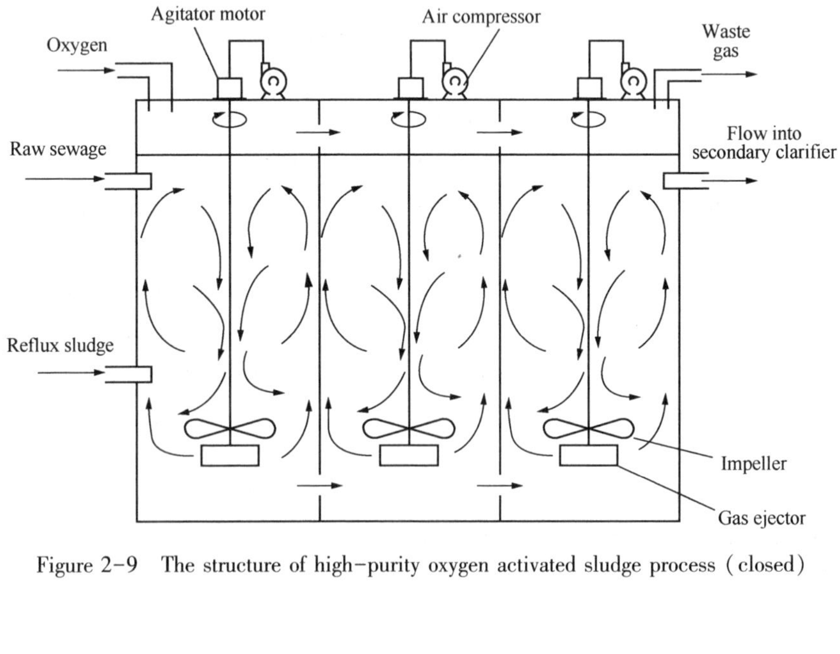

Figure 2-9 The structure of high-purity oxygen activated sludge process (closed)

The advantages of the process are as followed. The efficiency of oxygen is up to 80% ~ 90%, while the efficiency is only 10% for the blast aeration. The content of MLSS is up to 4,000 ~ 7,000 mg/L in the aeration tank, indicating that the volume load increased marked. SVI is less than 100, indicating that the possibility of sludge bulking is very small. The residual sludge amount is very small.

The closed aeration tank is used widely in the high - purity oxygen activated sludge (Figure 2-9). There are lots of small rooms in the tank and the rooms are in series. The flowing in the room is completely mixing. The pressure in the room is lightly higher than the outside of the room to prevent from the air penetration. At the same time, the waste air is discharged outside of the room. The modification is showed in the Figure 2-10. Its advantage is the low construction investment.

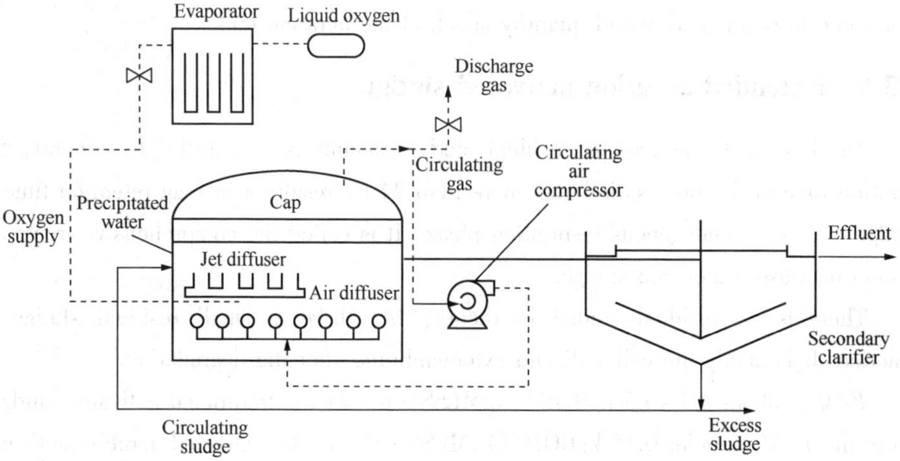

Figure 2-10 The structure of high-purity oxygen activated sludge process (modified)

2.3.5 Step-feed activated sludge

The step-feed activated sludge process is showed in the Figure 2-11.

The process is proposed according to the problems in the traditional activated sludge. The characteristics are as followed. The waste water was introduced into the aeration tank along the length of the aeration tank. The organic matter load and the oxygen demand is balanced because the difference between the organic matter load and the oxygen demand decreases. Therefore, the typical problem in the traditional activated sludge is conquered. The advantage is to decrease the energy consumption and to exhibit sufficiently the degradation of organic matter by the activated sludge. Increasing the

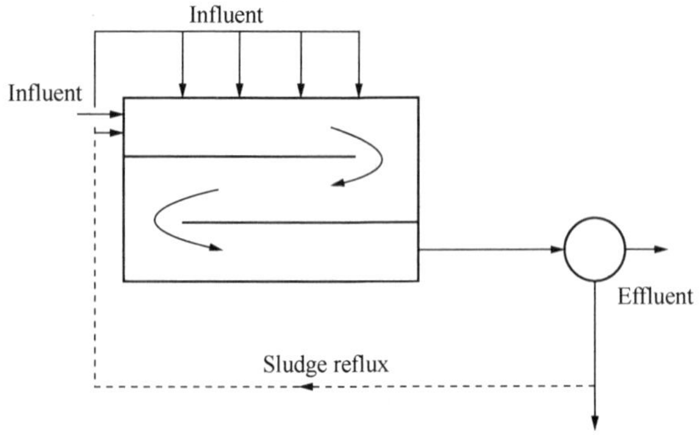

Figure 2-11 Step-feed activated sludge process

resistance to water quality and quantity shock of the aeration tank.

2.3.6 Extended aeration activated sludge

The key of the process is residual sludge amount is very little. In general, the aeration time of the process is long, more than 24 h, resulting in long retention time of sludge and be in endogenous respiration phase. It is called the comprehensive treatment system on waste water and sludge.

There is no residual sludge in theory, but there is small residual sludge in practice. It is manly the cell wall and cytomembrane after the degradation.

F/M is about 0.2~0.5 kgBOD/(kgMLSS · d) for the traditional activated sludge, while the *F/M* is under 0.15 kgBOD/(kgMLSS · d) for the extended aeration activated sludge. It is the ultralow load of organic matter.

The process is suitable for the waste water plants with the high discharge standard and the small waste water amount. The amount is less than 1,000 m^3/d.

2.3.7 High-rate activated sludge

The key of the process is *F/M* ratio is high and aeration time is short. Therefore, the effluent quality is bad and the removal of BOD is less than 70%~75%. It is called non-completely treated activated sludge. On the contrary, the removal of BOD is more than 90%, and the BOD_5 in the effluent is under 20 mg/L. The process is called completely treated activated sludge.

The advantage of the high-rate activated sludge is the removal of organic matter is quick, but the disadvantage of the process is the residual sludge is big. In practice, the

F/M is over 0.5 kgBOD/(kgMLSS · d). The F/M is over 1.5 kgBOD/(kgMLSS · d), so the process is called modified aeration process.

2.3.8 Selector activated sludge

To avoid of sludge bulking, selector activated sludge is developed. The designment is that there is a small reactor before the aeration tank (Figure 2-12 and Figure 2-13). The whole waste water and the backflow sludge come into the selector, resulting in a higher load area. The higher content of organic matter is favor of the growth of zoogloea and inhibiting the explosive growth of filamentous bacteria, resulting in the good settling of the sludge.

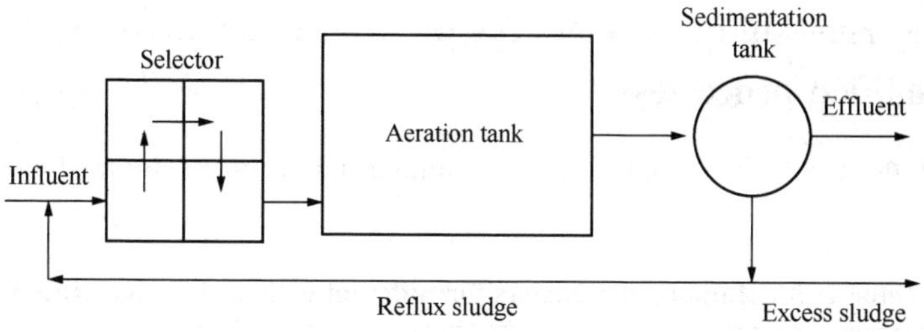

Figure 2-12 The flow diagram of selector activated sludge

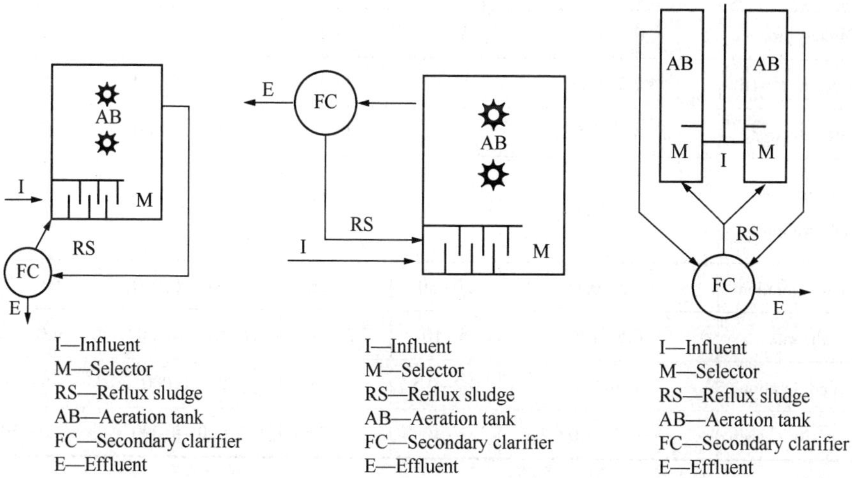

Figure 2-13 The layout of selector activated sludge

The selector is divided into aerobic selector, anaerobic selector and anoxic selector. For the aerobic selector, aeration is necessary to make the mixed liquid in the selector aerobic. The mechanism on the inhibiting the explosive growth of filamentous bacteria is to provide the suitable content of DO and enough substrates, and permit the zoogloea to uptake organic matter firstly. Therefore, the explosive growth of filamentous bacteria is killed. For the anaerobic selector, most of the filamentous bacteria are absolutely aerobic, while under the absolutely anaerobic conditions, the growth will be inhibited. On the contrary, most of the zoogloea are facultative, so under the absolutely anaerobic conditions they can conduct the breeding by the anaerobic metabolism. For the anoxic selector, most of the zoogloea can uptake the nitrate as the electron receiver and conduct the breeding. On the other hand, the filamentous bacteria have not the function. Therefore, its breeding speed is less than that of the zoogloea.

2.3.9 The running parameters for the traditional activated sludge and modified processes

Based on the above analysis, the running parameters are as followed in the Table 2-3.

Table 2-3 Running parameters for different activated sludge process

Item / Process	$F/M/$[kgBOD/(kgMLSS · d)]	SRT/d	Ta/h	MLSS/(mg/L)	$R/\%$
Traditional process	0.2~0.5	5~15	4~8	1,500~3,000	25~75
Completely Mixed	0.2~0.5	5~15	3~5	2,500~4,000	25~100
Step-feed	0.2~0.5	5~15	3~5	2,000~3,500	25~75
Adsorption-stabilization	0.2~0.5	5~15	—	—	50~150
Extended aeration	0.05~0.15	20~30	18~36	3,000~6,000	50~150
High-rate	0.5~1.5	5~10	2~4	4,000~10,000	100~500
Modified aeration	1.5~5.0	0.2~0.5	1.5~3	200~1,000	5~50
High-purity oxygen	0.25~1.0	3~10	1~3	2,000~5,000	25~50

In the adsorption-stabilization activated sludge, for the adsorption tank, the Ta is about 0.5~1.0 h and MLSS is 4,000~10,000 mg/L, while for the stabilization tank, the Ta is 1.5~3 h and MLSS is 1,000~3,000 mg/L.

2.4 The cultivation and running management of the activated sludge

2.4.1 The cultivation of the activated sludge

The bacteria are the key in the activated sludge. After the construction of the waste water plant is finished, the cultivation of the activated sludge is the most important. Before the practice of the waste water treatment, the content of the activated sludge should be at the special level in the aeration tank. The steps of the cultivation are as followed.

① Before aeration, the aeration tank is filled with the domestic waste water. Aeration time is 22 h without influent and then settling time 2 h.

② After 2~3 days, stopping the aeration and settling 1 h. The part of the overlying water is drained and then the fresh waste water is filled into the aeration tank. Each day, there are 3 steps including aeration, settling and decanter. It is noticeably that the influent into the aeration tank increased and the aeration time decreased, compared with the before.

③ In practice, the index (MLSS, SV) should be measured. During the cultivation, if the pH, COD, NH_3-N change little, indicating that the bacteria is little by the cultivation.

④ After 14 days, the color of the sludge is lightly black, and the interface between the overlying water and sludge has clear margin. Under the microscope, there are the *paramecium*, *amphieptus Sp*, *litonotus SP*, *suctorial*, et al in the activated sludge. It is indicating that the biofacies is better and better. By the measurement of COD and NH_3-N, if the removal is not high, the cultivation keeps going on.

⑤ From now, two periods are needed for the cultivation. A period is aeration time with 10 h and settling time with 2 h.

⑥ The flocculation and settling property is good 30 days later. After 30 min settling, the overlying water is clean and the interface of water and sludge is clear. The color of the sludge is yellow and brown. Under the microscope, there are lots of close-grained zoogloea. The active *vorticella* is observed. By the measurement of COD and NH_3-N, if the removal is over 90% and 30%, respectively, the cultivation is end.

2.4.2 The running of the activated sludge

(1) Pilot run After the cultivation of the activated sludge, the pilot run is

necessary. The aim is to be sure the optimal running conditions. For example, the MLSS, aeration amount, waste water injection, and so on.

(2) Normal operation After the pilot run, the normal operation starts. During the normal operation process, the measures are needed to keep the higher and stable content of activated sludge, and to meet the demand of the effluent quality. The widely used measures are regulation of aeration amount, regulation of backflow sludge and drainage of residual sludge.

① Regulation of aeration amount. The consumption on power for aeration is the main part (50% ~ 60%) for the whole power consumption in waste water treatment plant. The measures are divided into four parts, such as constant aeration amount, the control according to the influent amount, the content of DO, the optional aeration amount.

The content of DO in the exit of the aeration tank should be controlled at the range of 1.5 ~ 2 mg/L, although in summer. Another request is to meet the demand on the mixing to the mixed liquid. The disturbance degree is that the content of the activated sludge is uniform at the surface, in the middle, at the bottom of the aeration tank.

If the control is constant aeration amount, the aeration amount should be regulated in the morning and evening. For the big amount waste water treatment plant, the constant aeration amount should be regulated each week according to the content of DO in the aeration tank.

② Regulation of backflow sludge. The measure is to keep the constant content of activated sludge (MLSS) in the aeration tank.

The methods for the control include constant backflow sludge, the control according to the influent amount (the constant R), the constant MLSS, the constant F/M.

③ Drainage of residual sludge. The measure is to keep the constant MLSS in the aeration tank. The activated sludge grows and the MLSS increases and the SVI increases, resulting in the value of MLSS up. Therefore, drainage of the residual sludge is necessary. Generally, the drainage of the residual sludge is similar as the multiplication of the activated sludge. But too big or too small drainage makes the MLSS changing.

(3) The measurement on the efficiency of the running The regular measurement on the process of the aeration tank and the second sedimentation tank is necessary to keep the good efficiency and accumulate the experience and to forecast the running. The measurement items are as followed.

① About the efficiency. The total BOD, COD and the dissolved BOD, COD in the influent and the effluent. The total and volatile SS in the influent and the effluent. The toxic matter in the influent and the effluent.

② About the sludge. SV, MLSS, MLVSS, SVI, microscopic examination of the bacteria in the activated sludge.

③ About the nutrient and the environmental conditions. N, P, pH, DO, water temperature.

In practice, SV and DO should be measured one time per 2 ~ 4 h to immediately regulate the backflow sludge and aeration amount, at least one time for each shift. The microscopic examination of the bacteria should be measured one time per shift to find the abnormity of the activated sludge. The N, P, MLSS, MLVSS and SVI should be measured at regular. The other items should be one time per day. The water sample should be the mixed liquid. The DO, pH, redox potential can be measured by the on-line monitoring system. Nowadays, the NH_3-N, total N, total P, suspended solid, turbidity, the inflow and water discharge amount can be measured by the on-line monitoring system.

In addition, the inflow, water discharge amount, the backflow sludge, the residual sludge should be recorded. The trend of the residual sludge, the working status of the aeration equipment, the aeration amount and the consumption of power are recorded. The content of the residual sludge should be measured regularly.

2.5 The abnormal phenomenon during the running of the activated sludge

During the running of the activated sludge, there are some abnormal phenomena, resulting in the efficiency decrease and sludge runoff.

2.5.1 Sludge bulking

The sludge bulking is the biggest problem in the running. During the bulking, the settling ability of the sludge is down, and the normal solid-liquid separation in the secondary sedimentation tank disappeared. It is indicating that the supernatant liquid is reduced and the sludge goes away easily. The content of the suspended solid in the effluent is exceeding and the microorganism is reduced in the aeration tank. When the SVI of the activated sludge is at about 100, the sedimentation ability is the best. When the SVI is over 150, it is suggesting that the activated sludge is bulking and the necessary measures are in practice.

The sludge bulking is divided into two kinds, such as filamentous bacteria bulking and non – filamentous bacteria bulking. For the former, the main reason is that the excessive breeding of the filamentous bacteria in the activated sludge. As for the later, the zoogloea absorbed much polysaccharides, resulting in the density of the activated sludge down. In real waste water treatment plants, the activated sludge bulking is attributed to the filamentous bacteria bulking. In addition, the industrial waste water is more liable to be bulking, compared to the domestic waste water, while the mixed activated sludge is more liable to be bulking, compared to the plug flowing.

The experience in many waste water treatment plants show that the activated sludge bulking is appeared under these conditions.

① The content of the carbohydrate or the soluble organic matter is high.

② Corrupt or early digested waste water, the waste water with high content of the sulfuretted hydrogen.

③ The content of N and P is not balanced.

④ The waste water with toxic matters.

⑤ The waste water with high pH or low pH.

⑥ Low dissolved oxygen in the mixed liquid.

⑦ The waste water being short of microelement.

⑧ Shocking load for the mixed liquid.

⑨ The sludge age is too long or the organic load is too low.

⑩ Higher organic load under anoxic conditions.

⑪ The temperature is excessively high or low.

The measures are used as following, temporary measures, regulating measures according to the process running, environmental regulations.

(1) Temporary measures The measures include settlingassistance and killing bacteria. The former is that adding coagulant into the sludge to increase the density when it is bulking. It is favor of the separation of solid and liquid in the second sedimentation tank. The widely used coagulants are Polymerization Ferric Chloride (PFC), ferric sulfate, aluminum sulfate, polyacrylamide. The clay or the kieselguhr can be used in small waste water treatment plant. But the dose of the coagulant is not too big, the activity of the microorganism will be inhibited.

The latter is that adding the chemical matters into the sludge and killing the filamentous bacteria. The wildly used bactericide is NaClO, ClO_2, Cl_2, H_2O_2 and bleaching powder. Generally, the Cl_2 is used wildly as the bactericide to control the bulking, since the Cl_2 is used to kill the bacteria in the effluent. However, the Cl_2 is not only to kill filamentous bacteria but also to kill the other bacteria. Therefore, the content of Cl_2 should be controlled strictly at the adding point.

In a word, settling assistance is often used to control non-filamentous bacteria, while the bactericide is often used to control filamentous bacteria.

(2) Regulating measures The measures are used when the running is improper. For example, increasing DO into the waste water if the bulking is shortage of DO. Increasing pH or controlling the drainage of the industrial waste water in upstream if pH is too low. Increasing DO into the waste water if the reason is the corrupt waste water. Increasing nutrient (N and P) if the waste water is shortage of nutrient. Increasing F/M due to the low organic load.

(3) Environmental regulations The mechanism is to change the ecology in the aeration tank to create the environment favor of the bacteria. Therefore, it is to increase the competitiveness of the bacteria to inhibit the excessive breeding and growth of the filamentous bacteria.

2.5.2 Sludge disintegration

The effluent quality is cloudy and the sludge is fine in the activated sludge system, resulting in the effluent quality worsen. We called sludge disintegration.

When the running is improper or the toxic matter is mixed into the waste water, they are easy to cause sludge disintegration. For example, the aeration is excessive, the balance of the nutrient for the bacteria is destroyed. The results are that the content of the bacteria is reduced and its activity decreased. The adsorption ability for pollutants is down and the flocculant is fine. It is showing that the sludge is formed into the feather sludge, and the effluent quality is cloudy and the SVI is down. When there is toxic matter in the waste water, the bacteria are inhibited and damaged. The activated sludge disintegrates due to the shortage of activity, and then the clarification ability is disappeared, and even stop.

Once the sludge disintegration appeared, the amount of waste water, the back flow of sludge, the aeration, the drainage of sludge, SV, MLSS, DO, the sludge load should be checked to look for the reason and regulate it. The toxic matter should be removed if the reason is the toxic matter in the waste water.

2.5.3 Sludge corruption

The sludge stayed in the second sedimentation tank too long and corrupted to produce the H_2S or CH_4, resulting in the sludge floating. This is called sludge corruption. The corrupted sludge is general black and stench.

The solving measures can increase the slope of the second sedimentation tank or improve the drainage equipment, to strengthen the drainage of sludge.

2.5.4 Sludge floating

The denitrification appeared at the bottom of the second sedimentation tank, so that the nitrogen was adsorbed on the sludge, resulting in the density of the sludge down and the sludge floating. This is called sludge floating.

The control measures are as followed. It is to increase the sludge backflow and timely drainage of the residual sludge to avoid the denitrification. It is to decrease the content of the mixed liquid sludge, to reduce the sludge age and to decrease the content of DO, avoiding of the nitrification. It is to strengthen the denitrification.

In addition, the aeration is excessive in aeration tank, and then the mix in the sludge is too strong. Lots of fine air bubbles were adsorbed onto the flocculants. Another reason is the influence of lots of fat and oil, resulting in the floating of the sludge.

2.5.5 Bubbles

The bubbles are the common phenomenon in activated sludge treatment plant. The bubbles can be divided into two kinds, such as chemical bubbles and biological bubbles. The former is milk white. The latter is brown. The bubbles are flowing with the effluent to decrease the water quality. In winter, the bubbles are icy and it is difficult to remove. In summer, the bubbles can bring the stench. The biological bubbles decrease the content of dissolved oxygen due to the surface aerator. The biological bubbles come to the sludge area, worsen the running of the thicken tank.

Chemical bubbles generally come from the detergent in the waste water and the surfactant in the industrial waste water. The biological bubbles mainly come from Nocard's bacillus.

For the control on the chemical bubbles, spraying or adding antifoam agent is wildly used. For the control on the biological bubbles, it is difficult to disappear, although there are many plants using the methods, such as adding Cl_2, increasing drainage of sludge, decreasing SRT. The problem is not radically solved. Therefore, the prevention is key for the biological bubbles.

2.5.6 Abnormal biofacies

When the running is improper or the influent water quality changes sharply, it is resulting in the abnormal biofacies. The microfauna in the activated sludge are the vorticella when the running is proper. The quantity and the activity of the vorticella is favor of the forecast of the status and the development of the activated sludge. The suitable DO is 1 ~ 3 mg/L, while the DO is too high or too low, the vorticella are abnormal. Therefore, it is necessary to regulate the DO. When the content of DO is nearly zero, the vorticella die.

When there are lots of non-biodegradable organic matters or toxic matters, some particles that are not digested accumulate in the body of the vorticella. It is necessary to measure the specific oxygen uptake rate (SOUR).

The values of pH change, and they are not in the range of 6 ~ 9. The vorticella is not active, and the cilium stops to swing. It is necessary to measure the pH in the influent.

2.5.7 Insufficient nitrification

This is referred to the transformation from ammonia nitrogen to nitrate decrease, resulting in the ammonia nitrogen removal down.

The main reason maybe as followed.

① The sludge age is shorter than the generation of nitrobacterium, resulting in the nitrobacterium does not stay in the aeration tank and the content of the nitrobacterium is not enough to remove ammonia nitrogen.

② The retention time in the aerobic stage is short, resulting in the insufficient time for nitrification.

③ the insufficient alkalinity and DO.

In practice, the measures can be used to regulate the insufficient nitrification.

① The sludge age is more than 10 days in the nitrification reactor.

② Hydraulic retention time in the nitrification reactor is enough to make the nitrification completely.

③ adding sodium bicarbonate to replenish the insufficient alkalinity.

④ Increasing the aeration amount or improving the aeration efficiency to increase the content of DO in the nitrification reactor. The aim is to increase the activity of the nitrobacterium and improve the nitrification efficiency.

2.5.8 Insufficient denitrification

Insufficient denitrification is referred to the transformation from nitrate to nitrogen decrease, resulting in the total nitrogen removal down.

The main reason maybe as followed.

① Insufficient organic matter in the anaerobic stage, resulting in the insufficient electron donor, handling the reduction of the nitrate. Therefore, the removal of nitrate is low and the denitrification is not complete.

② The content of DO in the anaerobic stage is higher, resulting in the destruction of the anaerobic environment and decreasing the activity of the denitrifying bacteria.

In practice, the measures can be used to regulate the insufficient nitrification.

① When the content of the organic matter is not enough, the measures can be as followed. adding the external carbon source (such as methyl alcohol). Sufficient use on the organic matter in the raw water (step-feeding process). Uses on the organic carbon source from the primary sludge and residual sludge.

② Decreasing the content of DO in the effluent from aerobic stage.

2.6 The designment of the activated sludge

The designment of the activated sludge includes the aeration tank, the secondary settling tank and the backflow of sludge.

2.6.1 Aeration tank

(1) The volume of aeration tank　Generally, the volume of the aeration tank is calculated by the organic matter load. The load includes BOD load (N_s) and volume load (N_v). The content of the activated sludge (MLSS) depends on the experience, in general, 3,000~6,000 mg/L. Therefore, the numbers of the aeration tank, the depth, the length and the width are sure by the calculation on the aeration tank.

(2) The aeration amount and the demand of the oxygen　The demand of the oxygen (ΔO_2) to remove 1 kgBOD$_5$ by the activated sludge is calculated according to the following equation.

$$\Delta O_2 = a' + \frac{b'}{N_s}$$

In the equation, a'—The demand of the oxygen rate for the degradation of the organic matter by the bacteria. That is the demand of the oxygen to remove 1 kgBOD$_5$, kg;

b'—The demand on the oxygen used for the self-metabolism of the activated sludge, kg/d;

N_s—The BOD load, kgBOD/(kgMLSS · d).

After the demand on the oxygen and thesecurity coefficient are sure, the real demand on the oxygen is transferred to the standard demand on oxygen.

$$R_0 = \frac{RC_{sm(20)}}{\alpha \times 1.024^{T-20}[\beta \rho C_{sm(T)} - C_L]}$$

In the equation, $C_{sm(20)}$—The saturated content of oxygen in the water under the water temperature of 20℃ and the barometric pressure of 1.013×10^5 Pa, mg/L;

$C_{sm(T)}$—The mean of the saturated content of oxygen in aeration tank at temperature of T, mg/L;

α—The ratio of the oxygen transfer rate in the waste water to the oxygen transfer rate in the clean water. In general, the value is 0.5~0.95;

β—The ratio of the saturated oxygen in the waste water to the saturated oxygen in the clean water. The value is 0.90~0.97;

ρ—The real air pressure, Pa;

T—Water temperature, ℃;

C_L—The content of oxygen in the mixed liquid, mg/L. In general, 2 mg/L.

According to the aerator, the aeration amount is calculated by the following equation.

$$G_s = \frac{R_0}{0.3E_A} \times 100\%$$

In the equation, G_s—The air amount;

E_A—The aeration efficiency.

2.6.2 The secondary settling tank

(1) The area of the secondary settling tank The area of the secondary settling

tank is calculated by the equation.

$$A=\frac{Q_{max}}{q}=\frac{Q_{max}}{3.6v}$$

In the equation, A—The area of the settling tank, m^2;

Q_{max}—The biggest inflowing of the waste water, m^3/h;

q—Hydraulic surface loading, $m^3/(m^2 \cdot h)$;

v—The settling speed during stratified settling, mm/s.

(2) The volume of the sludge bucket The function of the sludge bucket is to store and thicken the residual sludge. The volume of the sludge bucket is too big, and it is easy to corrupt for the sludge. In general, the retention time of the sludge in the bucket is 2 h.

$$V_s=\frac{4(1+R)QX}{24(X+X_r)}=\frac{(1+R)QX}{6(X+X_r)}$$

In the equation, V_s—The volume of the sludge bucket, m^3;

Q—The inflowing of the waste water, m^3/h;

X—The content of the mixed liquid, MLSS, mg/L;

X_r—The content of the backflow sludge, mg/L;

R—The backflow ratio of the sludge.

(3) The discharge of the residual sludge The residual sludge is discharged in the secondary settling tank. The discharge amount is equal to the growth of the activated sludge (ΔX).

$$Y_{obs}=\frac{Y}{1+K_d\theta_c}$$

$$\Delta X=Y_{obs}Q(S_0-S_e)$$

In the equation, Y_{obs}—The real sludge yield, kgMLVSS/kgBOD_5;

K_d—The self-metabolism of the activated sludge, d^{-1};

θ_c—The average retention time, SRT, d;

S_0—The content of BOD_5 in the influent, mg/L;

S_e—The content of BOD_5 in the effluent, mg/L.

2.6.3 The design on the backflow of the sludge

$$R=\frac{X}{X_r-X}$$

In practice, the R is an important index for the running of the waste water treatment plant. The R depends on the content of the mixed liquid and the content of the backflow sludge. The content of the backflow sludge is closely related to the SVI.

Questions

1. Taking 1 L mixed liquid from the aeration tank in a waste water treatment plant, adding it into a cylinder, the scale on the cylinder of the surface of the sludge is 500 mL after 30 min. Please tell us the SV. If the MLSS in the aeration tank is 4, 000 mg/L, what is the SVI? Based on the value of SV and SVI, please evaluate the settling property of the sludge.

If the MLSS in the aeration tank is reducing to 2, 000 mg/L, what happened.

2. For the designment of a waste water treatment plant, the influent amount of the waste water is 10, 000 m^3/d, the BOD is 220 mg/L in the influence.

If the N_s is 0.2 kgBOD/(kgMLSS · d), please give the volume of the aeration tank under the conditions of MLSS with 4, 000 mg/L.

If you selected high rate activated sludge, how to get the volume of the aeration tank under the same MLSS with 4, 000 mg/L.

If we increase the MLSS, what happened, please give us the reason.

3. In a waste water treatment plant, the volume of the aeration tank is 10, 000 m^3, and SVI is 100 mL/g, and R is 50%, and r is 1.2, and V_c is 8 days, please calculate the MLSS and the drainage amount of the residual sludge each day.

If V_c is increased, please make a forecast about the change of MLSS and the drainage amount of the residual sludge each day.

4. In a waste water treatment plant, the influence amount of waste water is 20, 000 m^3/d, the volume of the aeration tank is 11, 000 m^3, the MLSS is 3, 000 mg/L, the BOD of the influence is 180 mg/L, the BOD of the effluence is 10 mg/L, Give the need on the oxygen.

5. What difference between tapered aeration process and step - feed activated sludge.

6. In your opinion, what relationship between the N_s and the volume of aeration tank.

7. What effect of temperature on the growth of activated sludge?

8. The characteristics of the activated sludge?

9. What is sludge bulking? Once meeting the abnormal phenomenon, What measures can be chosen to solve the problem?

10. What are the differences between sludge disintegration, sludge corruption and sludge floating?

11. Why the activated sludge with selector can avoid of sludge bulking?

12. What is the function of sludge age affecting N removal?

13. In your opinion, the notice for the cultivation of activated sludge.

14. The advantages and disadvantages for SBR.

15. If AAO is selected as the process for the waste water treatment plant, please tell us the range of DO in different unit.

16. Mechanism of N and P removal in activated sludge.

Chapter 3 The biofilm

> **Emphasis:** The basic mechanism on waste water clarification by biofilm. The designment of biofilm process. The types and structure of biofilm reactor. The running and management of biofilm.
>
> **Nut:** The designment of biofilm process. The matrix migration. Biological characteristics of biofilm. The application of biofilm process. The difference between biofilm and activated sludge.

The biofilm and the activated sludge are the aerobic treatment process, which remove the organic matter in the water by the bacteria, for the waste water. However, the bacteria in the activated sludge are suspended in the aeration tank by the form of zoogloeal. It is attributed to the suspended growth system. The bacteria in the biofilm grow on the filter or the carrier as the membrane sludge. It is attributed to the attached growth system. The bacteria and the protozoa ,and the metazoa grow and breed on the filter or the carrier, forming the membrane sludge. This is called biofilm. When the waste water contacts with the biofilm, the bacteria can uptake the organic matter in the waste water for the metabolism of the bacteria. The waste water is clarified. In practice, the biofilm is mainly aerobic process. Therefore, in the chapter, only the aerobic biofilm is discussed.

The basic process of the biofilm is showed in the following figure (Figure 3-1). The waste water comes into the biofilm reactor, and then the pollutants are removed by the bacteria. The clean water is discharged after the second sedimentation tank. The primary sedimentation tank is to remove the suspended solids to avoid of the blocking of the biofilm reactor. This is very important for the filter with small pore. The function of the second sedimentation tank is to remove the deciduous biofilm. When the content of the organic matter in the influent is big, the growth of the biofilm is very quick. It is not good for the running of the reactor. It is to reduce the content of the organic matter in the influent by the backflow of the effluent. The mechanism is to flush the biofilm using the backflow to avoid of the accumulation of the biofilm. It is noticeably that the backflow from the second sedimentation tank depends on the content of the organic matter in the influent.

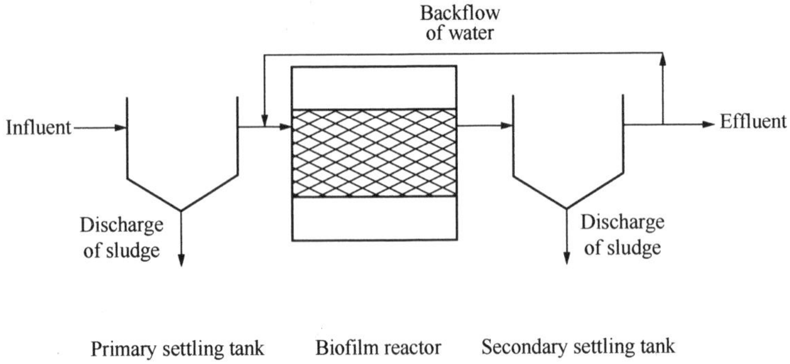

Figure 3-1 The basic process of the biofilm

3.1 The basic concepts

3.1.1 The formation of the biofilm and the clarification

The mechanism of thebiofilm is to degrade the organic matter by the bacteria under the exchange of the matter between the solid phase and the liquid phase when the waste water contacts with the biofilm. The waste water is clarified, and the bacteria quantity increases by breeding. The key of the methods of the biofilm is to form the excellent biofilm.

(1) The structure and the clarification mechanism The structure is showed in the Figure 3-2.

The biofilm is attached on the surface of the filter. The biofilm is the hydrophilic and sticky substance. With the waste water renew, there is an adhesive water layer on the surface of the substance. Therefore, there are lots of bacteria and microfauna on the surface of the biofilm and in the biofilm. The food chain is formed by the pollutants, bacteria and microfauna.

Based on the Figure 3-2, the transfer processes of many materials appear between the biofilm and water layer. The oxygen is dissolved in the flowing water layer and the DO is transferred into the biofilm by the adhesive water layer. The organic matter is transferred into the biofilm by the adhesive water layer. Under the metabolism of the bacteria, the organic matter is degraded. The waste water is clarified. The metabolite,

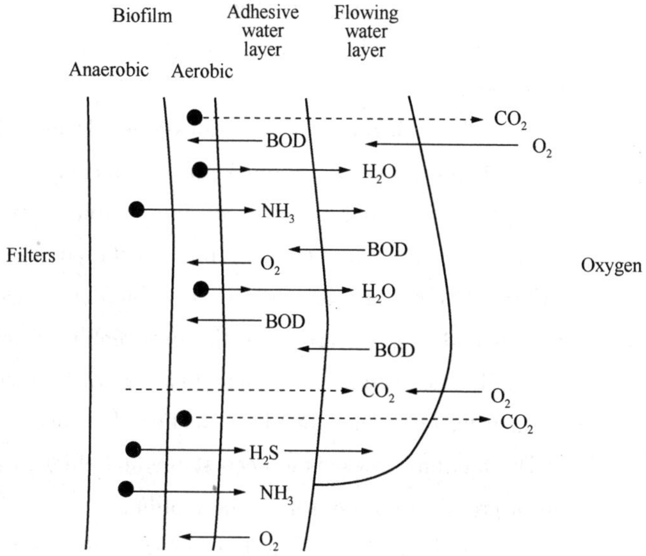

Figure 3-2 The structure of the biofilm

such as H_2O, comes into the flowing water layer by the adhesive water layer and then it is discharged, while CO_2, H_2S, NH_3 and CH_4 go to the atmosphere passing the water layer.

(2) The formation, renew and maturity of the biofilm The formation of the biofilm is the combined action under the physical, chemical and biological processes, according to the research.

① The organic matter is transferred to the surface of the biofilm.

② The irreversible adsorption of the bacteria on the surface of the carrier.

③ The usage and the oxygenolysis on the organic matter by the enough bacteria in the biofilm.

The bacteria on the surface of the carrier breed when there is enough organic matter, microelement and DO. With the growth of the bacteria, the quantity of the bacteria increases and the thickness of the biofilm increases. When the thickness is up to some degree, the DO penetration is impossible, so the inner of the biofilm is under anaerobic conditions. Therefore, the biofilm consists the aerobic layer and anaerobic layer. The aerobic layer of the biofilm consists the aerobic bacteria and facultative bacteria, and the thickness of the layer is about 2 mm. The growth and the breeding of the bacteria are limited due to the insufficient organic matter and DO inside of the biofilm closing to the carrier. The aerobic bacteria do not survive and the metabolism of the facultative bacteria is anaerobic metabolism. The facultative bacteria are transferred into the anaerobic bacteria. Therefore, the biofilm, closing to the carrier, are facultative bacteria and anaerobic bacteria.

With the increase of the thickness of the anaerobic layer, the metabolite increases. Because the metabolite must pass through the aerobic layer, the ecology is destroyed, resulting in the loss of the balance between the aerobic layer and anaerobic layer. In addition, the gaseous metabolite must pass through the aerobic layer, it is to reduce the adhesion of the biofilm on the carrier. The biofilm is called the aged biofilm. The biofilm is aged and dropped and then the new biofilm is formed. But the new biofilm is able to clarify the waste water some time later. In practice, it is to postpone the biofilm age and faster the new biofilm formation. It is necessary to avoid of the whole biofilm drop. The biofilm process is to keep stable and efficient removal on the pollutants by the period of growth-drop-growth of the biofilm.

The sign of the maturity of the biofilm is that the ecology system of the bacteria and different kinds of microorganisms and its degradation on the organic matter is in the balanced and stable state, along the water flowing. In practice, the time from the formation to the maturity of the biofilm is about 20~30 d under the 20℃ conditions for the municipal waste water.

3.1.2 The carrier of the biofilm

The material to provide the adhesive surface for the bacteria is called filter or carrier. The filter is very important because of the development and the characteristics of the biofilm.

3.1.2.1 The types of the filter

(1) Inorganic filter The carbonates, zeolites, ceramics, carbon fiber, slag, activated carbon are called the inorganic filter. The characteristics of the inorganic filer are as followed. ① High mechanical strength. ② Stable chemical characteristic. ③ Huge specific surface area. The main shortage is high density.

(2) Organic filter The organic filter is the main carrier in the biofilm process, including PVC, many kinds of resin, plastic, fiber and gelatin, et al.

Nowadays, the new filters appear continually, such as elastic packing combination packing.

3.1.2.2 The principle for the choose of the filter

The efficiency of the biofilm is related closely to the filters. The principles are as followed.

① Enough mechanical strength.

② Excellent stability, including physical, chemical and biological stability.

③ Good electrification on the surface.

④ No toxicity or inhibition.

⑤ Excellent physical shape.

⑥ Reasonable price.

3.1.3 The characteristics of the biofilm

Generally, the biofilm can replace the traditional activated sludge under some conditions. In addition, there are some special characteristics, such as easy running, little residual sludge, strong resistance on the water shock. The characteristics from the microorganism phase and process phase are as followed.

(1) Microorganism phase:

① Microorganism diversification and the long food chain. The environment conditions are suitable for the growth and breeding of the microorganisms. It is possible for the appearance of the filamentous bacteria, eelworm, rotifera and oligonucleotides caterpillar class. The comparison between the biofilm and the activated sludge is as Table 3-1.

Table 3-1 The comparison of the kinds and quantity between the biofilm and the activated sludge

Kinds of the microorganism	Activated sludge	Biofilm	Kinds of the microorganism	Activated sludge	Biofilm
Bacteria	++++	++++	The other ciliates	++	+++
Fungus	++	+++	Rotifera	+	+++
Algae	—	++	Eelworm	+	++
Flagellate	++	+++	Oligonucleotides Caterpillar class	—	++
Sarcodina	++	+++	Other metazoa	—	+
Peritrich	+++	++++	Insects	—	++
Suctoria	+	+			

Based on the above table, the food chain for the biofilm is longer than that for the traditional activated sludge. The residual sludge is little due to the main reason.

② Partition running and dominant fungi. The retention time for the bio-solid is long. Therefore, the bacteria with long generation time and specific increment rate can survive, such as nitrification bacteria. It is to run he biofilm process at different subsections, so there are special bacteria suitable for the water quality at the subsections, improving the formation of the dominant fungi. This is favor of the metabolism of microorganism and the degradation of the organic matter.

(2) Treatment process phase:

① Strong resistance on the shock load. The water content of thebiofilm is low, the amount of the bacteria in unit volume in the reactor is larger than that of activated sludge. Therefore, the treatment ability is bigger for the biofilm reactor and the clarification increases. The effect of the water quality and quantity on the biofilm is little. There is no fatal effect on the clarification of the biofilm.

② Good settling and no sludge bulking. The density of the dropped sludge from the biofilm is big and the size of the particle of the sludge is big, so the settling of the sludge is excellent. It is easy to be separated. In addition, although there is lots of filamentous bacteria, no sludge bulking appears due to the adherent growth. On the other hand, using the strong degradation on the organic matter by the filamentous bacteria is to improve the efficiency of the waste water treatment.

③ Treatment on the waste water with low content. The biofilm is able to clarify the waste water with low content and the treatment efficiency is good. If the running is proper, the content of the BOD_5 is lower as 20~30 mg/L, and the BOD_5 in the effluent is reduced to 5 ~ 10 mg/L. On the contrary, the traditional activated sludge is not suitable for the treatment on the waste water with low content. If the BOD_5 in the raw water is less than 50~60 mg/L, the formation and breeding of the activated sludge are affected and the clarification decreases and the effluent quality is down.

④ Easy running and management. The quantity of the biofilm is big, and there is no need for the backflow of the sludge. It is necessary to regulate the MLSS and residual sludge. It is indicating that the consumption on the power is low, so the consumption on power for removal unit BOD_5 is lower, compared to the activated sludge.

(3) The disadvantages of the biofilm:

① The biofilm is limited by the mass transfer. Although the biomass is big for the biofilm, the active bacteria is limited and focusing on the surface of the biofilm. In practice, the DO with 2~3 mg/L is enough for the activated sludge, but for the biofilm it is not enough.

② The investment on the construction is bigger than that of the activated sludge.

③ Too high BOD content is not good. In practice, the content of BOD is 30 ~ 50 mg/L. If the BOD is too high, resulting in the fast growth of the biofilm. This causes the blocking of the biofilm system. On the contrary, the activated sludge is favor of the treatment on the BOD with high content.

④ There is strict requirement on the SS in the influent.

⑤ It is difficult to control the biomass for the biofilm, so the running of the biofilm is not flexible.

3.1.4 The biofilm reactor

If the filter or the carrier is introduced into the waste water treatment process and there are the bacteria growth on the surface of the filter or carrier, the process is called biofilm reactor.

At 1893, the waste water was sprayed to the coarse filter for the clarification and the good efficiency was gotten for the removal of pollutants in UK. This is the origin of the biofilter.

During the 1920s and the 1930s, lots of biofilters were built. During the 1940s and the 1950s, the biofilters were replaced by the activated sludge due to the low water load and low BOD load. The main filter or carrier is the broken stones, pebbles, slags and cokes. The disadvantages of the above filter or carrier are the small specific surface area and low pore. At 1960s, lots of organic synthetic fillers appeared, which have the huge specific surface area and well-developed pores. At 1970s, the other biofilm reactors were researched and applied, such as rotating biological contactor, biological aerated filters, submerged biofilm reactor, fluidized-bed bioreactor. Nowadays, more and more research pay attention to the biofilm reactors. Lots of new biofilm reactors come forth, such as membrane biofilm reactor, air-lifts biofilm reactor, moving bed biofilm reactor, hybrid activated sludge-biofilm reactor, sequencing batch biofilm reactor, attached-growth ponds.

3.2 Some important parameters

There are many models and evaluation tools to describe the mass transfer process and the dynamics of substrate usage by the bacteria for the biofilm. But it is difficult to describe the physical parameters and the model coefficient due to the complex of the biofilm, the experience equation was used for the designment of the biofilm reactor.

3.2.1 Specific increase rate of biomass

The specific increase rate of biomass (μ) is used widely to describe the breeding of the biofilm. The parameter reflects the activity of the bacteria. The definition is as following.

$$\mu = \frac{dX/dt}{X}$$

In the equation, X—Bacteria content, [mass][volume]$^{-1}$;

μ—Specific increase rate of biomass, [time]$^{-1}$.

(1) The biggest specific increase rate of biomass (μ_0) The μ_0 is described by the equation.

$$\frac{dM_b}{dt} = \mu_0 M_b$$

After integral, the equation is as followed.

$$\ln M_b = \mu_0 t + C$$

In the equation, M_b—Biofilm biomass, [mass][area]$^{-1}$;

C—Constant.

According to the above equation, the curve graph with t—$\ln M_b$ is gotten and the μ_0 can be calculated by the graphical method.

(2) The average specific increase rate of biomass ($\overline{X}$) The average specific increase rate of biomass is calculated by the following equation.

$$\overline{X} = \frac{\dfrac{M_{bs} - M_{bo}}{t}}{M_{bs}}$$

In the equation, M_{bs}—Biofilm biomass when the biofilm are at the stable state, [mass][area]$^{-1}$;

M_{bo}—The initial biomass, [mass][area]$^{-1}$.

3.2.2 Specific removal rate of the substrate

$$q_{obs} = \frac{Q(S_0 - S)}{A_0 M_b}$$

In the equation, q_{obs}—Specific removal rate of the substrate, [time]$^{-1}$;

Q—The inflowing amount, [volume][time]$^{-1}$;

S_0—The content of the substrate in the influent, [mass]

[volume]$^{-1}$;

S—The content of the substrate in the effluent, [mass] [volume]$^{-1}$;

A_0—The surface of the carrier, [area].

The higher of the specific removal rate of the substrate, the higher of the activity of the biofilm.

3.3 Bio-filters

The bio-filters are the origin of the biofilm reactors, and the history has been more than one hundred years. Since 1893 the waste water was sprayed to the coarse filter for the clarification, and the efficiency to remove pollutants is good. At 1900, the process was accepted and the process was called bio-filter method and the tank was called bio-filters.

However, for the early bio-filters, the hydraulic load and the BOD load are low. Although the clarification is very good, the occupied area is big and it is easily blocking. After that, the backflow of the effluent was taken, and the hydraulic load and the BOD load increased obviously. That is the high-rate filter. At 1951, the engineer, Schurz, created the tower biofilter.

3.3.1 The definition of the bio-filter

3.3.1.1 The principle

The waste water was sprayed to the surface of the coarse filter by the water distributor in the bio-filter. Under the gravity, a part of waste water was adsorbed on the surface of the filters, forming the skinny adhesive water layer, and the other part of the waste water penetrate the filters by the skinny, forming flowing water layer. Finally, the water arrived to the discharge system and is drained out of the tank. When the waste water passes the filters, the suspended solids and the colloid and the organic matter are hold back and adsorbed by the filters. The soluble organic matter is utilized by the bacteria for growth and breeding. And then the bacteria adsorb the suspended solids and the colloid and the organic matter in the waste water, forming the biofilm. Once the biofilm is in maturity, the bacteria use the organic matter for metabolism and then the waste water is clarified.

The clarification process is very complex in the bio-filter, including the mass transfer, DO diffusion and adsorption, organic matter degradation and the microorganism degradation.

3.3.1.2 The affecting factors

(1) The height of the filter tank Along the depth of the filter bed, the mass, the types of the bacteria and the removal rate of the organic matter are different completely. In practice, upper layer in the filter bed, the content of the organic matter is high and the breeding of the bacteria is fast, but they are mainly the bacteria. The biomass is high and the removal rate of the organic matter is high. With the depth of the filter bed, microorganism are from low level to high level and the types of the microorganism increase and the biomass decrease.

(2) Load The load is an important parameter affecting the running of the biofilters, similar as the height of the biofilters. The load includes the hydraulic load and the BOD load.

The hydraulic load is referred to the treated waste water by the unit area of the biofilter or the unit volume of carriers each day. The unit is m^3 (waste water)/[m^2(filter · d)] or m^3(waste water)/[m^3(carrier · d)]. The parameter reflects the contacting time and the flushing of the water. The hydraulic load is too big, and the contacting time is short. The clarification is not good. On the other hand, the hydraulic load is too small, the carriers are not used completely. In practice, the hydraulic load is 1~4 m^3(waste water)/[m^2(filter · d)] for the tricking filters. The hydraulic load is 5~28 m^3(waste water)/ [m^2(filter · d)] for the high-rate filters.

The BOD load is referred to the organic matter amount provided by the unit volume carriers in unit time. The unit is $kgBOD_5$/[m^3(carrier · d)]. The volume of the carried represents the surface of the biofilm and the biomass, so the BOD load is the F/M. In practice, the BOD load is 0.15~0.3 $kgBOD_5$/[m^3(carrier · d)] for the tricking filters. The hydraulic load is about 1.1~28 $kgBOD_5$/[m^3(carrier · d)] for the high-rate filters.

(3) Backflow There is the obvious effect on the running of the high-rate filters by the backflow. The advantages are as followed. Increasing the hydraulic load. Promoting the drop of the aged biofilm. Decreasing the BOD load. Promoting the growth of the biofilm. Increasing the DO amount. The disadvantages are as followed. Decreasing the content of BOD in the influent. Decreasing the mass transfer and removal of organic matter. Decreasing the water temperature in winter. Increasing the consumption on energy.

The backflow should be taken at the following conditions.

① The content of the organic matter is high in the influent.

② The waste water amount is too small, and the hydraulic load is lower than the smallest hydraulic load according to the experience.

③ The content of some organic matter is high, resulting in the inhibition on the bacteria.

(4) Aeration It is necessary to provide enough O_2 to the biofilm. In general, the O_2 comes from the atmosphere, depending on the natural ventilation. The main factors on the natural ventilation are the stack effect and wind power. The stack effect depends on the difference of the temperature inside and outside of the tank and the height of the tank.

3.3.2 Trickling filter

The trickling filter is the first generation of bio-filter. The hydraulic load is 1 ~ 4 m^3(waste water)/[m^2(filter · d)], while the BOD load is 0.15 ~ 0.3 $kgBOD_5$/[m^3(carrier · d)].

3.3.2.1 The structure

The component of the trickling filter is tank body, carriers, water distributor and drainage system (Figure 3-3).

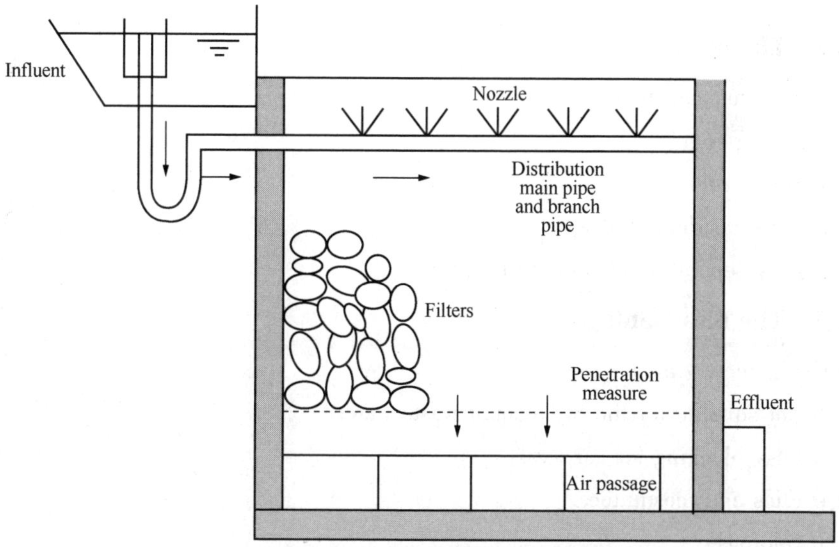

Figure 3-3 The schematic diagram of the trickling filter

The flat shape of the tank body is square, rectangle or circle. The main materials are bricks or stones, resulting in the tank can resist the pressure from the carriers. The function of the tank body is to protect the carriers. Generally, there are small holes on the tank wall to promote the ventilation inside the filter. The tank bottom is to support the filters and to decant the effluent. There are ventilation openings at the bottom, but the area is less than 1% of the surface area of the tank.

(1) Carriers The carriers are important because the bacteria growth on the surface of them. For the trickling filters, the size of the carriers is about 25 ~ 100 mm. The effect of the carriers is obvious for the clarification of the tank.

(2) Water distributor The function of the water distributor is to spray the waste water to the surface of the carriers. In addition, the water distributor has the function to adapt the change of the water quantity. The other characteristic is no-blocking.

(3) Drainage system The drainage system is at the bottom of the bio-filter. The functions include the drainage of the effluent and the excellent ventilation. The drainage system includes the penetration equipment, the ditch for collecting, the total drainage ditch and the interspace for ventilation.

(4) The application and the characteristics Generally, the trickling filter is suitable for the small amount of waste water treatment plant and the waste water amount is less than 1, 000 m^3/d.

3.3.2.2 The advantages

① Easy management.

② Saving energy.

③ Stable running.

④ Little residual sludge.

⑤ Easy separation of solid and liquid.

3.3.2.3 The disadvantages

① Big occupy area.

② Not suitable for the big amount waste water.

③ Easy blocking for carriers.

④ Flies and mosquitoes.

⑤ Odorous.

Nowadays, the trickling filters have been gradually eliminated due to the above disadvantages.

3.3.3 High-rate filter

The high-rate filter is the second generation of product in the bio-filters. It comes from the trickling filter by improvement of the clarification and running.

3.3.3.1 Structure

There are some differences between the high-rate filter and the trickling filter.

(1) The flat shape of the tank body is circle The carriers are granular with the size of 40 ~ 100 mm. The pores of the carriers are big, up to 70%. The height of the carriers is 2.0 m.

Nowadays, the artificial carriers are used widely in the high-rate filter. Generally, the materials of the carriers are polyvinyl chloride, polystyrene and polyamide.

(2) Rotating distributor There are many kinds of rotating distributors. Figure 3-4 is a kind of wide use and simple structure. The waste water with pressure comes into the fixed standpipe in the center and then comes into the horizontal pipe. The horizontal pipe rotated the standpipe. There are lots of small holes with the different interval at the same of the horizontal pipe. The characteristic is that the interval is big at the center and the interval is small at the rim. The interval should be sure by the calculation. Waste water is sprayed from the porthole, producing counterforce and driving the horizontal pipe rotation.

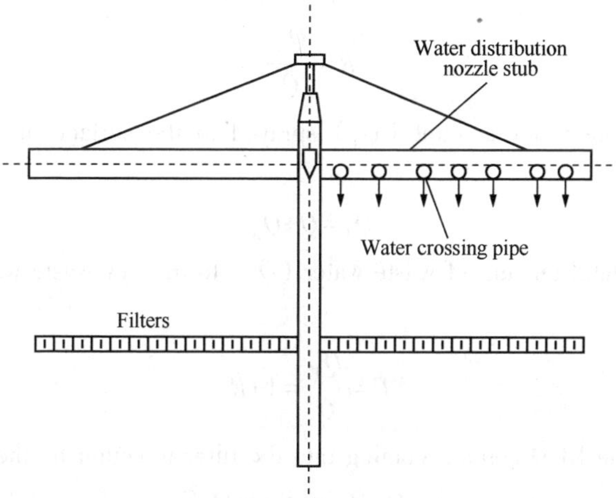

Figure 3-4 The schematic diagram of the rotary distributor

3.3.3.2 Process characteristics

The high-rate filter increases the load of the filter obviously. The BOD load is up to 0.5~2.5 $kgBOD_5/[m^3(filter) \cdot d]$ and the hydraulic load is up to 5 ~ 40 m^3(waste water)/[m^2(filter) · d]. The high load is gotten by the limit on the BOD_5 in the influent and backflow of effluent. Therefore, the BOD_5 should be less than 200 mg/L. On the contrary, it is necessary to dilute the influent by the backflow of effluent. The process is showed in the Figure 3-5.

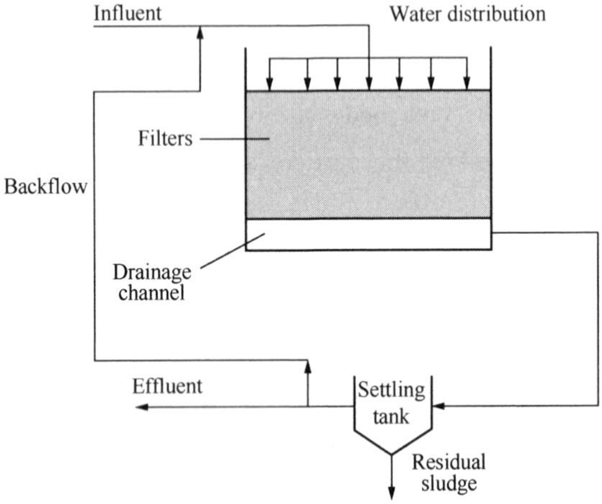

Figure 3-5 The schematic diagram of high-rate filter

The ratio of the backflow amount (Q_R) to the raw waste water (Q) is backflow ratio (R).

$$R=\frac{Q_R}{Q}$$

The total waste water amount (Q_T) sprayed to the surface of the carriers is as followed.

$$Q_T=Q+Q_R$$

The ratio of total amount of waste water (Q_T) to the raw waste water (Q) is cycle ratio (F).

$$F=\frac{Q_T}{Q}=1+R$$

Therefore, the BOD content coming into the filter is gotten by the equation.

$$S_a(Q+Q_R)=S_0Q+Q_RS_e$$

$$S_a=\frac{S_0+RS_e}{1+R}$$

In the equation, S_a—The BOD content sprayed to the carriers, mg/L;

S_0—The BOD content in raw waste water, mg/L;

S_e—The BOD content in the effluent, mg/L;

R—Backflow ratio.

3.3.3.3 Process system

There are many process systems for the high-rate filter due to the backflow of effluent. Figure 3-6 shows two typical processes. For the process a, the effluent from the bio-filter directly returns to the filter to avoid of the big volume of the primary settling tank. The sludge returns to the primary settling tank to promote the inoculation of the biofilm and to fast the renewal of the biofilm. The settling efficiency increases due to the backflow of the sludge in the primary settling tank. For process b, the principle is similar as the process a. The difference is the backflow water. The effluent comes from the secondary settling tank.

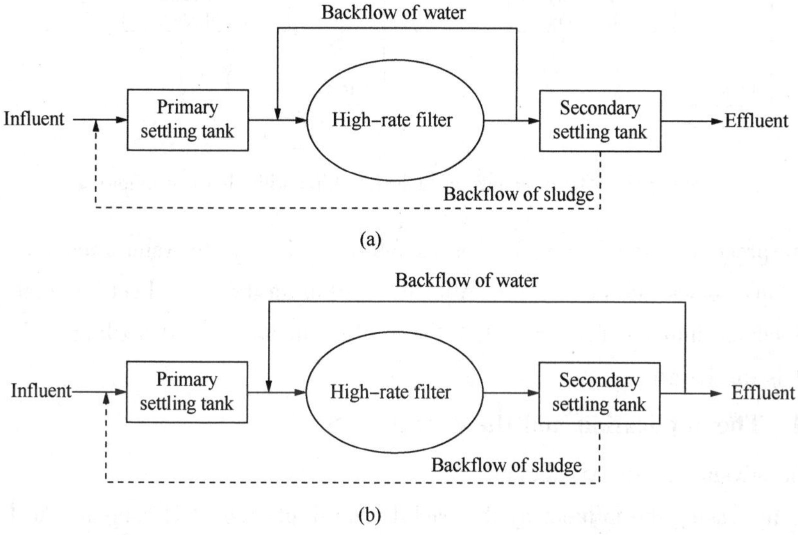

Figure 3-6 The typical processes of the one-stage high-rate filter

When the BOD load in raw waste water is high or the requirement on the effluent is strict, the two-stage bio-filters are used. For the series two-stage bio-filter, the advantages are as followed.

① Reduction of the depth of the filter.

② No-blocking.

③ Excellent ventilation.

④ The removal of BOD is up to 90%.

The disadvantages are mainly the uneven load. Generally, the load is high for the first stage of the bio-filter and the load is low for the second stage of the bio-filter. The alternant inflowing is chosen to avoid of the disadvantage (Figure 3-7).

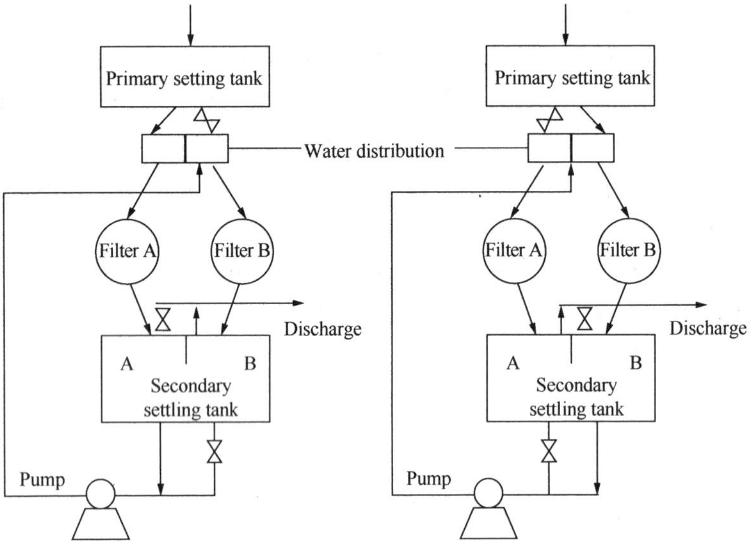

Figure 3-7 The two-stage of high-rate filter with alternant inflowing

The process in the Figure 3-7 is as followed. The waste water from the primary settling tank comes into the filter A. After the settling in the second settling tank A, the effluent comes into the filter B. After the settling in the second settling tank B, the effluent is discharged.

3.3.3.4 The application and the characteristics

The advantages are as followed.

① Increasing the influent by the backflow to dilute the BOD content. Flushing the biofilm is to keep the activity of the bacteria. It is to avoid of the blocking of the carriers and to inhibit the over growth of the filter-fly and the odor.

② Increasing the size of the filter to avoid of the blocking.

③ Increasing the hydraulic load and the BOD load.

④ Occupying area is small.

The disadvantages are as followed.

① The quality of the effluent is worse than the trickling filter. The content of BOD_5 is more than 30 mg/L. No obvious nitrification.

② The color of the sludge is brown. The oxidation is not sufficient.

3.3.4 Tower biofilter

Tower biofilter is anew kind of high-rate biofilter. The height of the tower biofilter is big and the ventilation is good to avoid of the bad ventilation due to the small size of carriers.

3.3.4.1 Structure

The sketch diagram of the tower biofilter is showed in the Figure 3-8. The height of the tower is up to 8 ~ 24 m, and the diameter is up to 1 ~ 3.5 m. The ratio of the diameter to the height is (1 : 6) ~ (1 : 8). It is like a tower. The flat shape of the tank body is circle. It includes tank body, carriers, water distributor and ventilation and drainage system.

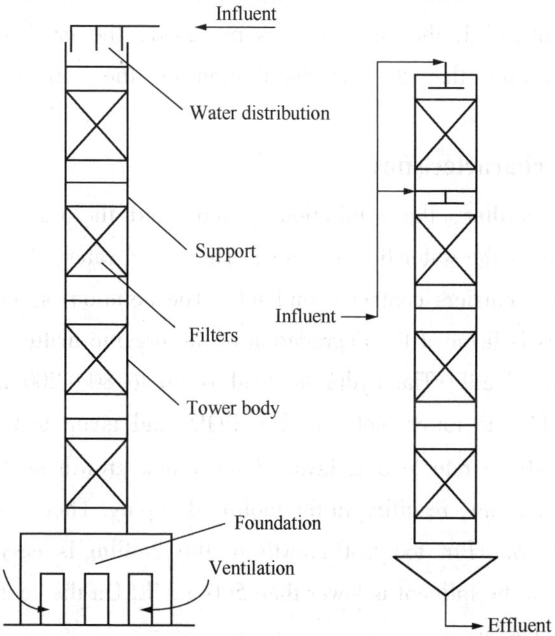

Figure 3-8 The sketch diagram of the tower biofilter

(1) Tank body The tank body is to protect the carriers. Generally, the tank body is built in layers. The screen is built at the layer to support the carriers. The height of the layer is less than 2.5 m. There is inspection hole for the change of the carriers and thermometer hole for the monitoring on temperature and pothole for the observation of the biofilm growth and the water distribution at each layer. The superior border is higher 0.5 m than the carriers on the top layer.

The experiment and running experience showed that the efficiency increases with the height of the tower increase under the same volume load.

(2) Carriers The light material filters are necessary for the tower biofilter. In China, the glass cloth honeycomb filter with the epoxy resin is common. The specific surface area of the carrier is big and the structure is even. It is favor of the even distribution of waste water and the ventilation.

(3) Water distributor The water distributor of the tower biofilter is same as the trickling filter. For the big or middle tower biofilter, the rotating water distributor is common with electric machine. For the small tower biofilter, the fixed spray nozzle is used wildly.

(4) Ventilation The natural ventilation is used wildly for the tower biofilter. Generally, there is space with a certain height and there is ventilation opening around the tower at the bottom. If the ventilation is not good, the mechanical ventilation is chosen. It is noticeably that the air distribution on the surface of the carrier is important.

3.3.4.2 Process characteristics

In the tower biofilter, the ventilation is good and the waste water drops. The turbulent fluctuation of the water flow is strong. The waste water, the air and the biofilm on the surface of the carriers contact completely. The aeration is very good. The mass transfer is fast. This is favor of the degradation of the organic matter.

(1) High-rate load The hydraulic load is up to 80 ~ 200 m^3 (waste water)/ [m^2(filter) · d] for the tower biofilter. The BOD load is up to 0.5 ~ 2.5 $kgBOD_5$/ [m^3(filter) · d]. High-rate load is favor of the quick growth of the biofilm and the strong flush on the biofilm, resulting in the biofilm dropping. Therefore, it is favor of the activity of the bacteria. But too fast growth of the biofilm is easy blocking for the carriers. The BOD in the influent is lower than 500 mg/L. On the contrary, the backflow of the effluent is necessary.

(2) The layer of the carriers In the tower biofilter, the layer is obvious. At different layer, the species of the bacteria are different, which are suitable for the characteristics of the waste water to the layer. This is favor of the growth, breeding and metabolism of the bacteria, promoting the degradation and removal of the organic matter.

3.3.4.3 Application and characteristics

Generally, the tower biofilter is suitable for the small amount of waste water treatment plant and the waste water amount is less than 1, 000 m^3/d.

The advantage of the tower biofilter is saving occupied area and the strongadaptive capacity on the sharp change of the water quality and quantity. The disadvantage is that the pumping fee is big and the running and management is not easy.

3.3.5 Biological aerated filter

Biological aerated filter (BAF) is a process with the biological degradation and separation of solid and liquid. The process combined the submergedbiofilm reactor with the filtration, and no secondary settling tank. The filter is running at period by the backwashing.

3.3.5.1 The principle and the characteristics

The complete pretreatment on the influent is necessary to remove the suspended solids. If the content of the suspended solids is too high, it is necessary to renew the filtration bed and increase the backwashing times. In practice, the requirement on the SS is less than 50 ~ 60 mg/L. If the BAF is the advanced treatment, the effect of the content of SS on the efficiency of BAF is neglected. If the BAF is the main biological treatment, the content of SS is hardly reduced to under 50 mg/L by the traditional pretreatment. The combination of BAF with enhanced primary treatment is necessary to avoid of the blocking.

The process of BAF is showed in the Figure 3-9.

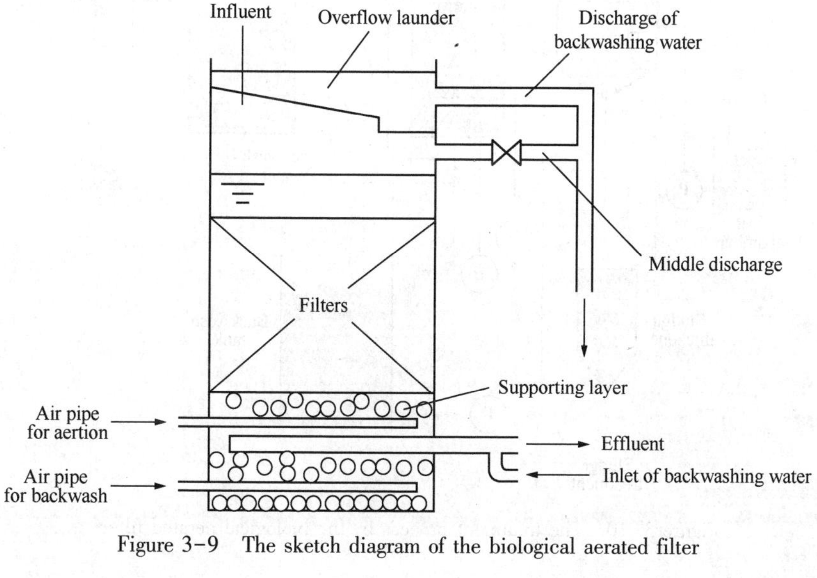

Figure 3-9 The sketch diagram of the biological aerated filter

The influent of the BAF is downflow and the air is upflow at the distance of 30 cm to the carriers. This is favor of the complete contacting between waste water and air to promote the O_2 transfer. It is favor of the degradation of biofilm on the carriers at the lower levels. The biomass is very big due to the small size of the carriers and the big specific surface area. The accommodation on the pollutants increased and the backwashing period is long.

The effluent goes to the drainage system at the bottom and goes to the collecting tank. After a long running time, the biomass and the impurities in the filters increased and the head loss increased too, resulting in the water table up. It is necessary to backwash the carriers.

3.3.5.2 The process of the BAF

The BAF process includes the pretreatment, BAF and backwashing system (Figure 3-10). The pretreatment includes the grit chamber and primary settling tank. The function of the pretreatment is to reduce the SS content. In practice, the influent SS is under 60 mg/L. The pretreatment can reduce the backwashing times to keep the normal running of the filter. The removal of organic matter and nitrification can be finished in the single stage BAF, while it can be finished in the multistage BAF.

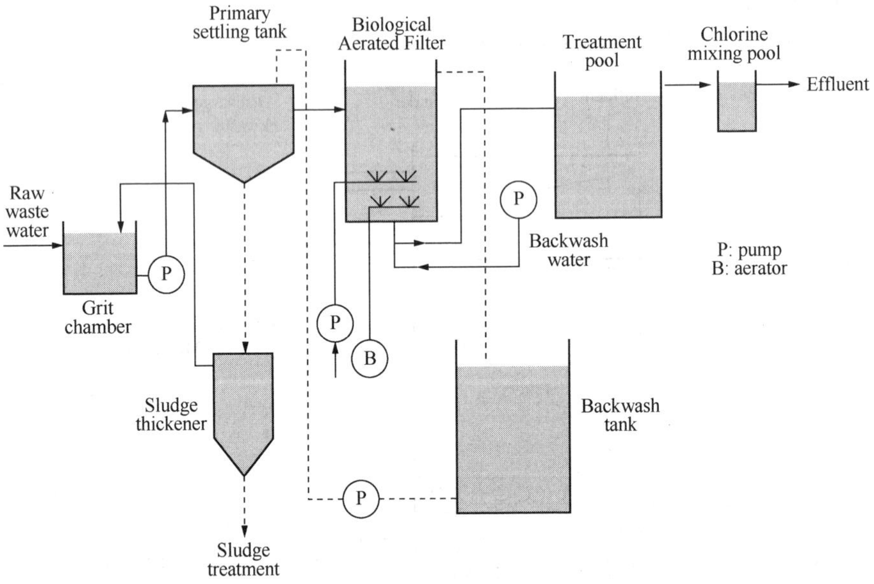

Figure 3-10 The treatment process by the biological aerated filter

The BAF is divided into downflow and upflow, according to the inflowing.

(1) BIOVARBONE The structure is showed in the Figure 3-11.

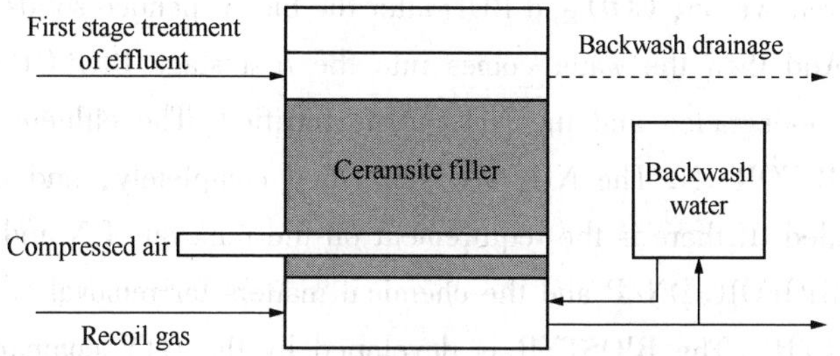

Figure 3-11 The sketch diagram of the BIOVARBONE

The inflowing is downflow. The aeration pipes are at the bottom of the tank and they are equipped at the 20~45 cm to the bottom. The biodegradation is upside the aeration pipes. SS is retained in the carriers and the bacteria adhere to the surface of the carriers forming the biofilm. It is resulting in the increase of head loss. When the head loss is to the designment, it is necessary to backwash the carriers. Generally, the combination of air and water is used for the backwashing. The disadvantages of the BIOVARBONE are the low BOD load and most of SS is hold back at the top of the carriers. In the latter of the running, there is negative head in the carriers.

(2) BIOFOR The BIOFOR is referred to the bio - filtration oxygenated reactor. The process is showed in the Figure 3-12. The inflowing is upflow. The waste water comes into the tank for the mixing of waste water and air. Under the distribution, the mixed liquid comes into the carriers and the BOD, COD, NH_3-N, SS are removed in the carriers. During the backwashing, the air and water come into the mixing tank and then go to the carrier layer by the water distributor. The effluent goes to the primary settling tank and mixed with the raw waste water. The advantage of the parallel flow for the influent and the air is as followed. The first is the even distribution of water and air. The second is to avoid of the jam of the carriers due to SS.

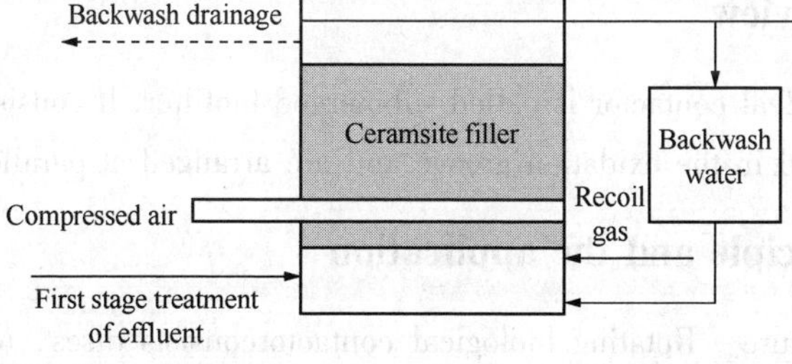

Figure 3-12 The sketch diagram of the BIOFOR

The BIOFOR is a kind of combination process. The influent comes into the primary settling tank to remove SS, COD and BOD after the big suspended solids are removed by the screening. And then the water comes into the first stage BIOFOR – C/N. Most of COD, BOD is biodegraded and the NH_3 – N is nitrified. The effluent comes into the second stage BIOFOR – N. The NH_3 – N is nitrified completely, and COD, BOD are farther biodegraded. If there is the requirement on the removal of N and P, it is to add the third stage BIOFOR–DN/P and the chemical matters for removal of P.

(3) BIOSTYR The BIOSTYR is developed by the OTV company in France to overcome the disadvantages of the BIOVARBONE. The biggest difference is the use of the light material carriers—biostyrene. The main material is polystyrene with the density lower than 1.0 g/cm^3.

3.3.5.3 Process characteristics

① Small occupied area. There is no secondary settling tank, saving the investment and the occupied area. The hydraulic load and the BOD load are higher than that in the traditional activated sludge. The HRT is short.

② The effluent quality is good. The content of SS is very low and is less than 10 mg/L.

③ The transfer efficiency of O_2 is high and the aeration amount is small.

④ The resistance on the shock is strong and it is suitable for the low temperature.

The problems of the process are as followed.

① The running management is complex.

② The requirement on the SS in the influence is higher.

③ The head loss is big.

④ The residual sludge is bigger, compared with the traditional activated sludge.

3.4 Rotating biological contactor

3.4.1 The overview

Rotating biological contactor is called submerged biofilter. It consists of many discs which are submerged in the oxidation groove and are arranged at parallel.

3.4.2 The principle and the application

(1) The structure Rotating biological contactorconsists discs, oxidation groove, spindle and drive unit. (Figure 3 – 13) The discs are arranged in series with the

spindle. 40% of the area of the discs are submerged in the waste water in the oxidation groove. The spindle is higher 10~25 cm than the water table in the oxidation groove.

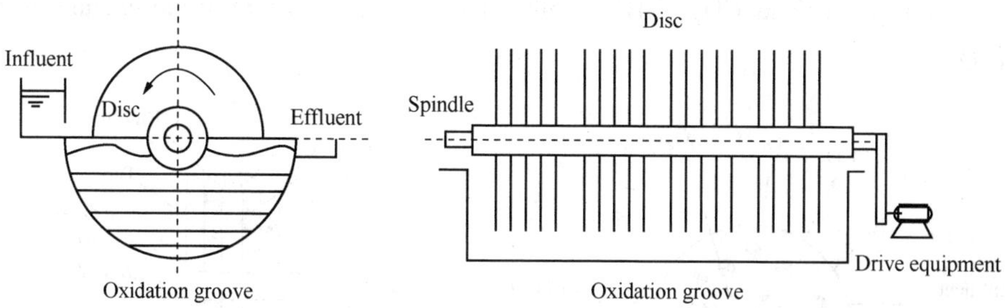

Figure 3-13 The structure of the rotating biological contactor

① Disc. The disc is the key of the rotating biological contactor. The characteristics of the disc is as followed: light material, strong intensity, corrosion resistance, ageing resistance, easy biofilm culturing, non - deformation, high specific surface area, handling ease.

In practice, the shape of the disc is the circle. Nowadays, the corrugate disc is used to increase the surface area. The diameter of the disc is 2.0~3.6 m. The biggest is up to 5.0 m. The bigger surface area of the disc is favor of the reduction of the area of the oxidation groove. The gap of the discs is 30 mm in standard. If the multistage discs are used, the gap is 25~35 mm for the former stages, the latter is 10~20 mm. The material of the disc is plastic, such as polyvinyl chloride.

② Oxidation groove. The depth of the oxidation groove is more than the 35% of the diameter of the disc. The shape of the groove is generally arch. The structure and the construction of the groove is different depending on the amount of the waste water and the place of the site.

③ Spindle. The spindle is the key part for the rotating biological contactor. The both ends are fixed on the bearings at the both ends for the oxidation groove.

④ Drive unit. The drive unit includes power equipment, reduction gear and transmission chain. For the large-scale discs, one drive unit is to drive one disc. For middle-scale and small-scale discs, one drive unit is to drive 3~4 stage discs.

(2) The clarification The disc rotates in the oxidation groove at low linear velocity. The disc alternately contacts with air and waste water. The formation of biofilm is formed on the surface of the disc after a long time. The species of the microorganism keep stable and the organic matter is degraded by the bacteria in the biofilm. When the disc contacts with the air, the oxygen is adsorbed by the fixed water layer on the biofilm

and then it is transferred into the bacteria and the water in the oxidation groove (Figure 3-14). The content of the dissolved oxygen increases. There is another transfer of the matters, such as CO_2, NH_4^+, et al (Figure 3-15), except of the organic matter and O_2.

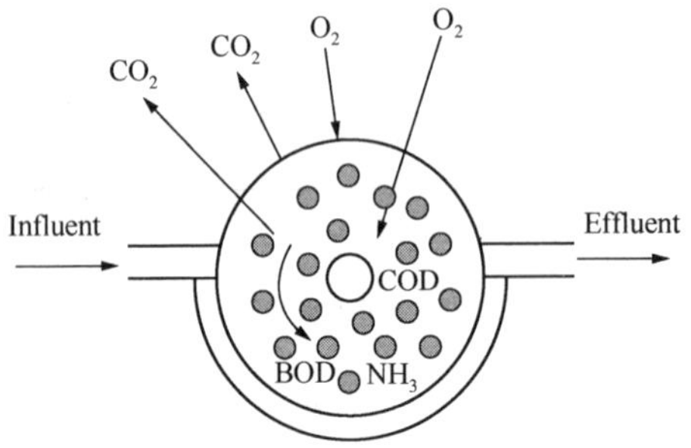

Figure 3-14 The clarification of the rotating biological disc

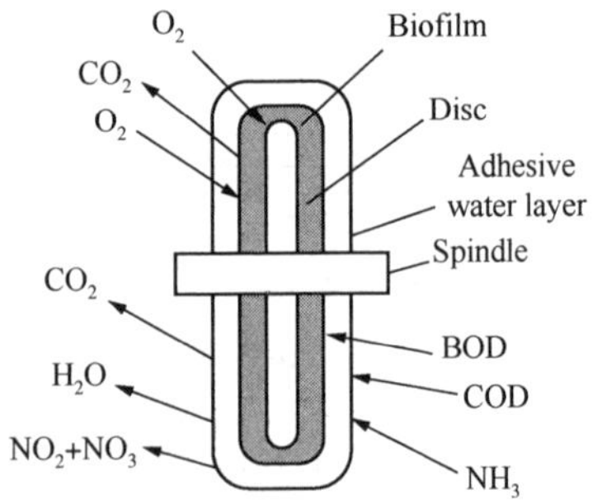

Figure 3-15 The mass transfer on the rotating biological disc

During the treatment, the biofilm continuously grows and the surplus of the biofilm is dropped under the shear force due to the rotating of the disc. The dropped biofilm comes to the secondary settling tank.

(3) Typical processes The basic process is showed in the Figure 3-16. The processes include the discs, the pretreatment and the secondary settling tank. The function of the secondary settling tank is to remove the dropped biofilm. It is appropriate to use the multistage discs. In practice, it is favor of the improvement of the effluent under the same area of the discs for the same waste water. The biofacies show that the thickness of the biofilm on the first stage disc is big and then the thickness decreases at the latter stage discs with the organic matter decrease.

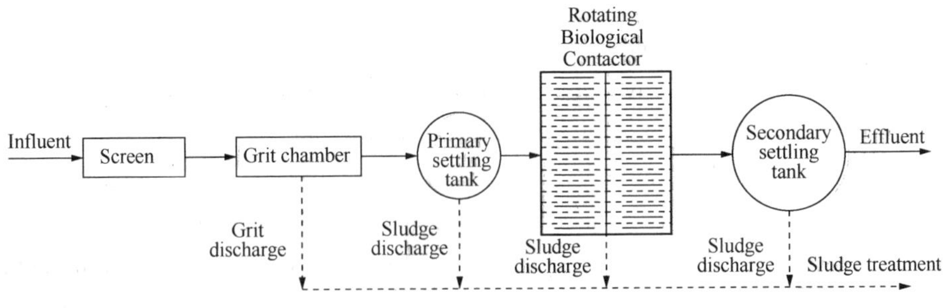

Figure 3-16 The basic process of the rotating biological process

(4) The characteristics of the process The advantages of the disc are as followed, compared with the biofilters.

① No-blocking.

② Classification of the biofacies.

③ Longer SRT. It is favor of the nitrification and denitrification.

④ Strong hydraulic and BOD load resistance.

⑤ Saving energy.

The disadvantages of the disc are as followed.

① The investment is huge.

② Volatile organic matters come into the air.

③ The application is affected by the environmental factors. At north, the disc should be in the room to avoid of the freezing. For the disc outside of the room, it is necessary to build the rain cover to avoid of the flushing of the rain.

3.4.3 The designment of the rotating biological contactor

The designment of the rotating biological contactor is as followed.

① The total area of the discs:

$$A=\frac{Q(S_0-S_e)}{F_A}$$

In the equation, A—The total area of the discs, m^2;

Q—The amount of waste water, m^3/d;

S_0—The content of BOD_5 in the influent, g/L;

S_e—The content of BOD_5 in the effluent, g/L;

F_A—BOD loading, $g/(m^2 \cdot d)$.

② The number of the discs:

If the disc is the circle, the number is gotten by the following equation.

$$M=\frac{2A}{\pi D^2}$$

In the equation, A—The total area of the discs, m^2;

M—The total area of the discs;

D—The diameter of the disc, m.

③The effective length of the oxidation groove:

$$L=m(d+b)k$$

In the equation, L—The effective length of the oxidation groove, m;

m—The number of the discs for each rotating disc;

d—The interval of the disc, m;

b—The thickness of the disc, mm;

k—The coefficient.

④ The effective volume of the oxidation groove:

$$V=(0.294\sim0.335)\times(D+2\delta)^2L$$

In the equation, δ—The distance between the edge of the disc and the inside wall of the oxidation groove, m.

⑤ The rotating speed:

$$n_{min}=\frac{6.37}{D}\times\left(0.9-\frac{1}{F_s}\right)$$

In the equation, n_{min}—The smallest revolution, r/min;

F_s—The hydraulic loading in the oxidation groove, $m^3/(m^2\cdot d)$.

⑥ The power of the motor:

$$N_p=0.385\,R^4n_{min}ma\beta$$

In the equation, N_p—The power of the motor, kW;

R—The diameter of the disc, m;

m—The number of the discs on a spindle;

a—The number of the spindles;

β—The coefficient of the thickness of the biofilm.

⑦ The retention time:

$$t=\frac{V'}{Q}$$

In the equation, t—The retention time, h;

V'—The effective volume of the oxidation groove, m^3.

3.5 Submerged biofilm reactor

The submerged biofilm reactor comes from the biofilter and the aeration tank. At 1930s, the similar equipment appeared with the filters. The filter can be broken stones, bamboo and wood, metal products. The application is to treat the waste water with low content of organic matter. The advantage of the equipment is to avoid of the loss of the activated sludge. At 1970s, the application of the submerged biofilm reactor is wide because of the appearance of the honeycomb tube filters and stereoscopic corrugated filters with big pores and high specific surface area.

The tank is full of the filters with the density. The filters are submerged by the waste water. The aeration pipe is at the bottom of the tank. Under the metabolism of bacteria, the organic matter is removed. The waste water is clarified. The process has the advantages of thebiofilm and the activated sludge.

In these days, the process is applied widely for the treatment on the domestic waste water and some industrial organic waste water.

3.5.1 The structure

The submerged biofilm reactor includes the tanks, the filters, the support, the aeration equipment, the influent equipment, the effluent equipment, the discharge for the sludge (Figure 3-17).

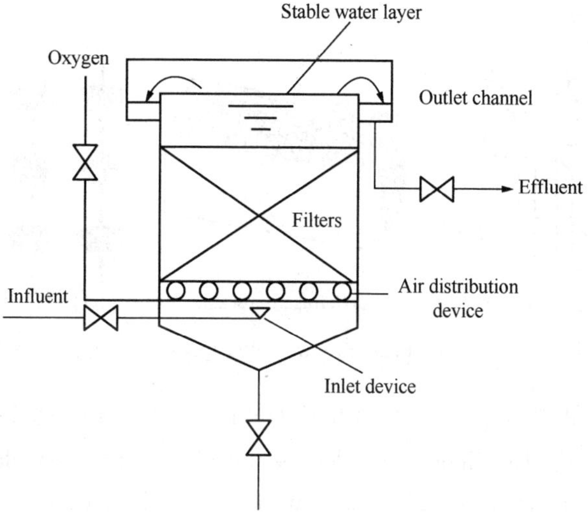

Figure 3-17 The basic structure of the submerged biofilm reactor

(1) Tank body The aeration tank is the major structure, which is the main tank for clarification of waste water by the microorganism. The flat shape is circle, square, or rectangle. The material of the tank body is generally the steel plate, reinforced concrete or brick.

(2) Filter The bacteria main inhabit on the filter, which is the carrier for the biofilm. In addition, the filter has the function to remove the suspended solids. The characteristics of the filter affects the efficiency of the treatment. The most important is the investment on the filter is very big in the total investment. Therefore, the choice of the appropriate filter is very important on technology and economy.

The filters can be divided according to the shapes, characteristics and materials. According to the shape, the filter can be honeycomb, cylindrical, corrugated, shield, annulus radiation, ball, irregular granularity. According to the characteristics, the filter can be hard, soft, semi-soft. According to the material, the filter can be plastic, fiber, glass fiber reinforced plastics. In China, the following is in practice.

① Honeycomb filter (Figure 3-18a). The material is plastic, glass fiber reinforced plastics. The advantages are as followed: high specific surface area, high pores, light material, strong intensity. The aged biofilm is easy drop. The disadvantages are as followed. When the size of the filter is not suitable for the BOD load, there is no balance between the growth and the drop of the biofilm. The filter is easy blocking. When the aeration mode is not good, the speed in the Honeycomb filter is not even.

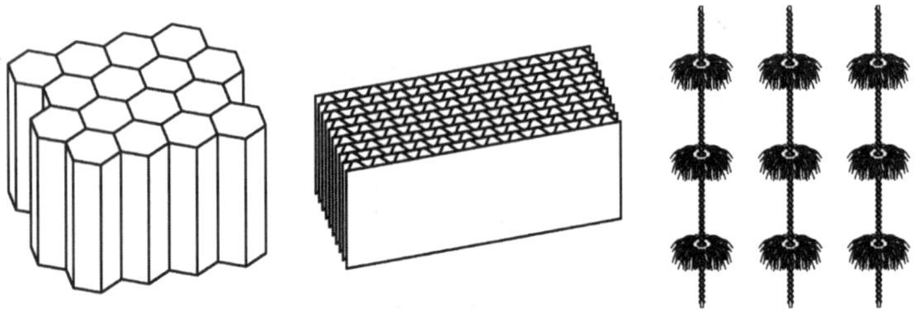

(a) Honeycomb packing　　(b) Corrugated packing　　(c) Soft fiber packing

Figure 3-18　Filter types

② Corrugated filter (Figure 3-18b). In China, the corrugated filter is based on the flocor from UK. The filter is bonded between the corrugated plate and the hard polyvinyl chloride tablets. The advantages are as followed: big pores and no-blocking, the structure is simple, it is easy to be transported and installed, light material and strong intensity, anticorrosion. The disadvantages are as followed. The speed is not even.

③ Soft filter (Figure 3-18c). Generally, the nylon, vinylon, dacron or acrylic was knitted into the bunch and connected by the center rope. The advantages are as followed: high specific surface area, light material and strong intensity, stable chemical and physical property, easy transportation, easy installment. The disadvantage is that it is easy to be agglomeration for the filter, while in its center it is anaerobic.

3.5.2 The form of the aeration tank

The form can be divided into the divided flow and straight-flow according to the plate of the aeration. In foreign countries, the divided flow is used widely (Figure 3-19).

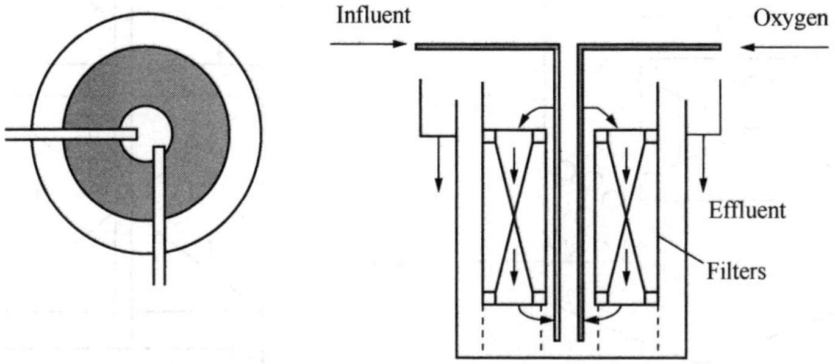

Figure 3-19 The divided flow of submerged biofilm reactor with center aeration

The divided flow aeration tank is that the aeration is in a separated room for the waste water and there is drastic aeration and the transfer processes of the oxygen. After aeration, the waste water goes slowly to the other room full of filters, contacting with the biofilm completely. Under the cycling, the waste water has an experience of repeated aeration and contacting. This is favor of the growth and breeding of the bacteria. However, the flowing speed is slow and the flushing is small, resulting in the slow renew of the biofilm. It is easy to form anaerobic layer and blocking. The mode is not suitable for the high BOD load. According to the aeration site, the divided flow aeration tank is divided into center aeration (Figure 3-19) and unilateral aeration.

The unilateral aeration is showed in the Figure 3-20. The filter is at the one side of the tank, while the other side is aeration area. The raw waste water comes into the aeration area. After the aeration, the waste water goes to the filter. The waste water cycles between the filter and the aeration area repeated. The effluent passes the gap outside the aeration area and goes to the sedimentation tank.

In China, the straight-flow is used widely (Figure 3-21). The characteristic is that the aeration site is at the bottom of the tank, promoting the water up flowing. Under the flushing and mixing of the air and water, the renew and the drop of the biofilm is quick, keeping the high activity of the biofilm. In addition, there is no blocking. The contact probability between the air and the water increases due to the small size of the air bubble.

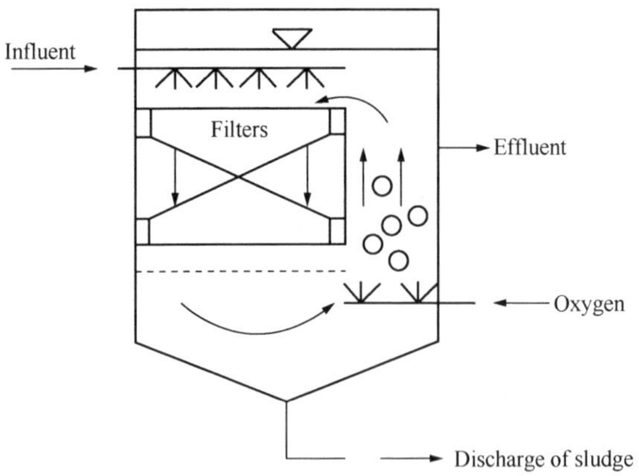

Figure 3-20 The divided flow of submerged biofilm reactor with unilateral aeration

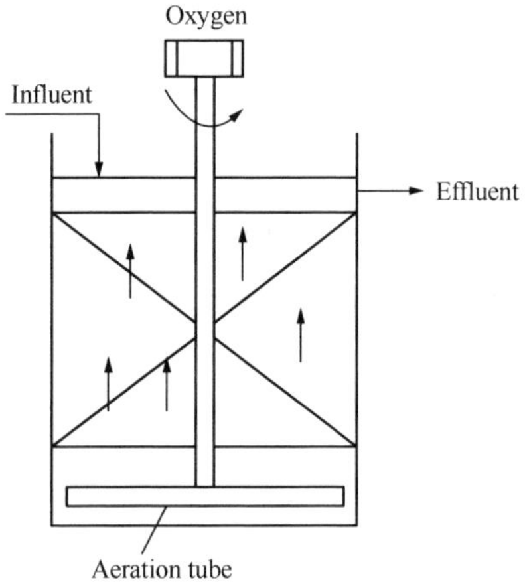

Figure 3-21 The divided flow of submerged biofilm reactor with blast aeration

3.5.3 The characteristics of the submerged biofilm reactor

(1) Process Lots of filters can be used. The bacteria are rich in thebiofilm. There are filamentous bacteria in the biofilm, except of bacteria, protozoa and metazoan.

(2) Function There are many functions of the submerged biofilm reactor. Denitrification is feasible, except of the removal of organic matter.

(3) Running High resistance on the hydraulic and BOD load. The efficiency is good under the intermittent conditions. Easy running. Easy management. No backflow of sludge. No sludge bulking. No filter-fly. Small residual sludge. High particle size sludge and easily settling.

The problems of the process are as followed. The removal efficiency is lower than that of the traditional activated sludge. The investment is higher, too. It is easy blocking under improper running conditions. The water distribution and the air distribution are not even. Lots of metazoan may promote the instantaneous drop of biofilm, resulting in the deterioration of the effluent.

3.5.4 The processes of the submerged biofilm reactor

The main process of the submerged biofilm reactor is including one-stage treatment process, two-stage treatment process, multi-stage treatment process. In practice, the characteristics of the three processes are obvious and they are suitable for different conditions.

The one – stage treatment process is showed in the Figure 3 – 22. After the pretreatment by the primary settling tank, the raw waste water comes into the submerged biofilm reactor. And then the effluent is discharged after the separation by the secondary settling tank. The dropped biofilm is discharged from the treatment process as the residual sludge.

The two–stage treatment process is showed in the Figure 3–23. The removal of pollutants is stable, compared with the one–stage treatment process. In general, there is a middle settling tank or no.

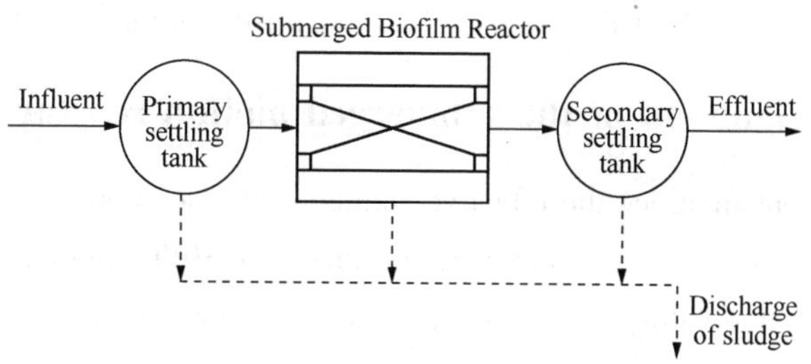

Figure 3–22 The one–stage treatment process of submerged biofilm reactor

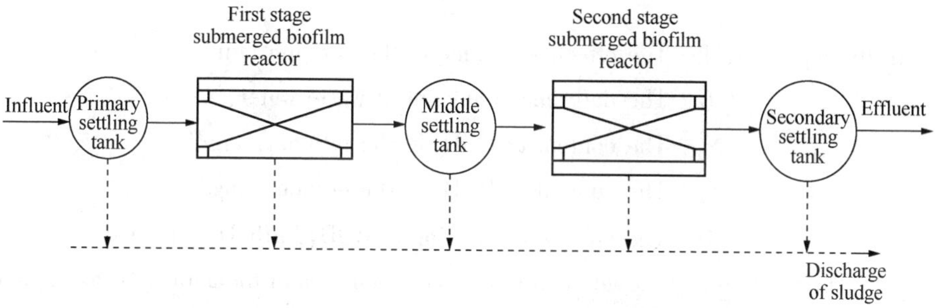

Figure 3–23 The two–stage treatment process of submerged biofilm reactor

The multi–stage treatment process is showed in the Figure 3–24. It includes three or more submerged biofilm reactors.

In theory, after the treatment by the primary settling tank, the raw waste water comes into the aeration tank. The bacteria are at the end of increased logarithmic phase or the begging of the deceleration proliferative stage. The growth and breeding are fast for the bacteria due to the high BOD load. The removal rate of organic matter degradation is high. For the series submerged biofilm reactors, the bacteria are at the end of deceleration proliferative stage or the begging of the endogenous respiration phase in the

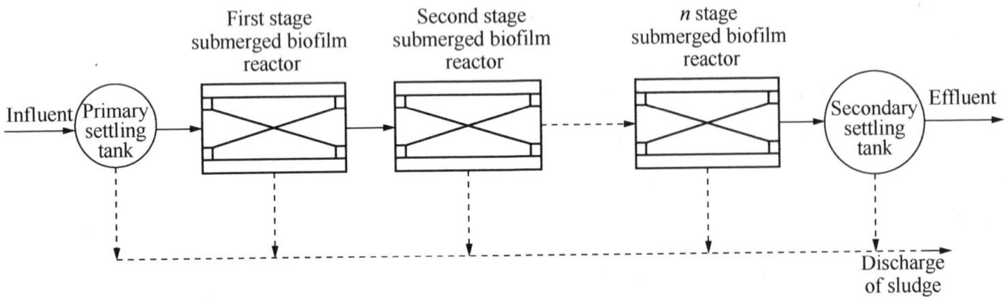

Figure 3-24 The multi-stage treatment process of submerged biofilm reactor

latter reactors. The growth of the bacteria is slow and the effluent quality is excellent.

3.5.5 The designment of the submerged biofilm reactor

The designment includes the effective volume of the aeration tank, the total area of the aeration tank, the number of the aeration tank, the HRT, the oxygen demand.

(1) The effective volume of the aeration tank The effective volume of the aeration tank is gotten by the equation.

$$V=\frac{Q(S_0-S_e)}{M}$$

In the equation, V—The effective volume of the aeration tank, m^3;

Q—The daily mean amount of waste water, m^3/d;

S_0—The content of BOD_5 in the influent, mg/L;

S_e—The content of BOD_5 in the effluent, mg/L;

M—The volumetric loading of BOD_5, $gBOD_5/(m^3 \cdot d)$.

(2) The total area of the aeration tank The total area of the aeration tank is gotten by the equation.

$$F=\frac{V}{H}$$

In the equation, F—The total area of the aeration tank, m^2;

H—The height of the filter, m. In general, it is 3 m.

(3) The number of the aeration tank The number of the aeration tank is gotten by the equation.

$$n=\frac{F}{f}$$

In the equation, n—The number of the aeration tank, count, $n \geqslant 2$;

f—The area of an aeration tank, m^2, $f \leqslant 25m^2$.

(4) HRT The contacting time is gotten by the equation.

$$t = \frac{nfH}{Q} \times 24$$

In the equation, t—The contacting time, h.

(5) The height of the aeration tank The height of the aeration tank is gotten by the following equation.

$$H_0 = H + h_1 + h_2 + (m-1)h_3 + h_4$$

In the equation, H_0—The height of the aeration tank, m;

h_1—The super elevation, m, $h_1 = 0.5 \sim 0.6$ m;

h_2—The water depth on the filter, m, $h_2 = 0.4 \sim 0.5$ m;

h_3—The gap height of the filter layer, m, $h_2 = 0.2 \sim 0.3$ m;

h_4—The height of the water distribution area, m, People come into the area for overhaul, $h_4 = 1.5$m. No people come into the area for overhaul, $h_4 = 0.5$m;

m—The layers of the filter.

(6) The oxygen demand The oxygen demand is gotten by the following equation.

$$D = D_0 Q$$

In the equation, D—The oxygen demand, m^3/d;

D_0—The oxygen demand for 1 m^3 waste water, m^3 O_2/m^3 waste water.

3.6 Biological fluidized bed

The definition of the biological fluidized bed is that the bed is full of the filters which are covered with the biofilm. Under the waste water flushing, the filters are fluidized due to the small intensity of the filter under. Generally, the filters are sand, activated carbon, coke, et al. The process solves the problem of blocking.

3.6.1 The structure

The biological fluidized bed includes the bed, the filters, the aeration equipment, the water distribution, the removing biofilm equipment.

(1) The bed The flat shape is in general circle. It is made by the steel plate.

(2) The filters They are the key parts in the biological fluidized bed. The

common filters are sand, anthracite, coke, granular activated carbon, polystyrene sphere. When the speed of the influent is close to critical velocity, the filters are loose and the filters are starting to fluidize. When the speed of the influent is bigger than the critical velocity, the height of the filter bed increases and the degree of the fluidization increases. Once the gravity of the filters is the same of the floating force, the bed is in fluidization stage. The expansion rate of the bed is about 20%~70%. The whole surface of the filters is contacting with the waste water, providing more surface for the bacteria to contact with the organic matter.

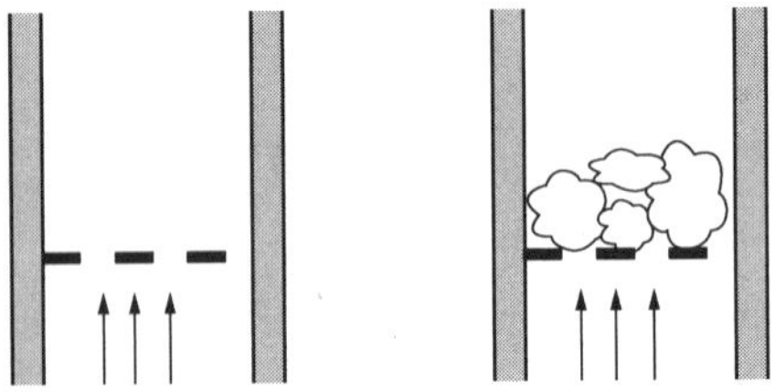

Figure 3-25 The water distribution of the biological fluidized bed

(3) Water distribution The even distribution of water is very important for the clarification of the bed, especially for the two-phase fluidized bed. The equipment for distribution is the support of the filters. The filters are not lost when there is no water in the bed. The equipment is showed in the Figure 3-25.

(4) The removing biofilm equipment The equipment is mainly used in the biological fluidized bed. The function of the equipment is to remove the aged biofilm and to keep the activity of the biofilm.

3.6.2 The characteristics

In theory, the removal of organic matter is mainly depending on the biofilm on the surface of the filters. In the form of the reactor, it is different from the rotating biological contactor, biological filters. In the biological fluidized bed, the carriers, which are covered with the biofilm are suspended in the water. In addition, there is free biofilms and zoogloea in the reactors. Therefore, the process has the characteristics of the activated sludge and the biofilm. The biological fluidized bed has special advantages such as bacteria content, mass transfer, biological reaction rate. In addition, the process has the other advantages, except of the advantages of the biofilm.

(1) High activity of the bacteria The thickness of the biofilm is less than 0.2 μm and the biofilm is even due to the collision and friction between the carriers which are covered with the biofilm. The researcher reported that respiratory rate for the biofilm is higher 2 times than that of activated sludge for the same waste water. It is indicating that

the reaction rate is so fast. It is certified that the BOD load is high in the biological fluidized bed.

(2) Good mass transfer The running of fluidization is favor of the mass transfer in the reactor. The renewal of the interface of the liquid and the air promotes the oxygen and substrate transfer. This is favor of the adsorption and degradation of the bacteria, promoting the biological reaction rate.

Although there are many advantages for the biological fluidized bed, showing the use and the scale of the process increasing, the popularizing rate is less than activated sludge and submerged biofilm reactor. The investment and the occupying area are only 70% and 50% of the traditional activated sludge, respectively. But the running cost is higher, due to the consumption on the energy.

3.6.3 The process type of the biological fluidized bed

The biological fluidized bed can be divided into two-stage biological fluidized bed, three-stage biological fluidized bed and mechanical mixing biological fluidized bed according to the power. The biological fluidized bed can be divided into aerobic and anoxic bed according to the aeration conditions. The types, aeration and the function are showed in the following table (Table 3-2).

Table 3-2 The types, aeration and the function of the biological fluidized bed

The types	The removal objects	Fluidization mode	Aeration
Aerobic fluidized bed	BOD, COD, NH_3-N	Two-stage biological fluidized bed	Surface aeration, blast aeration, pressurized aeration
		Three-stage biological fluidized bed	Blast aeration
		Mechanical mixing fluidized bed	Blast aeration
Anoxic fluidized bed	Nitrate, nitrite	Two-stage biological fluidized bed, mechanical mixing fluidized bed	

(1) Two-stage biological fluidized bed The flowing process is showed in the Figure 3-26. The waste water is the power to make the carriers fluidization for the two-stage biological fluidized bed. In the bed, the waste water contacts with the carriers, while the aeration is in the other equipment.

(2) Three-stage biological fluidized bed The flowing process is showed in the Figure 3-27. The air is the power to make the carriers fluidization for the three-stage biological fluidized bed. In the bed, there is air, liquid and solid. The air comes into the bed from the pipes at the bottom of the bed. The biofilm is dropped due to the mixing and friction among the carriers. The running consumption of the three-stage biological fluidized bed is lower than that of the two-stage biological fluidized bed. The use of the three-stage biological fluidized bed is widely, compared with the two-stage biological fluidized bed.

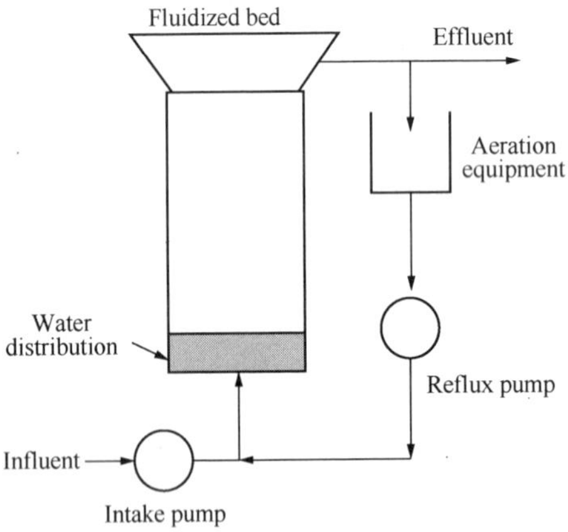

Figure 3-26 Two-stage biological fluidized bed

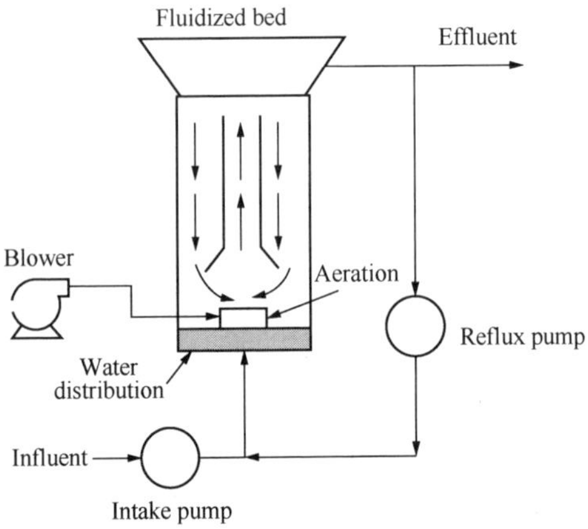

Figure 3-27 Three-stage biological fluidized bed

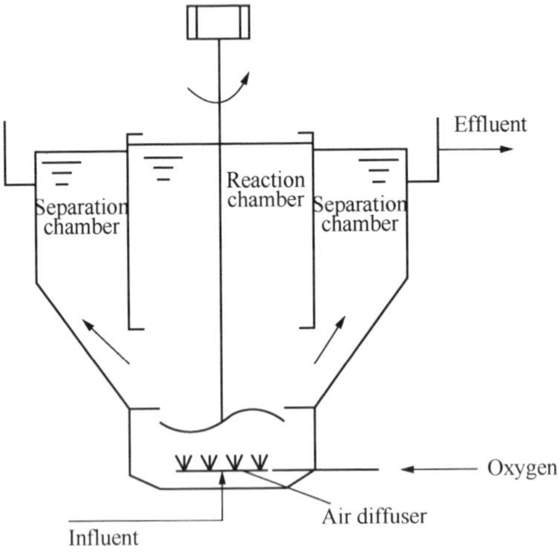

Figure 3-28 Mechanical mixing fluidized bed

(3) Mechanical mixing fluidized bed The flowing process is showed in the Figure 3-28. The bed includes the reaction room, liquid and solid separation room. There is blender at the center at the bottom of the bed. The carriers are fluidized under the mixing conditions. The carriers are main the sand, coke or activated carbon, and their size is less than 0.1~0.4 mm.

3.6.4 The designment of the biological fluidized bed

The designment of the biological fluidized bed includes the kinds of the filters, the parameters of the filters, the thickness of the biofilm and the biological particles, the settling characteristics of the biological particles, the expansion of the bed, the effective volume of the reactor.

(1) The parameters of the filters The mean size of the filter is 0.3 ~ 1.0 mm for the sand or the activated carbon. The size of the artificial filter is sure by the experiment.

(2) The thickness of the biofilm Generally, the thickness of the biofilm is 0.1 ~ 0.2 mm. The thickness is related with the BOD in the influent.

(3) The settling characteristics:

$$U_t = \sqrt{\frac{40(\rho_p - \rho_L) g d_p}{3 \rho_L C}}$$

In the equation, U_t—The settling speed of the biological particles under static conditions, cm/s;

ρ_p—The density of the biological particles, g/cm^3;

ρ_L—The density of the waste water, g/cm^3;

g—The acceleration of gravity, 9.8 m/s^2;

d_p—The mean size of the biological particles, mm;

C—The constant.

(4) The effective volume of the reactor:

$$V_s = \frac{M_s}{\rho_s} \times 10^{-3}$$

In the equation, V_s—The real volume of the filter, m^3;

M_s—The total quantity of the filters, kg;

ρ_s—The density of the filters, g/cm^3.

$$V = \frac{\left(\frac{d_p}{d_s}\right)^3 V_s}{1-\varepsilon}$$

In the equation, V—The effective volume of the reactor, m^3;

d_s—The mean size of the filters, mm;

ε—The pore rate of the layer;

ρ_s—The density of the filters, g/cm^3.

(5) The sludge loading:

$$F_s = \frac{(S_0 - S_e)Q}{\left[\left(\frac{d_p}{d_s}\right)^3 - 1\right]\rho_f V_s (100 - P) \times 10^4}$$

In the equation, F_s—The sludge loading, In general, 0.1~0.3 kg/(kg · d);

d_s—The mean size of the filters, mm;

S_0—The content of BOD_5 in the influent, mg/L;

S_e—The content of BOD_5 in the effluent, mg/L;

Q—The amount of waste water, m^3/d;

P—The water content of the biofilm, %. In general, 95%;

ρ_f—The density of the wet biofilm, g/cm^3, In general, 1.02 ~ 1.04 g/cm^3.

(6) The size of the reactor:

$$A = \frac{QHR}{864U_L}$$

$$H = \frac{V}{A}$$

In the equation, A—The area of the bed, m^2;

R—The backflow ratio, %; In general, 100% ~ 200%;

U_L—The upflow speed of the waste water in the bed, m/s;

H—The height of the bed, m.

3.7 The new biofilm reactor and combination process

3.7.1 Combined biofilm reactor

Nowadays, more and more researchers pay attention to the combined biofilm reactor. These reactors combine the advantages of the single reactor to obviously increase the clarification. The widely used reactors are as followed.

(1) Combined activated sludge and biofilm reactor The flowing process is showed in the Figure 3-29. The carriers are added into the aeration tank in the activated sludge process and the bacteria are attached on the surface of the carriers. Therefore, the removal of the organic matter is finished by the suspended activated sludge and the adhesive biofilm. The filters or the carriers can be powdered activated carbon, anthracite, porous foam, et al. This is favor of the increase of the bacteria in the

aeration tank. The fixed MLSS is up to 2, 000 ~ 19, 000 g/m^3 on the base of the suspended MLSS. The thickness of the biofilm depends on the aeration degree or the shear force due to the aeration.

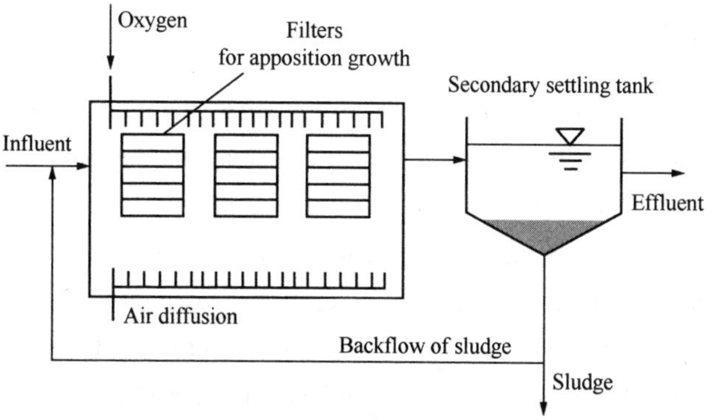

Figure 3-29 Combined activated sludge and biofilm reactor

(2) Sequencing batch biofilm reactor The flowing process is showed in the Figure 3-30. On the base of the sequencing batch reactor, the biofilm is introduced. The new process is called SBBR. The new reactor has the characteristics of the SBR and the advantages of the biofilm. According to the running, the SBBR is similar as the SBR. The SBBR includes five steps, inflowing, aeration, sedimentation, decanting, idle. The filters can be soft fiber filter, Polyethylene packing, activated carbon, et al. According to the clarification, the process has the function of removal of non-biodegradable organic matter, N and P and resistance of hydraulic load.

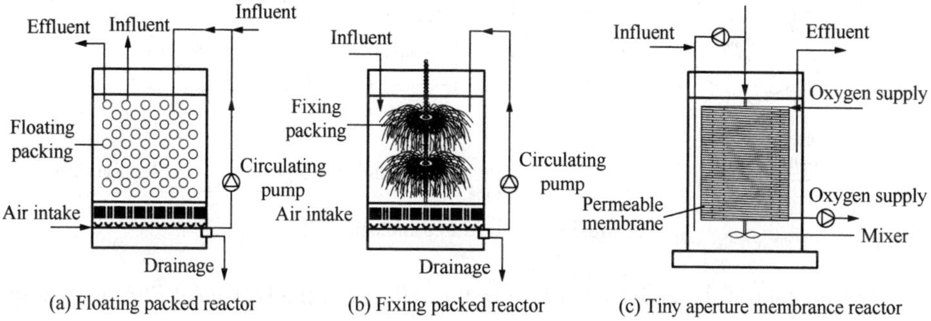

Figure 3-30 Sequencing batch biofilm reactor

3.7.2 Union process of biofilm and activated sludge

The union process originates from 1970s. The BOD load in the biological filter with the new carriers is higher 10 ~ 15 times than that with the traditional carriers. In addition, the biological filter has no odor and blocking. The union of the two processes has the advantages of biofilm and activated sludge, such as resistance of hydraulic and BOD load and easy running and management, the excellent quality of the effluent and nitrification.

There are two kinds for the union process. The first is combined biofilm reactor. That is to add the suspended or fixed carriers into the aeration tank in the activated sludge process. The second is the series connection between the biofilm and the activated sludge. The following is focusing on the second.

(1) Activated biological filter The activated biological filter is the series connection between the biological filter and the aeration tank in the activated sludge. The characteristic is the backflowing of the part of the effluent from the biological filter with the activated sludge from the secondary settling tank. The process is showed in the Figure 3-31.

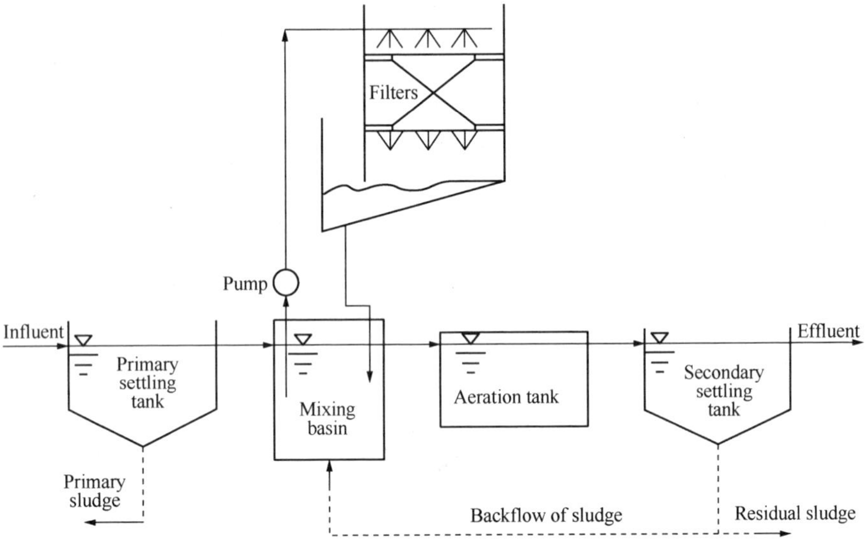

Figure 3-31 Activated biological filter

The content of the activated sludge in the influent of the biological filter is high due to the backflow from the secondary settling tank. This is favor of the adsorption and biodegradation by the microorganisms. However, the retention time of the waste water and sludge is short in the biological filter. It is difficult to degrade the whole adsorbed organic matter in the biological filter. It is necessary to aerate and degrade the residual organic matter in the aeration tank to meet the demand of the effluent quality. But the

BOD load is reduced in the aeration tank due to the adsorption and biodegradation by the biological filter. The experiments showed that the resistance on the BOD load is big in the biological filter. The BOD load maybe higher, but the fluctuation of the effluent quality is even. In addition, the union of the biological filter and the activated sludge is favor of the improvement on the running of the aeration tank, such as, avoiding of the sludge bulking. It is indicating that the running of the whole process is stable.

(2) Trickling filter/Activated sludge The characteristic is to design a middle settling tank between the trickling filter and activated sludge. The function of the middle settling tank is to remove the dropped biofilm. The process is as following (Figure 3-32).

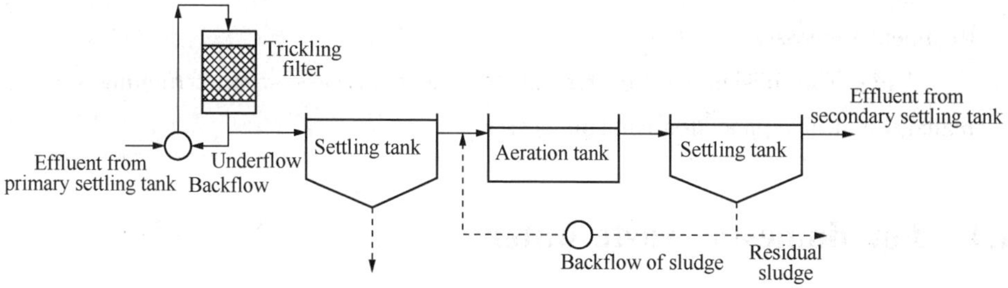

Figure 3-32 Trickling filter/Activated sludge

Questions

1. What is biofilm? What difference between the biofilm and activated sludge?

2. Please describe the formation process of biofilm?

3. The formation mechanism and the clarification mechanism of biofilm.

4. The principles of the biofilm.

5. What are the characteristics of biofilm waste water treatment system in terms of microbial cultivation and treatment process?

6. What are the factors that affect the treatment efficiency of biofilter? How do they affect treatment?

7. The distinction of biofilters, rotating biological disc, biological contactor and submerged biofilm reactor?

8. What is the importance of backflow of effluence in the bio-filters?

9. What conditions we can modify activated sludge process with biofilm?

10. In the one-stage high rate biofilter, the backflow of water can be the effluence of secondary sedimentation tank or high rate biofilter, please give the explanation.

Chapter 4 Typical treatment processes on domestic and industry waste water

Emphasis: The designment of typical treatment processes. The types and structure of treatment processes. The running and management of typical treatment processes.

Nut: The designment of typical treatment processes. The running and management of typical treatment processes.

4.1 For domestic waste water

4.1.1 Oxidation ditch

4.1.1.1 The overview

The oxidation ditch (Figure 4 - 1), also named continuous loop reactor, was developed in the Netherlands with the first full scale plant installed in Voorschoten, Holland, in 1954. An oxidation ditch is an extended aeration activated sludge treatment process that has been used in small towns, isolated communities, and institutions. Some oxidation ditches are named according to manufacturer's trade names such as Pasveer, Orbal, or Carrousel. Typical oxidation ditch (Figure 4 - 2) consists of a single or multichannel of an oval or ring shape equipped with aeration equipment called rotors or brushes for generating a water flow and stirring water in the channel to supply oxygen. The only pretreatment typically used in an oxidation ditch system is the bar screen (primarily settling before an oxidation ditch is sometimes practiced). After passing through the bar screen, waste water flows directly into the oxidation ditch. Ditches may be constructed of various materials, such as concrete, gunite, asphalt, and impervious membranes. Concrete is the most common. L-and horseshoe-shaped configurations have been constructed to maximize land usage. The waste water moves through the ditch at 0.3 to 0.5 m/s. The cross-sectional area of the ditch is commonly 0.9 to 1.5 m deep, which helps to prevent anaerobic conditions from occurring at the bottom of the ditch. Oxidation ditch systems with depths of 2.5 m or more with vertical sidewalls and vertical shaft aerators may also be used. Activated sludge is added to the oxidation ditch so that the microorganisms will digest the BOD in the water. Longer retention time within the ditch will allow for a greater amount of organic matter to be broken down by the aerobic bacteria. After treatment, the mixed liquor flows to a secondary clarifier where the sludge and the water will be allowed to be separated. Some of the sludge is returned to the oxidation ditch to maintain a desirable MLSS concentration. The MLSS concentration in the oxidation ditch generally ranges from 1, 500 to 5, 000 mg/ L. The rest of the sludge is sent to waste.

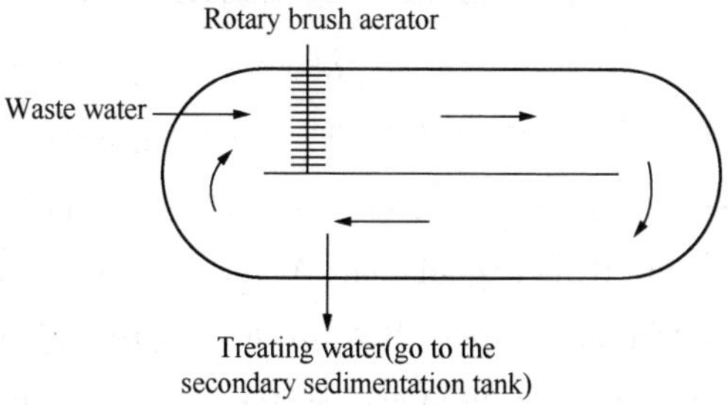

Figure 4-1 Plane graph of oxidation ditch

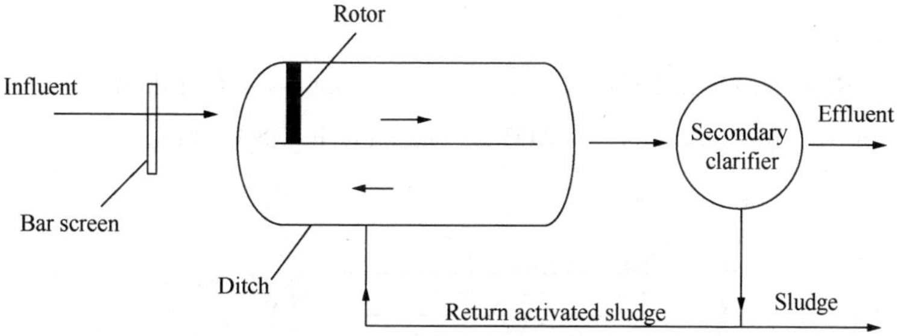

Figure 4-2 Typical oxidation ditch flow diagram

4.1.1.2 The typical oxidation ditch

(1) Carrousel oxidation ditch At 1960s, the carrousel oxidation ditch system is invented by the DHV company in Holland. It included the series oxidation ditch, the secondary settling tank and the backflow of sludge (Figure 4-3).

① Common carrousel oxidation ditch. The common carrousel oxidation ditch is showed in the Figure 4−3. The raw waste water passes the screen and grit chamber, and then goes to the oxidation ditch. The biodegradation of BOD, nitrification, denitrification appears in the one tank. In fact, the system is similar as the A/O process.

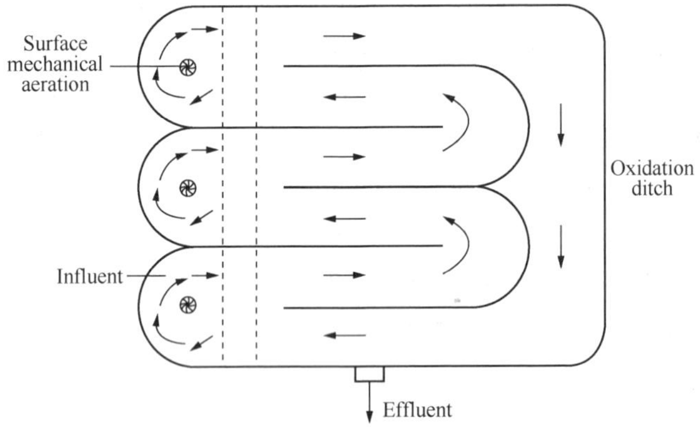

Figure 4−3 The sketch of the common carrousel oxidation ditch

In the ditch, the content of the DO is about 2 ~ 3 mg/L due to the surface aerator. It is enough for the BOD degradation. At the same time, the ammonia nitrogen is transferred into the nitrite and nitrate. But, at the downstream of the surface aerator, the mixed liquid is under anoxic conditions, so there is denitrification. After this, the mixed liquid is back to the upstream of the surface aerator (aerobic conditions).

Generally, the removal of BOD, COD, N and P is 95%, 90%, 75% and 65%, respectively.

② Carrousel 2000 oxidation ditch. Carrousel 2000 oxidation ditch is a process with the front denitrification function by EIMCO company in USA (Figure 4−4).

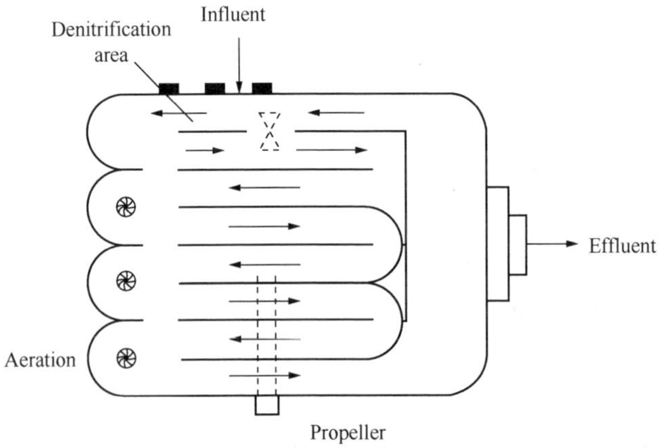

Figure 4−4 The sketch of the carrousel 2000 oxidation ditch

During the process, the mixed liquid is recycled to the front area by the propeller and the amount of recycled mixed liquid is regulated by the cartridge valve. The denitrification reaction appears in the front area under the conditions of denitrifying bacteria. The removal of N is strengthened in the carrousel 2000 oxidation ditch due to the lots of mixed liquid backflow and the excellent nitrification. In addition, the biological removal of P is realized under the conditions of phosphorus-accumulating bacteria. In practice, the volume of the front area is only 10% of the total volume in the oxidation ditch.

Generally, the removal of BOD, COD, N is 98%, 95%, 95%, respectively. The P in the effluent is reduced to 1~2 mg/L. However, the surface aerator limits the water depth of the oxidation ditch, less than 4.5 m. The energy efficiency is only 1.8 $kgO_2/(kW \cdot h)$, indicating that the energy consumption is high. If the microporous aeration is used in the carrousel 2000 oxidation ditch, resulting in the increase of the water depth, 5.8 m. In addition, the energy efficiency is about 2.5~3.0 $kgO_2/(kW \cdot h)$.

③ Carrousel 3000 oxidation ditch. Carrousel 3000 oxidation ditch is called deep carrousel oxidation ditch. The water depth is up to 7.5~8.0 m. The special wrap-around designment of the carrousel 3000 oxidation ditch is favor of the saving of the occupied area and the investment (Figure 4-5). The component of the oxidation ditch is as followed. At the center, there are series cycle units which are used for the inflowing and the backflow of sludge. Outside of the center, there are two circles which include the four stages. The circles are used for the selector and anaerobic tanks, respectively. Outside of the tanks, there are three surface aerators and the main carrousel 2000 reaction tank for pre-denitrification.

The biggest advantage of the oxidation ditch is the pre-denitrification tank, which is favor of the sufficient denitrification using the biodegradable organic matter.

(2) Alternating oxidation ditch The oxidation ditch is invented by the Kruger company in Demark. Alternating oxidation ditch includes two ditches (Figure 4-6) and three ditches (Figure 4-7). The former is called D mode, and the latter is called T mode. For the two mode, there are no secondary settling tank. The water depth is 2~3.5 m.

①D mode. The volume of the two ditches in the D mode is the same and there is the connection pipe between the two ditches, resulting in the same water depth. There is no secondary settling tank in the D mode. The running process is showed in (Figure 4-8). Each period includes inflowing, aeration and sedimentation.

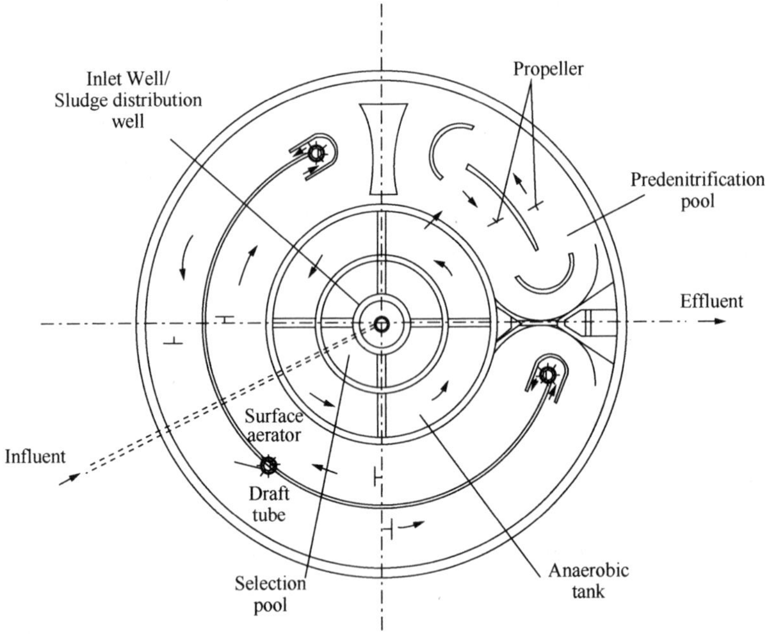

Figure 4-5　The sketch of the carrousel 3000 oxidation ditch

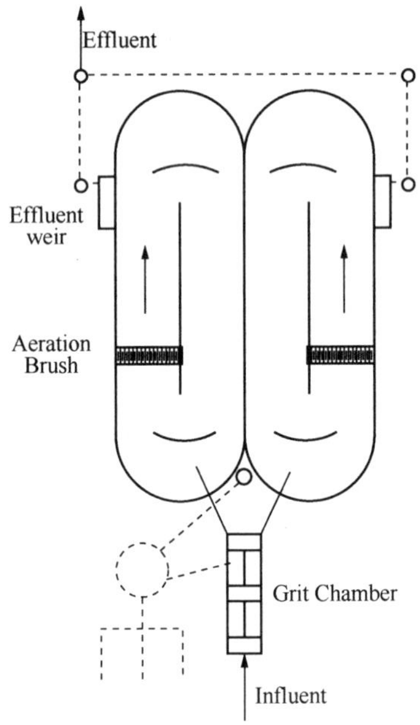

Figure 4-6　Alternating oxidation ditch with two ditches

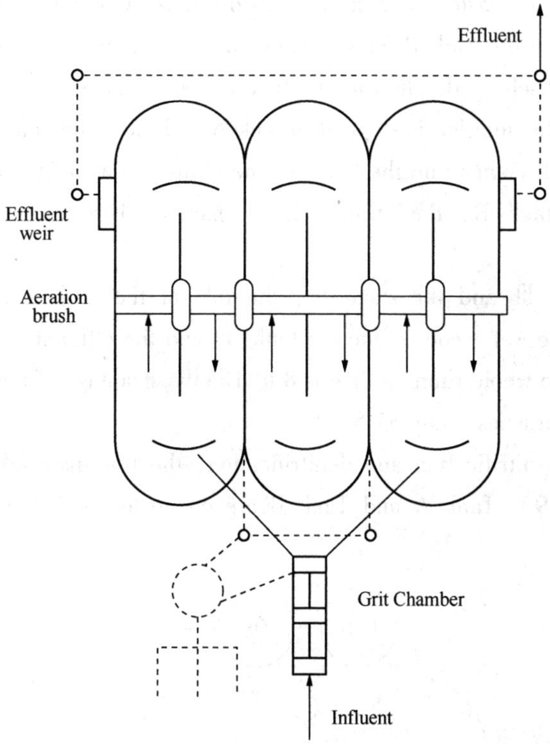

Figure 4-7 Alternating oxidation ditch with three ditches

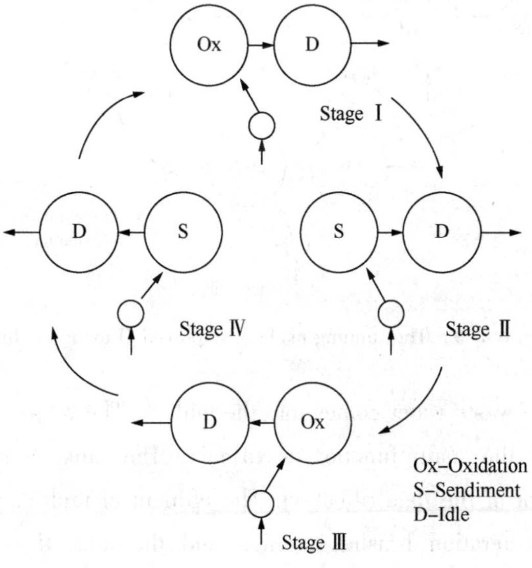

Figure 4-8 The running mode of D oxidation ditch

At stage Ⅰ, the waste water comes into the tank A and is aerated and then the mixed liquid goes to the tank B for sedimentation to separate the solid and liquid. The clarified water is discharged. The running time is 3 h in stage Ⅰ.

At stage Ⅱ, the aeration is stopped in tank A and the sedimentation is starting, but the waste watcr still comes into the tank A. The effluent of tank A goes to tank B. After the treatment in tank B, the effluent is discharged. The running time is 1 h in stage Ⅱ.

For the stage Ⅲ and the stage Ⅳ, the role of the tank A and the tank B is changed. The waste water comes into the tank B, and the effluent is in tank A.

Therefore, the whole running time is 8 h. The disadvantage of the process is the low usage of aeration brushes, only 37.5%.

To realize the nitrification and denitrification, the running mode was improved at 1970s (Figure 4-9). Tank A and Tank B are connected with the secondary settling tank.

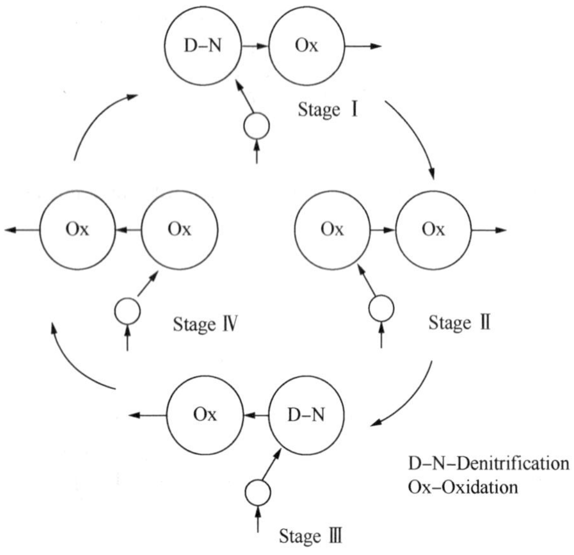

Figure 4-9 The running mode of improved D oxidation ditch

At stage Ⅰ, the waste water comes into the tank A. The rotate speed of the aeration brushes is low and the main function is to mix. The tank A is under the anoxic conditions. It is favor of the denitrification. The effluent of tank A goes to tank B. The rotate speed of the aeration brushes is high and the tank B is under the aerobic conditions.

At stage Ⅱ, the rotate speed of the aeration brushes is high in tank A and tank B. The BOD removal and nitrification occur at the same time.

For the stage Ⅲ and the stage Ⅳ, the role of the tank A and the tank B is changed. The waste water comes into the tank B, and the effluent is in tank A. The running is repeated similar as the stage Ⅰ and stage Ⅱ.

② T mode. The running of T mode is similar as D mode. The tank B is continuous aeration. The other tanks are oxidation and sedimentation alternating or denitrification and sedimentation alternating. There is no secondary settling tank. The advantage of the system is the increase of the usage of aeration brushes, up to 58.5%. Another advantage is to realize the nitrification and denitrification.

(3) Orbal oxidation ditch Orbal oxidation ditch was designed by Husiman in South Africa and then was transferred to the Envirex company.

Orbal oxidation ditch is composed by the three circle or oval ditches, and the ditches are connected by the holes. The raw waste water comes into the outside ditch and then comes into the other two ditches in turn. The process is similar as the series completed mixing aeration tank. The intake lies at the downstream of the exit of the effluent (Figure 4-10). Generally, the aerated disc was used in the Orbal oxidation ditch. The numbers of the discs depend on the content of the dissolved oxygen. The water depth is 2~3.6 m and the speed in the ditch is 0.3~0.9 m/s.

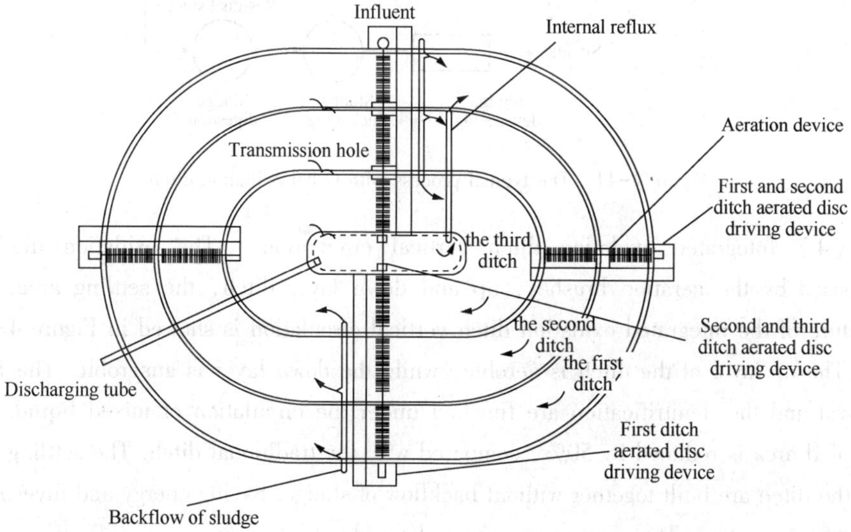

Figure 4-10 The sketch of Orbal oxidation ditch

For the three ditches, from outside to inside, the volume of the first ditch to the total volume is about 50% ~ 55%, and the second ditch is about 30% ~ 35%, and the third ditch is about 15% ~ 20%. During running, the content of the DO is 0 mg/L, 1 mg/L and 2 mg/L, respectively. In the first ditch, the nitrification and denitrification simultaneous occur, but the nitrification and BOD removal depend on the content of DO. The usage of the DO is very high in the first ditch. Although the oxygen supply is over 90% in the first ditch, the content of DO is close to 0. In the other ditches, the oxygen usage is lower, compared with the first ditch. The oxygen supply has some advantages.

① Saving energy and meeting the demand of nitrification. The oxygen supply not only meet the demand to degrade BOD but also keep at the level of 0 mg/L.

② It is favor of the P release from the phosphorus-accumulating bacteria. It is better to adsorb excessively P under the aerobic conditions.

The typical process with Orbal oxidation ditch is showed in the Figure 4-11.

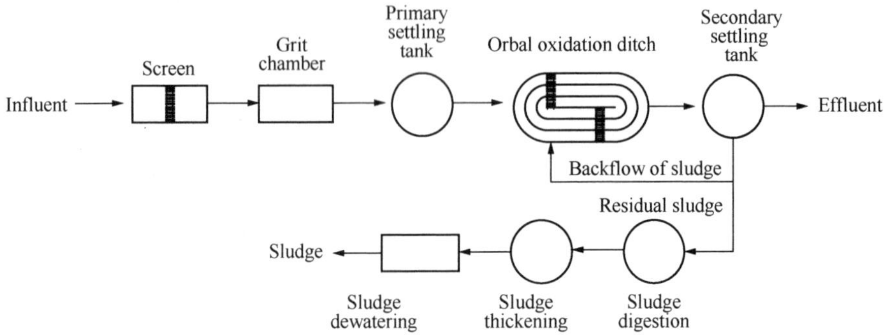

Figure 4-11 The typical process with Orbal oxidation ditch

(4) Integrated oxidation ditch vertical circulation The oxidation ditch is composed by the aeration brushes, up and down layer ditch, the settling area. The structure of the integrated oxidation ditch vertical circulation is showed in Figure 4-12.

The up layer of the ditch is aerobic, while the down layer is anaerobic. The BOD removal and the denitrification are finished under the circulation of mixed liquid. The occupied area is reduced by 50%, compared with the traditional ditch. The settling area and the ditch are built together without backflow of sludge. Saving energy and investment on the construction. The running is easy and simple.

HRT is the most important factor affecting the pollutants removal in the ditch. The COD removal is less than 75% when COD content in the influent is 480 mg/L and water

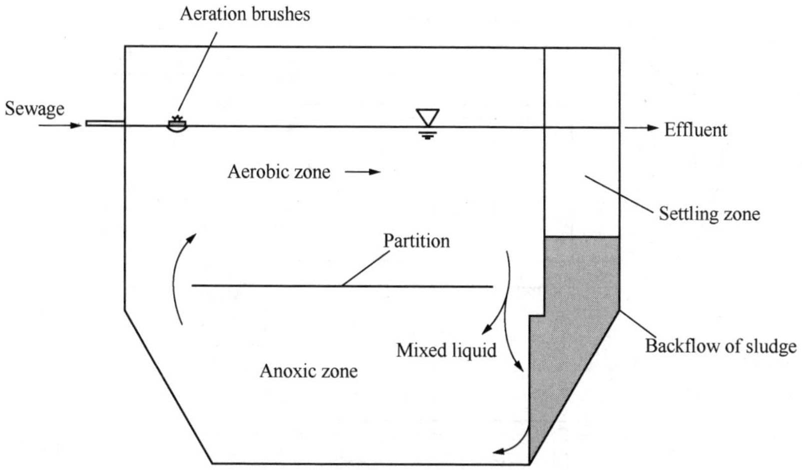

Figure 4-12 The integrated oxidation ditch vertical circulation

temperature is about 23℃, and HRT is less than 5 h. If HRT is more than 6 h, the COD removal is more than 90%. If HRT is more than 10 h and SRT is 30 d, the ammonia nitrogen removal is over 99% and the TN removal is over 90%.

There are some integrated oxidation ditches, such as oxidation ditch with lateral canal (Figure 4-13), BMTS oxidation ditch (Figure 4-14).

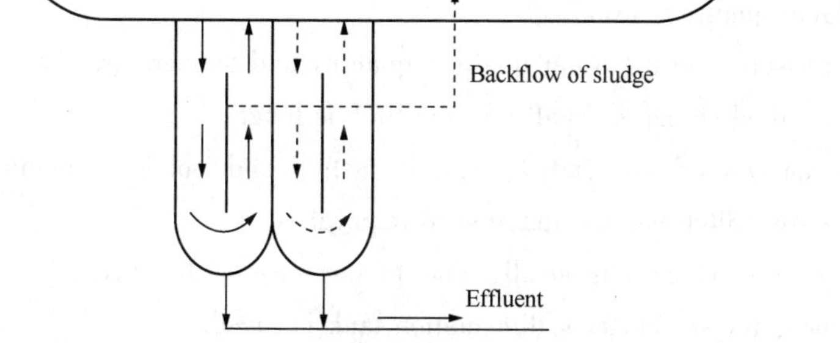

Figure 4-13 The sketch of oxidation ditch with lateral canal

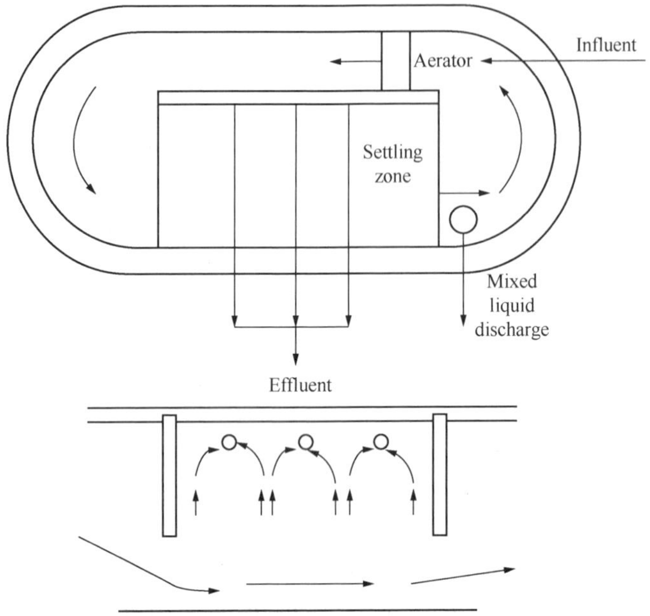

Figure 4-14 The sketch of BMTS oxidation ditch

4.1.1.3 The characteristics

① The ditch is long and narrow, up to tens of meters, even hundreds of meters. The depth of the ditch is shallow, in general 2 m.

② In practice, the surface blasting machine is used widely. The equipment of the influent and effluent is simple.

③ The flow in the ditch is completely mixed-plug flowing. The load of BOD is low. The effluent quality is excellent.

④ The resistance to the water quality, quantity and temperature shock is strong.

⑤ The residual sludge is small and the SRT is long.

⑥ The removal of suspended organic matter and soluble organic matter is complete. The oxic ditch has the function to removal N.

⑦ The occupying area is small, due to the save of the primary sedimentation tank. Sometimes, the secondary sedimentation tank is saved.

4.1.1.4 The shapes

There are many shapes for the oxic ditch. The typical shape is carousel oxidation ditch, orbal oxidation ditch, phased isolation ditch (VR、DE and T), aeration - precipitation integrated oxidation ditch, et al. The ditch maybe circle or ellipse and can be single ditch or several ditch.

4.1.1.5 The aeration

Surface aerators, such as brush rotors, disc aerators, draft tube aerators, and fine bubble diffusers, are used to circulate the mixed liquor. The mixing process entrains oxygen into the mixed liquor to foster microbial growth and the motive velocity ensures contact of microorganisms with the incoming waste water. The aeration sharply increases the dissolved oxygen concentration but decreases as biomass uptakes oxygen as the mixed liquor travels through the ditch. Solids are maintained in suspension as the mixed liquor circulates around the ditch. If design SRT is selected for nitrification, a high degree of nitrification will occur.

4.1.1.6 The designment

The designment parameters are as followed for the municipal waste water.

MLSS is 5, 000 mg/L. MLVSS is 2, 000~4, 000 mg/L. SRT is 5~8 d. HRT is 20~48 h. N_s is 0.03~0.07 kgBOD/(kg MLSS · d). N_v is 0.1~0.2 kgBOD/(m^3 · d). R is 50%~150%. The velocity of the mixed liquid in the ditch is about 0.4~0.5 m/s. The velocity at the bottom is 0.3 m/s.

(1) The volume of the oxic ditch.

① The volume of the oxic ditch. The volume is calculated by the following equation.

$$V_1=\frac{YQ(S_0-S_e)\theta_c}{X(1+K_d\theta_c)}$$

In the equation, V_1—The effective volume of the oxic ditch, m^3;

Y—Yield coefficient, kgVSS/kgBOD$_5$;

Q—Waste water amount, m^3/d;

S_0—The content of BOD$_5$ in the inflowing, mg/L;

S_e—The content of BOD$_5$ in the effluent, mg/L;

θ_c—The sludge age, d;

K_d—Endogenous respiration coefficient, d^{-1};

X—The content of MLVSS in the mixed liquid , mg/L.

In practice, the X is between 2, 000~4, 000 mg/L. The Y and k_d is gotten by the experienced data or according to the Table 4-1.

Table 4-1 The data for Y and k_d for the waste water

Parameters	Domestic	Defatted milk	Synthetic	Paper making	Municipal
$Y/(kgVSS/kgBOD_5)$	0.50~0.67	0.48	0.65	0.47	0.35~0.45
k_d/d^{-1}	0.048~0.050	0.045	0.18	0.20	0.05~0.10

② The volume of the anoxic area. If the oxic ditch has the requirement on the denitrification, the volume for the nitrification should be considered. Therefore, the volume of anoxic area (V_2) is calculated by the following equation.

$$V_2 = \frac{N_T}{q_{N,T} X}$$

$$q_{D,T} = q_{D,20℃}\theta_c^{T-20}$$

In the equation, V_2—The volume of anoxic area, m^3;

N_T—The amount of the nitrate, kg/d;

$q_{D,T}$—The ratio of the denitrification at temperature T℃, $gNO_3-N/(gVSS \cdot d)$;

$q_{D,20℃}$—The ratio of the denitrification at temperature 20℃, $gNO_3-N/(gVSS \cdot d)$;

θ_c—SRT, d.

③ The total volume of the oxic ditch:

$$V = V_1 + V_2$$

④ The retention time of the sludge. If the oxic ditch has the requirement on the denitrification, the nitrifying bacteria is the deciding factor on the retention time of the sludge. Therefore, the smallest SRT has relationship with the biggest specific growth rate of the nitrifying bacteria.

$$\theta_c^m = \frac{1}{\mu_N}$$

$$\mu_N = 0.47e^{0.098(T-15)} \times \frac{N}{K_N + N} \times \frac{DO}{K_O + DO}$$

In the equation, θ_c^m—The smallest sludge age, d;

μ_N—The specific growth rate of the nitrifying bacteria, d^{-1};

T—Water temperature, ℃;

N—The content of the NH_4^+-N, mg/L;

K_N—The saturation constant of the nitrosobacteria to oxidize NH_4^+-N, mg/L;

DO—The content of the dissolved oxygen, mg/L;

K_0—The constant of the DO, mg/L.

(2) The oxygen demand The oxygen demand includes the BOD_5 removal, ammonia nitrogen removal. On the other hand, the making of the oxygen by the denitrification process, the breeding of the activated sludge and the discharge of sludge resulting in the loss of BOD_u and ammonia nitrogen, for the above process, there is no oxygen consumption.

Generally, the nitrification is as followed.

$$NH_4^+ + 2O_2 \longrightarrow NO_3^- + 2H^+ + H_2O$$

The denitrification is as followed.

$$NO_3^- + 1.08CH_3OH + 0.2H_2CO_3 \longrightarrow 0.056C_5H_7O_2N + 0.47N_2 + 1.68H_2 + HCO_3$$

According to the nitrification and denitrification equation, the consumption of oxygen is 4.33 g O_2 to remove 1 g NH_3-N, and the consumption of the carbon source is 2.47 g methyl alcohol (about 3.7 mg COD) to reduce 1 g NO_3-N.

The transformation from the methyl alcohol to BOD is as followed.

$$2CH_3OH + 3O_2 \longrightarrow 2CO_2 + 4H_2O$$

According to the above equation, the consumption of oxygen is $1.05BOD_u$ to remove methyl alcohol. Therefore, the degradation on the BOD_u is 2.6 mg BOD_u to reduce 1 g NO_3-N. The removal on the BOD_u only consume the chemical oxygen in the NO_3-N.

As for the discharge of the residual activated sludge, the loss of BOD_u is 1.42 times on the VSS. In addition, it is assumed that the N content is 12.4% to the VSS. It is indicating that the two above parts do not consume oxygen. The equation is as following.

$$1.42P_x \frac{VSS}{SS} + 4.33 \times P_x \frac{VSS}{SS} \times 12.4\% = 1.42P_x \frac{VSS}{SS} + 0.54P_x \frac{VSS}{SS}$$

In the equation, P_x—The amount of the discharged residual sludge.

Therefore, the oxygen demand is calculated by the following equation.

$$O_2 = Q \times \frac{S_0 - S_e}{1 - e^{-kt}} - 1.42P_x \frac{VSS}{SS} + 4.33Q(N_0 - N_e) - 0.54P_x \frac{VSS}{SS} - 2.6Q\Delta NO_3$$

In the equation, O_2—The oxygen demand for the removal of BOD_5 and nitrate, kgO_2/d;

Q—The amount of the waste water, m^3/d;

S_0—The content of BOD_5 in the influent, mg/L;

S_e—The content of BOD_5 in the effluent, mg/L;

P_x—The amount of the discharged residual sludge, kg/d;

$\frac{VSS}{SS}$—The percentage of the volatile solids in the activated sludge, %;

N_0—The content of ammonia nitrogen in the influent, mg/L;

N_e—The content of ammonia nitrogen in the effluent, mg/L;

ΔNO_3—The content of nitrate which is reduced, mg/L.

(3) The alkalinity Generally, the removal of 1 mg BOD_5 is to produce 0.1 mg alkalinity ($CaCO_3$). The reduction of 1 mg NO_3-N to produce 3.0 mg alkalinity ($CaCO_3$). The oxidation of 1 mg ammonia nitrogen is to consume 7.14 mg alkalinity ($CaCO_3$). According to the above analysis, the residual alkalinity is calculated by the minus of the raw waste water and the above data. If the residual alkalinity is bigger than 100 mg/L, the value of pH is more than 7.2 and kept constant.

(4) The secondary settling tank and the backflow of sludge.

① The secondary settling tank. For the designment of the secondary settling tank, the advised parameters are as followed. The hydraulic surface loading is 12.6 ~ 21.0 $m^3/(m^2 \cdot d)$. The suspended solid loading is 20 ~ 100 $kgSS/(m^2 \cdot d)$. The overflow weir loading is 126 ~ 190 $m^3/(m \cdot d)$.

② The backflow sludge. The amount of the backflow sludge is calculated by the following equation.

$$QX_0+Q_rX_r=(Q+Q_r)X$$

In the equation, Q—The amount of the waste water, m^3/d;

Q_r—The amount of the backflow sludge, m^3/d;

X_0—The content of VSS in the influent, mg/L;

X_r—The content of VSS in the backflow sludge, mg/L;

X—The content of VSS in the mixed liquid, mg/L.

③ The backflow sludge. The amount of the backflow sludge (P_x) is calculated by the following equation.

$$P_x=Q\Delta S\left(\frac{Y}{1+K_d\theta_c}\right)+X_1Q-X_eQ$$

In the equation, ΔS—Theremovalcontent of BOD_5, mg/L;

X_1—The content of the inert solids in the influent, mg/L;

X_e—The content of the total suspended solids in the effluent, mg/L.

4.1.1.7 The advantages and disadvantages

The greatest advantage of the oxidation ditch is the efficiency of removal performance objectives with low operational requirements and operations and maintenance costs. Other advantages of oxidation ditches include ① producing less sludge than other biological treatment processes owing to extended biological activity during the activated sludge process, ② it is hard to be affected by load fluctuations, ③ it requires relatively little energy as the rotor is operated efficiently, and ④ it can be easily maintained. Disadvantages of the oxidation ditch include: ① as the tank is large and the depth is small it requires a large area, and ② effluent suspended solids concentration is relatively high compared to other modifications of the activated sludge process.

4.1.2 Sequencing batch reactor

The sequencing batch reactor activated sludge process (SBR) was put forward by British scholar Ardern in 1914. Because of low automation level, frequent switch operation of inlet and outlet valves brings inconvenience to operation and is hard to apply to production. With the improvement of the level of automation and the development of computers, SBR has the widespread application.

It is obvious that the principle of SBR is the same as the CFS (Continuous Flow System Activated Sludge Process). The key equipment in the SBR process is the SBR reactor (Figure 4-15). Compared with CFS (Figure 4-16), SBR process saves lots of structures. But the main difference is the running mode. For the CFS, The BOD removal is along the space by the continuous running, while for the SBR, all the reactions are observed in the SBR reactor. The BOD removal is along the time by the time control (Figure 4-17).

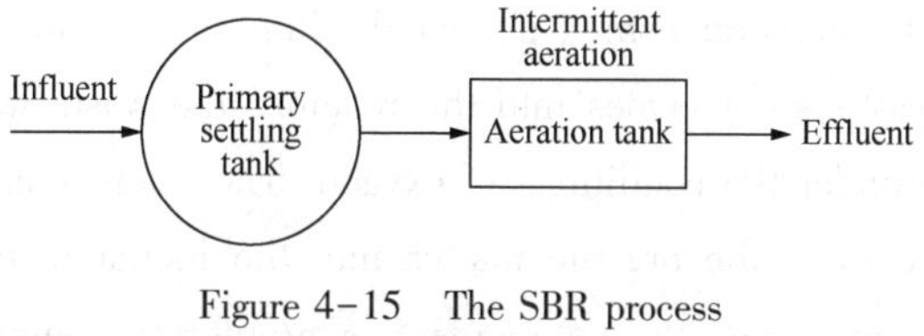

Figure 4-15 The SBR process

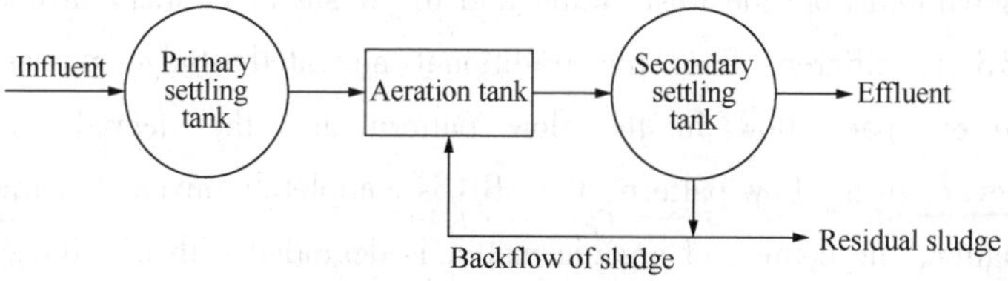

Figure 4-16 The CFS process

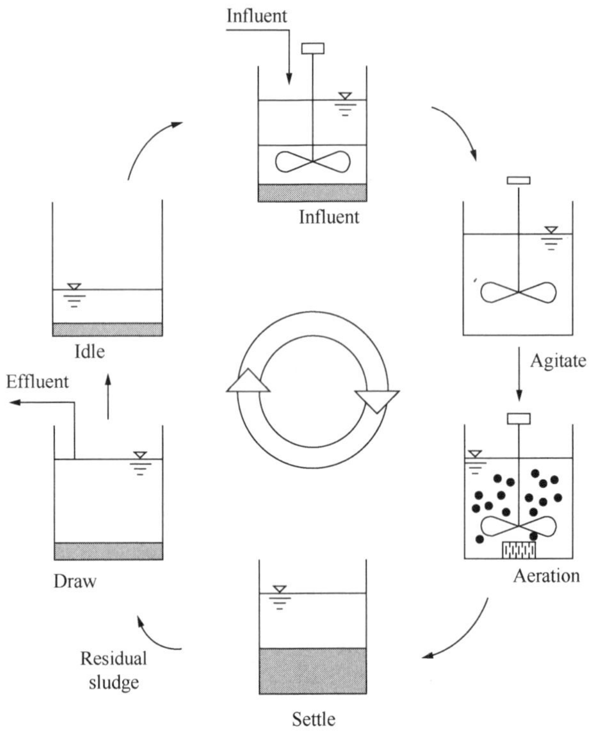

Figure 4-17 The running of the SBR process

4.1.2.1 The process of SBR

SBR consists aeration device, supernatant discharge device (decanter), reactor and other ancillary equipment. The removal mechanism of the organic matter is as followed. A part of microorganism (activated sludge) should be cultured in the reactor. When the waste water comes into the reactor, the waste water was mixed with the activated sludge under the conditions of oxygen. The microorganisms metabolize the organic matter and convert the organic matter into the inorganic compounds, such as CO_2, H_2O. At the same time, the microorganism proliferation appeared. The activated sludge was separated from the waste water and the waste water was clarified.

The SBR is different from the traditional activated sludge process. It is the characteristic of space flow in the flow pattern and the degradation of organic matter. However, in the flow pattern, the SBR is completely mixed. For the degradation of organic matter, the content of organic matter is degraded with the time.

The SBR reactor has 5 stages: water intake, reaction, sedimentation, effluent and idle. Batch type is the main characteristic for each stage. The sewage enters each reactor in succession according to the sequence. During the runtime, the relative order is orderly, and it is intermittent.

In the influent stage, the SBR tank is full of the high content of activated sludge, so the SBR is similar as variable volume reactor. The substrate concentration in the mixed liquid increases gradually on the base of the supernatant matrix concentration of the remaining sludge. At the end of the flushing stage, the aeration tank is full and the substrate concentration reaches the maximum. The adsorption, absorption and oxidation coexist during the biochemical reactions.

During the reaction period, the microorganisms are periodically in the environment of high and low substrate concentration under the conditions of mix and aeration. Therefore, there is the anaerobic, anoxic and aerobic process in alternation. The SBR has the function of removing organic matter, nitrogen and phosphorus.

During the sedimentation period, the settling of activated sludge was finished under relatively static conditions, avoiding that the mixed liquid continuously comes into the sedimentation tank by the pipes in the continuous activated sludge process, and avoiding that the flocculated activated sludge break up again. The sedimentation period is generally 1 ~ 2 hours.

During the period of drainage and sludge discharge, the treated water has been discharged to the lowest water level. Most of the activated sludge deposited at the bottom of the tank is used as the next cycle of reflux sludge, and excess sludge is discharged.

The role of the idle period is to restore the activity of the sludge by mixing, aerating and standing, and denitrification appeared to removing the nitrogen. At the same time, the microorganism per unit mass has a large adsorption surface area, and the activated sludge plays its adsorption capacity when it enters the next cycle.

4.1.2.2 SBR characteristics

① Flexible operation and strong adaptability.

② The occupied area is small and the investment is low.

③ The efficiency on the removal of pollutants is high.

④ The removal of N and P is excellent.

⑤ The sludge settle ability is good.

⑥ The adaptation is strong.

⑦ The safeguard and the management are easy.

4.1.2.3 The designment of SBR

The design formulas, the design parameters of the traditional activated sludge can be used for the designment of SBR.

① The volume of the SBR:

$$V=\frac{nQS_0}{N_v}$$

In the equation, V—The effective volume of the reactor, m^3;

n—The running period in a day;

Q—The waste water inflow amount in each period, m^3;

S_0—The average content of BOD_5 in raw waste water, $kgBOD_5/m^3$;

N_v—The volumetric loading, $kgBOD_5/(m^3 \cdot d)$.

Generally, the N_v of the SBR is 0.1~1.3 $kgBOD_5/(m^3 \cdot d)$. The N_v is about 0.1 $kgBOD_5/(m^3 \cdot d)$. When the BOD_5 in raw waste water is higher than 1,000 mg/L, the V goes to $2QS_0$. When the BOD_5 in raw waste water is lower than 1,000 mg/L, the V goes to $2Q$.

② The biggest waste water amount and the smallest waste water amount. The biggest waste water amount (V_{max}), which is referred to the amount during the reaction, is equal to the volume of the aeration tank (V).

The smallest waste water amount (V_{min}), which is referred to the amount during the sedimentation, is the residual water amount in the aeration tank. But, the interface of the sludge and the residual water is higher than the water table of the smallest waste water amount, the part of the activated sludge is lost. Therefore, the amount of the activated sludge is lower than the smallest waste water amount. The smallest waste water amount is gotten by the following equation.

$$V_{min}=V_{max}-Q$$

The relationship between SV_{30} (%), MLSS (mg/L) and SVI (mL/g) is as followed.

$$SV_{30}=\frac{\text{SVI}\times\text{MLSS}}{1000}$$

So,

$$\frac{SV_{30}\times V_{max}}{100}<V_{min}$$

The following equation is gotten.

$$\frac{V_{min}}{V_{max}} > \frac{SVI \times MLSS}{10^6}$$

Therefore,

$$\frac{Q}{V_{max}} \leqslant 1 - \frac{MLSS}{10^6} \times SVI$$

$$N_v < nS_0 \left(1 - \frac{MLSS}{10^6} \times SVI\right)$$

The period is calculated by the above equations for the waste water with high content of organic matter or the low content of organic matter.

In general, for the influent, if the MLSS is 2, 000 mg/L, and the SVI is 100 mL/g, the reactor can degrade the waste water amount with 80% of the volume of the reactor. If the SVI increases to 200 mL/g, the reactor can degrade the waste water amount with 60% of the volume of the reactor. if the MLSS is 5, 000 mg/L, and the SVI is 100 mL/g, the reactor can degrade the waste water amount with 50% of the volume of the reactor. if the MLSS is 5,000 mg/L, and the SVI is 200 mL/g, the running of the reactor is impossible.

4.1.2.4 The development of SBR

The technology of SBR is applied more and more in domestic water treatment plant, the processes are richer and richer based on the SBR, such as Intermittent Cycle Extended Aeration System (ICEAS), Cyclic Activated Sludge System (CASS), Demand Aeration Tank-Intermittent Aeration Tank (DAT-IAT), Modified Sequencing Batch Reactor (MSBR), UNITANK.

(1) Intermittent Cycle Extended Aeration System Intermittent Cycle Extended Aeration System (ICEAS) is the transformation of SBR. It was invented by The University of New South Wales in Australia and the ABJ company in USA. The running mode is the continuous inflowing and the intermittent discharge to resolve the intermittent inflowing problem in SBR.

Compared with SBR, there are two changes in ICEAS. The one is the change of running mode—the continuous inflowing and the intermittent discharge. During the settling stage and decanting stage, the water is inflowing, while there is no inflowing stage and idle stage for the reactor. The other is the change of the structure. There is a pre-reaction area in the reactor. The ratio of the pre-reaction volume to the main reaction area is about 1 : 30. The structure of ICEAS is showed in Figure 4-18.

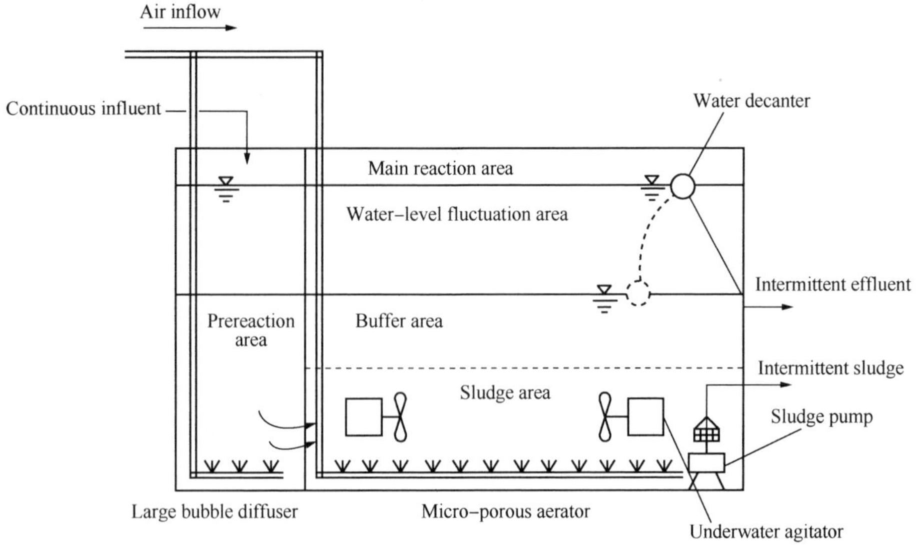

Figure 4-18 The structure of ICEAS

The waste water continuously comes into the pre-reaction area and then goes to the main reaction area by the small hole closing to the bottom. The most important is there is no effect of the coming of the waste water on the settling in the main reaction. Therefore, it is to realize the simultaneous settling and decanting without affecting the treatment on the waste water.

For the ICEAS, the BOD removal and nitrification and denitrification and P removal and the separation of solid and liquid and the decanting are finished in one tank. There is no backflow of sludge and mixed liquid. The typical running period includes aeration, settling and decanting due to the continuous inflowing.

After the grit chamber, the waste water comes into the pre-reaction area. The BOD content in the area is very high, resulting in the higher ratio of F/M. It is favor of promoting the adsorption on the organic matter by the bacteria and quickens the removal of dissolved organic matter. In addition, the pre-reaction area a biological selector, improving the growth of the flocculated bacteria and handling the growth of filamentous bacteria. The pre-reaction resists the shock of inflowing quality and quantity, pH and toxic matter. During the reaction, the waste water passes the repeated aerobic condition and anoxic condition. This is better to promote the BOD degradation, nitrification, denitrification and P removal. Generally, the period of ICEAS is 4 ~ 5 h. The biggest advantage of ICEAS is the little residual sludge due to the long SRT.

(2) Cyclic Activated Sludge System Cyclic Activated Sludge System (CASS) came from ICEAS. It was invented by Prof Goronszy by combination of biological selector and SBR at 1970s.

There are two changes for CASS, compared with ICEAS. The one is changing the two reactions (ICEAS) to three reactions (CASS). The other is the backflow of sludge in the main reaction respectively or simultaneously to the biological selector and facultative area.

CASS structure is showed in Figure 4 – 19. The reactor includes biological selector (A), facultative area (B) and main reaction area (C). Area A is a relatively independent area. Area B and area C are divided by a plate, but the two areas are connected by the small hole at the bottom. The ratio of the volume of the three areas is 5 : 10 : 85.

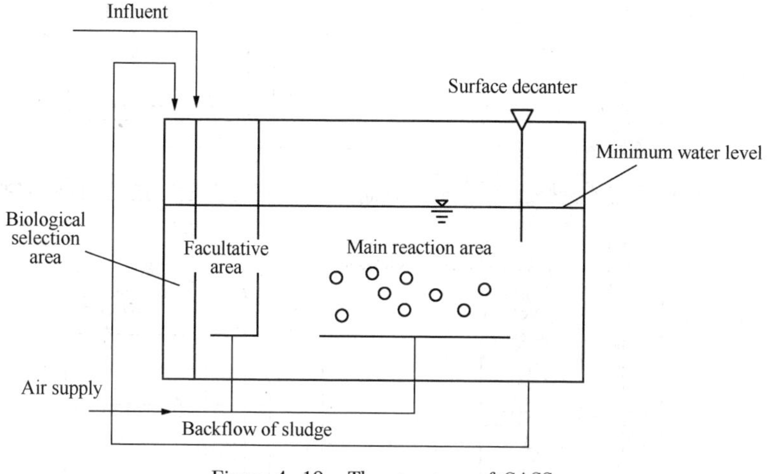

Figure 4-19 The structure of CASS

During the process, the raw waste water comes into the area A. The inflowing from area A to area B is controlled by the valve or the pump. Area C is the main reaction area. Its treatment period includes inflowing-aeration, inflowing-settling, decanting, inflowing-idle. The inflowing from area A to area B is stopped when area C is discharging. Area B is the pre-reaction area for area C. Most of the organic matter is removed by the adsorption in area A and area B to stabilize the influent of area C. After the stabilization in area C, the sludge is returned to the area A and area B.

The running of CASS is showed in the Figure 4-20.

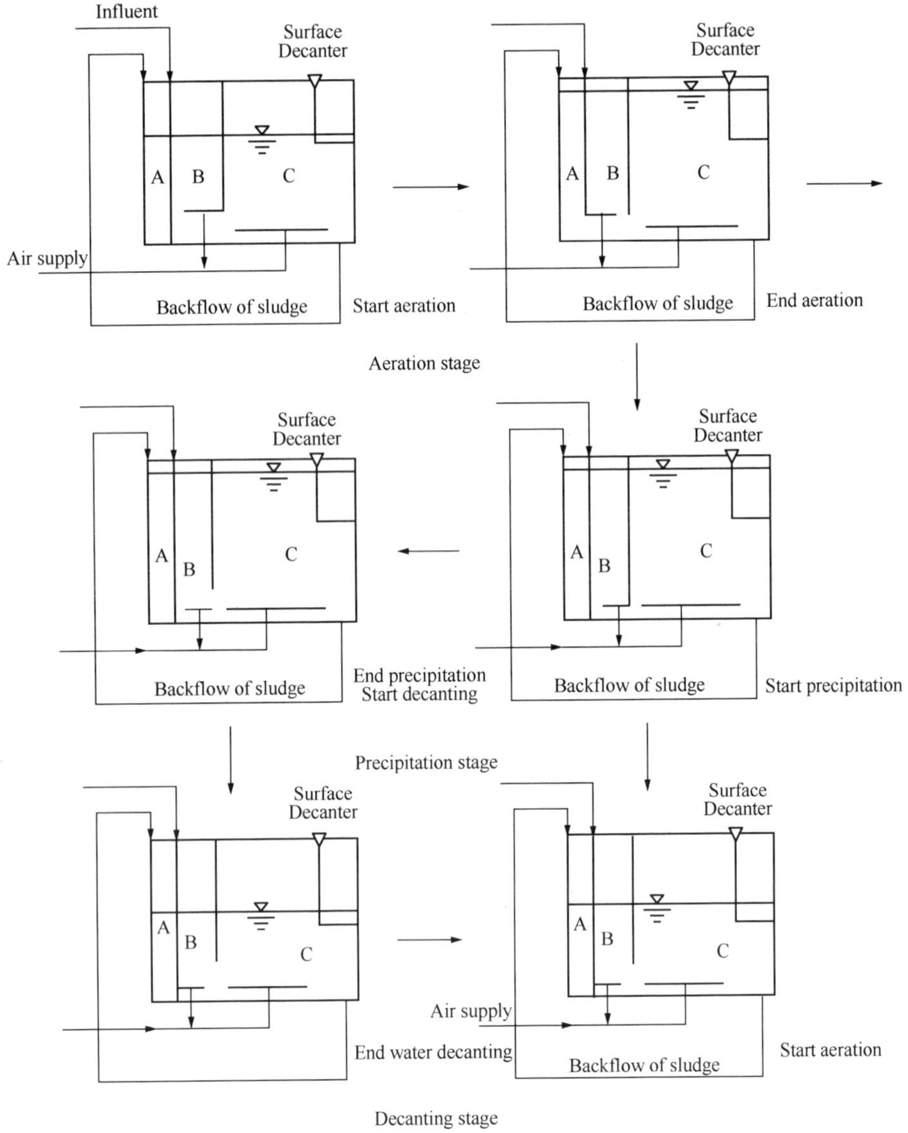

Figure 4-20 The running of CASS

A—Biological selection area; B—Facultative area; C—Main reaction area

At the starting, the water table is at the lowest level. The waste water in turn comes into area A, area B and area C. For B and C, it is inflowing and aerating simultaneously, and the sludge after stabilization in area C is returned to area A and area B. The backflow of the sludge is about 20% of the daily mean amount of the waste water. When the water table is at the designed level, the aeration stops and the settling starts. It is noticeably that the inflowing and backflow of sludge continues. During the discharge from area C, the inflowing of water from area A to area B stops. The discharge is finished when the water table is at the lowest level. This is idle stage. In this stage, the inflowing from area A to area B starts without aeration. The whole period is about 4 h and the half is used for aeration.

Area A is the biological selector which is a small area for the mixing of the waste water and sludge that comes from the main reaction (area C). This is favor of the inhibiting the growth of filamentous bacteria to avoid of the sludge bulking. In addition, the organic matter is quickly adsorbed by the bacteria in area A and the nonbiodegradable organic matter is hydrolyzed. The P release is promoted under the anaerobic conditions. For the facultative area B, the main function is the resistance on the shocking of water quantity and quality. In addition, the organic removal depends on the adsorption by the stabilized sludge, and the removal is up to 80%. Another function is to promote further the P release and strengthen the denitrification. The area C is the key area for the adsorption of organic matter. The realization of the simultaneous nitrification denitrification is under the control of DO.

(3) Demand aeration tank－intermittent aeration tank　Demand Aeration Tank－Intermittent Aeration Tank (DAT－IAT) is the combination of activated sludge and SBR. The main reactor is divided into two parts with the same volume by the plate (Figure 4－21). The one is demand aeration tank and the other is intermittent aeration tank.

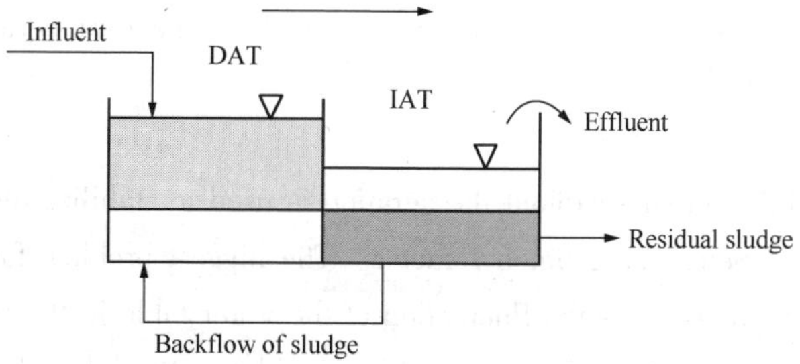

Figure 4－21　The structure of DAT－IAT

DAT is the pre－reaction tank, similar as the aeration tank in the activated sludge. The waste water continuously comes into the tank. Most of organic matter is degraded under the aeration and complete mixing. And then the effluent comes into the IAT passing the plate at laminar flow. IAT is similar as the SBR. After the aeration, settling, decanting and discharge of residual sludge, the whole period is finished.

DAT-IAT is period running and continuous inflowing. After the sedimentation, part of the sludge is returned to DAT. The following stages are showed.

Ⅰ Inflowing

The waste water comes into DAT after the pretreatment. And then the effluent goes to IAT. The designment is to not only simplify the inflowing but also resist the shocking.

Ⅱ Reaction

There is biological reaction in DAT and IAT.

DAT function is similar as the aeration tank in the CFS. The waste water is mixed with the returned sludge from the IAT. The adsorption of the organic matter is strengthened under the continuous aeration. The initial adsorption of the activated sludge is sufficiently used in DAT, resulting in most of organic matter removed.

IAT running and function is similar as SBR. Removing BOD, N and P removal depends on the aeration time.

Ⅲ Settling

Settling is only in IAT. When aeration and mixing stops in IAT, settling starts under the gravity.

Ⅳ Decanting

Decanting is only in IAT. The running is similar as SBR. When the water table is up to the designed level, the overlying water after the settling is slowly discharged by the decanter at the end of IAT. When the water table is down to the lowest level at the starting of the treatment period, the decanting is over. One part of settled sludge is returned to DAT by pump, and another part is used for the next period. The residual sludge is discharged.

Ⅴ Idle

During the idle, mixing without the aeration is used to stabilize the sludge.

(4) Modified Sequencing Batch Reactor The biggest problem for SBR, ICEAS, CASS, DAT-IAT processes is the fluctuation of the water table in the reaction tank and intermittent discharge of the effluent, resulting in higher head loss for the follow-up processes. So far, there are many modified SBR with not only the advantages of SBR, but also the same as the continuous activated sludge. A new Modified Sequencing Batch Reactor (MSBR) is introduced.

Generally, MSBR is the combination of A/A/O process with SBR. The key equipment is MSBR. MSBR is composed with two SBR reactors, aeration tank, anaerobic tank and anoxic tank (Figure 4-22).

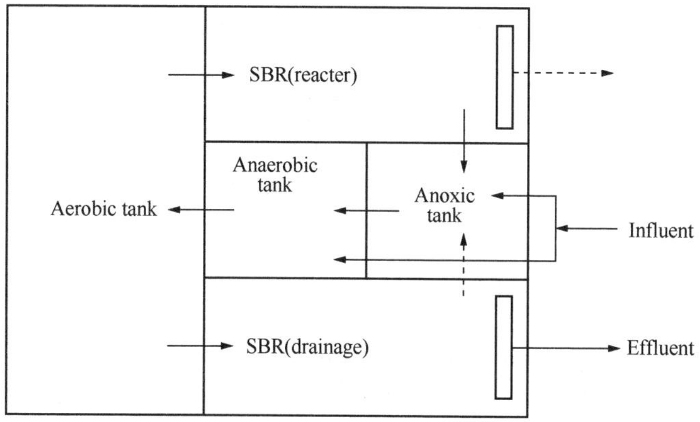

Figure 4-22 The structure of MSBR

The whole running process is observed from the Figure 4-23. Tank 1, tank 2 and tank 3 are anoxic tank, anaerobic tank and aerobic tank, respectively. The raw waste water continuously comes into tank 1 and tank 2. During the process, the mixed liquid flows from tank 1 to tank 3. Tank 4 and tank 5 are SBR reactors. Both tanks are alternate the discharge tank and reaction and settling tank. As the reaction tank, it includes three steps, such as backflow of mixed liquid, sequencing batch reaction and sedimentation. If tank 4 is the discharge tank, tank 5 is reaction tank. The running process is as followed. The first is the mixed liquid backflow, turning on the backflow pump, mixer and aeration equipment. The activated sludge and mixed liquid are returned from tank 5 to tank 1, at the same time, the activated sludge and mixed liquid are returned from tank 1 to tank 5. In addition, at the step, it is to regulate the aerator to create the best environment for denitrification. After the backflow, it is the second step, turning off the backflow pump, mixer and aeration equipment. In tank 5, the sequencing batch reaction occurs. And then the third step, turning off mixer and aeration equipment. The settling starts for the separation of solid and liquid. The whole period is end. The next is to change the function of tank 4 and tank 5.

It is noticeably that Figure 4-23 is a common mode for MSBR. According to the treatment aim and treatment object, MSBR can be changed, such as Figure 4-23.

A new process, which is called IDEA (Intermittent Decanted Extended Aeration), is developed on the base of CASS. The process of IDEA is showed in Figure 4-24. For IDEA, the biological selector is changed into the pre-reaction tank, compared with CASS. Part of the residual sludge is returned to the pre-reaction tank.

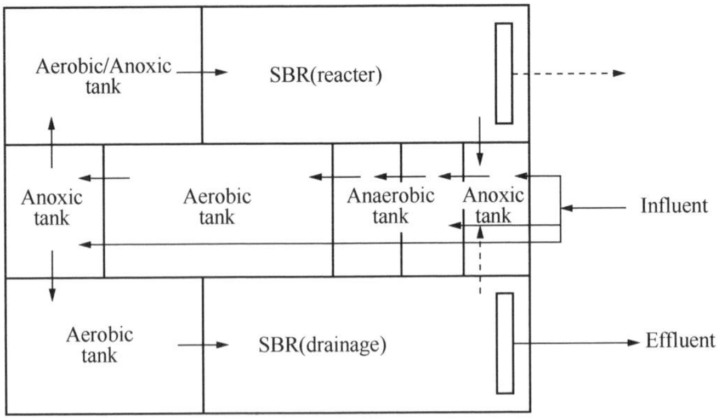

Figure 4-23 The structure of the improvement of MSBR

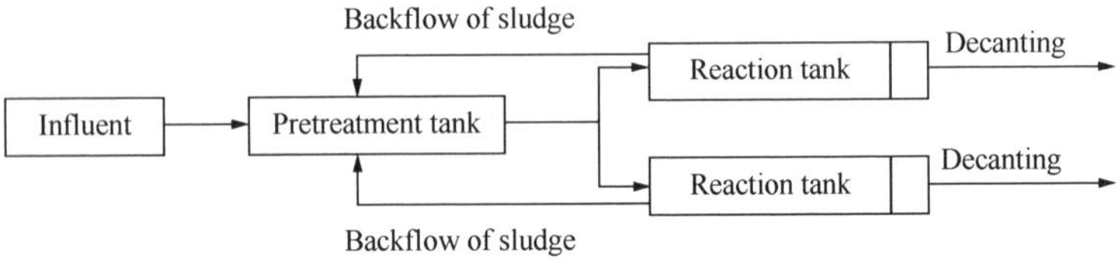

Figure 4-24 The structure of IDEA

(5) UNITANK UNITANK process was invented by Seghers engineering water NV at 1990s. The key part of UNITANK is a rectangle tank which is separated to some small rooms. The typical UNITANK is a rectangle tank with three rooms. Each room has aeration equipment, such as blasting aeration, mixing or mechanical surface aeration. The two rooms have effluent weir and discharge outlet of residual sludge, except of the middle room in UNITANK. Therefore, the two rooms are alternating aeration tank and settling tank, and the middle one is only as the aeration tank.

The running of UNITANK is continuous inflowing and periodic decanting, alternately. The basic running period includes two symmetrical running stage. The one is inflowing of the left room and discharging of the right room, and the other is inflowing of the right room and discharging of the left room. There is a short transition section between the above running stage.

The running process of UNITANK is showed in Figure 4-25.

The raw waste water comes into the left room with aeration. There are lots of activated sludge in the room because the room is the settling tank in the last running stage. After stabilization by the aeration, the activity of the activated sludge is recovered. Therefore, the organic matter is quickly biodegraded by the activated

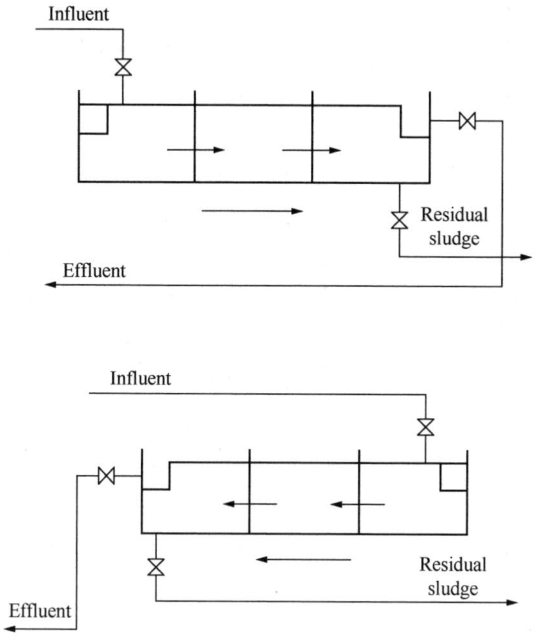

Figure 4-25 The running process of UNITANK

sludge. And then the water comes into the middle room, the organic matter is further biodegraded. After the removal of organic matter, the water goes to the right room for the separation of solid and liquid. The effluent is discharged by the effluent weir and the residual sludge is discharged too. During the flowing of the water from left to right, the sludge is divided again in the three rooms. And then it is the short transition section. After the transition section, it is the second running stage. The waste water comes into the right room and is discharged from the left room.

UNITANK has the function of removal N and P. It is necessary to design the mixer in the room, except of the aeration equipment. At the first stage, the waste water alternately comes into the left room and the middle room. The nitrate is transformed into N_2 by denitrification under the anoxic mixing in the left room, and the P is released from the activated sludge. When the waste water goes into the middle room with aeration, the organic matter is removed and the ammonia nitrogen is transformed into nitrate and P is adsorbed by the activated sludge. And then the waste water goes to the right room for separation of the solid and liquid to realize the P removal.

4.1.3 Membrane bioreactor

Membrane bioreactor (MBR) is a common way of modern sewage treatment. The technology is a combination of the biological technology and membrane separation technology. It has replaced the secondary sedimentation tank in the traditional process. It can efficiently carry out solid-liquid separation, and then the stable reused water can be gotten directly.

In the process of development, the first generation MBR is a split flow process (twentieth century 60 ~ 80 years). The first generation MBR used plate ultrafiltration membrane (polypropylene) and tube type ultrafiltration membrane (organic and inorganic). Because of the high cost of the membrane, the operation cost was large and the application was limited. The second generation of MBR is an integrated immersion process (after the end of 1980s). The characteristics of submerged MBR are as followed: submerged placement, low pressure (suction or gravity) effluent, gas-liquid two-phase flow disturbance, intermittent operation, stable operation time and low operation energy consumption. The types of submerged MBR include the internal immersion (with the aerobic reactor) and the external immersion (separated from the aerobic reactor).

4.1.3.1 The principle of MBR

Y. Magara et al found that there is straight line relationship between the flux of MBR and the logarithm of the activated sludge content at 1989.

$$J_v = -1.5711\lg(MLSS) + 7.84$$

In the equation, J_v—the flux of the membrane, $m^3/(m^2 \cdot d)$;

MLSS—the content of the suspended solid, (mg/L).

At 1991, Takeshi sato et al found the relationship is as followed using polysulfone flat membrane (molecular cut off 2,000,000).

$$J_v = 4.896(MLSS)^{-2.66}$$

$$J_v = 1.065 \times 10^{11}(COD)^{-3.93}$$

$$J_v = 0.7035\mu^{-0.936}$$

In the equation, J_v—the flux of the membrane, $m^3/(m^2 \cdot d)$;

MLSS—the content of the suspended solid, mg/L;

COD—the content of soluble COD in the mixed liquid, mg/L;

μ —viscosity coefficient, Pa · s.

4.1.3.2 The consistent of MBR

The basic structure of MBR consists of four parts, namely, influent system, biological reaction tank, membrane module and automatic control system. Because of the diversity of each part, MBR has different classifications. In a word, there are several categories of membrane bioreactors.

Membrane separation bioreactor (bioreactor) was used for solid-liquid separation in sewage treatment.

Membrane bioreactor, which is used for gas mass transfer, usually provides oxygen for aerobic process, can achieve bubble free aeration in bioreactor, and greatly improve the oxygen transfer efficiency in the reactor.

An extraction membrane bioreactor was used in the recovery of priority pollutants in industry. The selectively permeable membranes are used to extract specific pollutants. According to the placement of membrane components, it can be divided into a split type and an integrated membrane bioreactor.

Bioreactor and membrane were separated in the split type membrane bioreactor (Figure 4-26). The mixed liquid in the bioreactor comes into the membrane under the pressure. The clean water was gotten by the membrane under the pressure and the activated sludge was hold and backflow with the concentrated liquid into the biological reactor.

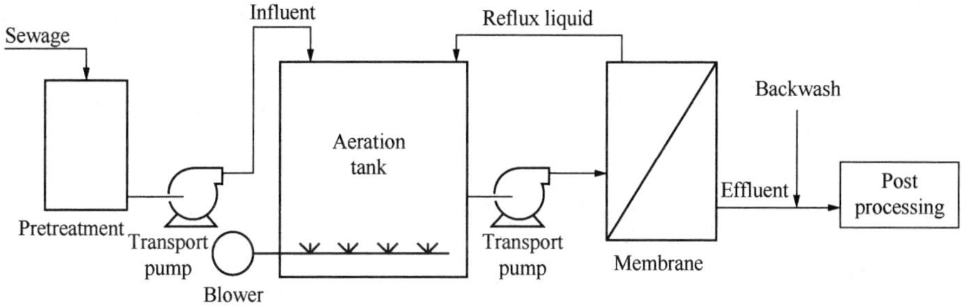

Figure 4-26 The separated MBR

The integrated MBR is that the membrane is placed into the reactor (Figure 4-27). The filtrate could be gotten by the suction. The cross flow for the membrane surface cleaning is generated by the air agitation. The aerator was placed below the membrane. The mixed liquid is up flow with the air and then shear stress on the membrane surface appeared on the surface of the membrane to reduce the membrane pollution. The reactor is the combination of sewage biological treatment technology and membrane separation technology.

4.1.3.3 The characteristics of MBR

Process characteristics:

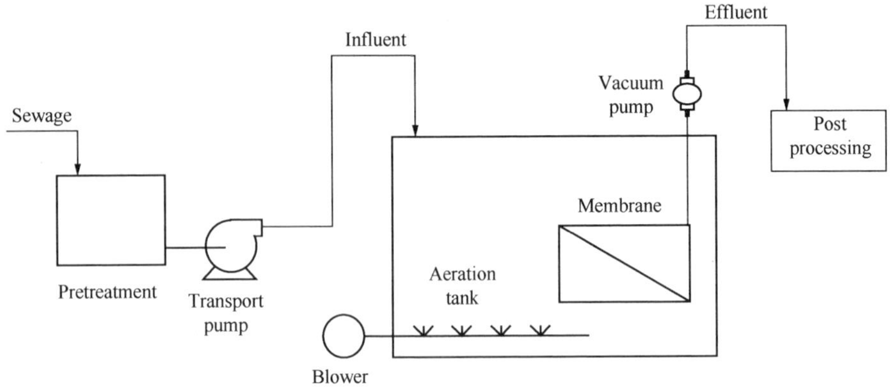

Figure 4-27 The integrated MBR

① The effluent quality is good, and the concentration of BOD, nitrogen, phosphorus and suspended matter is very low.

② The process is short, the area is saved, and there are no secondary sedimentation tanks.

③ The nitrifying bacteria with the long generation breed in the reactor due to the long SRT.

④ The amount of residual sludge is less 50% ~ 80% than that of conventional activated sludge process, which is beneficial to treatment.

4.1.4 Adsorption biodegradation (AB)

The AB is the abbreviation for the adsorption biodegradation process. The technology was applied by professor Bohnke of Germany in the early 1980s. The engineering practice shows that the AB has many outstanding advantages, such as high efficiency, low energy consumption and stable operation, and it is suitable for the various industrial waste water treatment. Therefore, the application of the technology is very wide in the field of waste water treatment.

4.1.4.1 The principle of AB

AB is a new technology on the base of the traditional activated sludge process and high load activated sludge process. Stage A is referred to the adsorption and stage B is referred to the biodegradation (Figure 4-28).

AB is made up of an aeration tank in stage A, middle sedimentation tank, an aeration tank in stage B, secondary sedimentation tank. In general, there is no primary sedimentary tank in the AB, and stage A is the primary treatment. Stage A and stage B

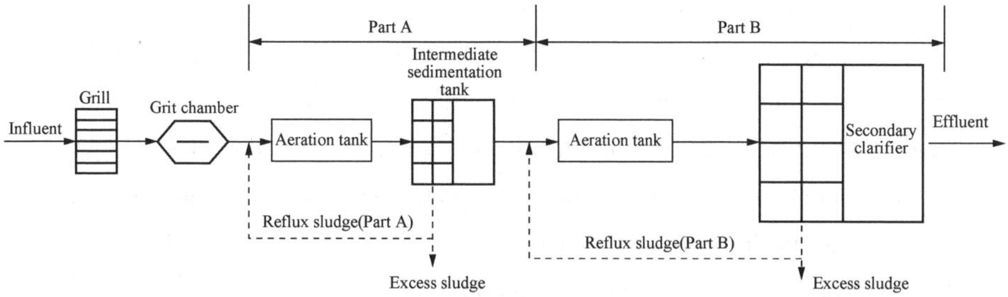

Figure 4-28 The adsorption biodegradation process

have the sediment backflow system, relatively. Therefore, stage A and stage B have the separate microorganism system. That is favor of the stability of the function.

For stage A, it receives lots of bacteria from the sewers continuously, and the bacteria will pass the breeding, adaptation, elimination and optimization. Therefore, the adaptation and the activity of the microorganism are very strong in stage A. It is indicating that the load is high in stage A, and it is favor of the breeding of the microorganism. On the other hand, only those microorganisms that can resist shock loading can survive. After the treatment in stage A, the removal of BOD is up to 40%~70%. Moreover, the bioavailability of the waste water increases and it is favor of the stage B.

The output of sludge in stage A is high, and the adsorption ability of the sludge is big. Therefore, heavy metals, nonbiodegradable organic matter, ammonia nitrogen and phosphorus can be removed by the adsorption of the sludge. It is noticeably that the biodegradation on the organic matter in stage A is only about 30%.

After the treatment of the stage A, the water quality and the quantity are stable for stage B. In addition, there is almost no condition on the shock loading. Since stage A can remove nitrogen, resulting in the ratio of BOD to N down, stage B has the condition for nitrification. The stage B runs at very low load, and it ensures the effluent quality by using the role of long-term microorganism. Compared with the traditional biological treatment, AB has obvious advantages in the processing efficiency, the operation stability, the investment and the operation cost of the project.

4.1.4.2 Basic process of AB

Stage A and stage B was strictly separated. Stage A runs under high road with the 2~6 $kgBOD_5/(kgMLSS \cdot d)$, while it is 10~20 times higher than the conventional activated sludge process. However, the sludge age is very short. The bacteria are the main component in the activated sludge. Biological adsorption is finished in 30 min and the removal of BOD is up to 40% ~ 70%. On the other hand, for stage B, the running parameters are as followed. The hydraulic retention time (HRT) is about 2 ~ 3 h. The sludge age is about 15 ~ 20 d. The sludge load is 0.15 ~ 0.30 $kgBOD_5/(kgMLSS \cdot d)$. The component of the microorganism is protozoon, zoogloea and metazoan. The organic matter can be further biodegraded in the stage B, compared to the stage A. The removal rate of nitrogen and phosphorus was 60% ~ 70% and 30% ~ 40%.

4.1.4.3 The characteristics of AB

① The efficiency on the removal of pollutants is high.

② The running of the process is stable, due to the high resistance on the shock loading.

③ The excellence on the removal of N and P.

④ The economy is very good. The main reason is that the load is reduced by 40% ~ 50% after stage A, so the volume of the aeration tank can be reduced by 40% ~ 50%. In a word, the running fee is reduced by 20% ~ 25% for the AB, compared to the traditional process.

4.1.4.4 The designing parameters of AB

For the domestic waste water, the designing parameters of AB are as followed.

For stage A: The sludge load is 2 ~ 6 $kgBOD_5/(kgMLSS \cdot d)$. The mixed liquid suspended solids (MLSS) is 1.5~2.0 g/L. The volume loading is 4~12 kg $BOD_5/(m^3 \cdot d)$. The HRT is 20 ~ 40 min. The sludge age is 0.3 ~ 1.0 d. The content of dissolved oxygen (DO) is 0.2 ~ 1.5 mg/L.

For stage B: The sludge load is 0.15 ~ 0.3 $kgBOD_5/(kgMLSS \cdot d)$. The HRT is 2 ~ 3 h. The sludge age is 15 ~ 20 d. The content of dissolved oxygen (DO) is 1 ~ 2 mg/L.

4.1.5 Moving bed biofilm reactor

The moving bed biofilm reactor (MBBR) was proposed and introduced at the end of 1980s and applied quickly in Europe. It takes the advantages of traditional activated sludge process and biological contact oxidation process, and then it becomes a new and efficient composite process. The key of MBBR is that the carrier of the bacteria is the suspended filler, and the density of the filler is close to water. In the aeration tank, the filler is in the fluidization state due to the aeration and water flow.

4.1.5.1 The principle of MBBR

Increasing the biomass and biological species in the reactors is favor of increasing the efficiency of MBBR by adding the suspended fillers in the aeration tank. The filler is completely mixed with the water due to its density close to water during the aeration. The collision and shearing of the filler in the water makes the air bubbles finer and increases the utilization of dissolved oxygen. In addition, each filler has different biological species inside and outside. Some anaerobic or facultative bacteria, in general, grow inside, and the outside is oxic bacteria. Therefore, each filler is a micro reactor, resulting in the coexist of nitrification and denitrification. It is favor of improving the efficiency.

4.1.5.2 The key part of MBBR

① Filler: its density is close to water. It is free to move under the slight stirring. The most important is that the effective surface area inside the filler is big, up to 500 ~ 1,200 m^2/m^3. The material of the filler is polyethylene or polypropylene plastic, and the service life is long, up to 20 years.

② Aeration system: a porous pipe system with middle and small holes was applied in the aeration tank. The characteristic is that the water distribution is uniform, and the gas volume can be adjusted and controlled. It is to ensure the irregular movement of the filler in the reactor, so that the filler can break up the upflow bubbles.

③ The effluent retention strainer: the strainer is to maintain a good flow pattern in the reactor and keep the fillers in the aeration tank. The form, the size of the pore, the location varied with the fillers. The outlet area depends on the unit discharging load of different pore diameter. The strainer has the porous plate type or winding welded pipe type (vertical or horizontal).

4.1.5.3 The advantages of MBBR

① The volumetric load is big. The efficiency increases by 2 ~ 3 times compared to the traditional activated sludge process.

② The resistance on the shocking load is strong. The running is stable.

③ The operation of the mixing and aeration system is easy. The maintenance is simple.

④ No blocking in the aeration tank.

⑤ The process is convenient and flexible. One is suitable for different tank types, such as circle, square, rectangle. The other is the filling rate of the filler is flexible.

⑥ The service life is long.

4.1.6 A/O process

4.1.6.1 Overview

The invention of A/O process (anerobic-aerobic) is related closely to the water eutrophication and the nitrogen and phosphorus pollution.

The nitrogen and phosphorus mainly come from the domestic waste water and the industrial waste water with the nitrogen and phosphorus. Another source is the runoff from the farms, such as the fertilizers, pesticide and animals faeces. However, the function of the traditional aerobic treatment is the removal of organic matters, while the removal of the N and P in the waste water is very low.

Wuhrmann process (Figure 4 - 29) is the earliest process for removal N at 1930s. The nitrification and denitrification are the main component in order in the process. The reactor is divided into two parts. One is aerobic tank and the other is anaerobic tank. The function of the aerobic tank is biodegradation of the organic matter and nitrification, while the anaerobic tank is denitrification. However, in the practice it is not practical. The main reason is that carbon source is necessary for the denitrification. In the process, the rate of the denitrification is very low due to the consumption of the carbon in the aerobic tank, since the aerobic tank is before the anaerobic tank. Therefore, the content of N in the raw waste water is very big, resulting in the huge anaerobic tank.

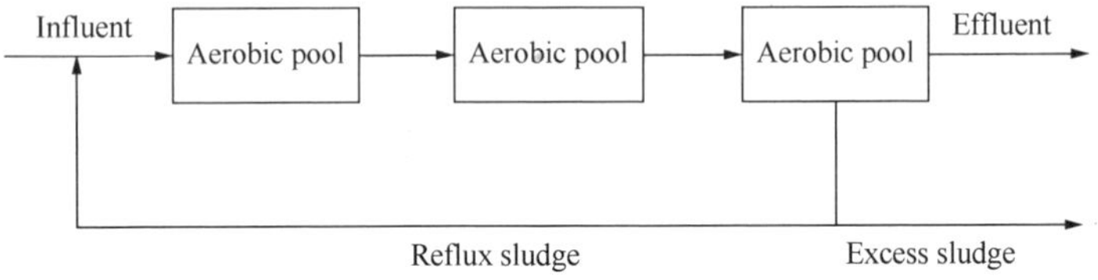

Figure 4-29 Wuhrmann process

At 1960s, Ludzack and Ettinger proposed a new technology - front denitrification (Figure 4-30). The process was that the anaerobic tank and aerobic tank were placed in order. Therefore, the carbon source for the denitrification can be the organic matters in the raw waste water. The problem is solved that the carbon source for the denitrification is not enough. For the process, there is a disadvantage that the nitrate goes to the sedimentation tank resulting in the effluent worsen.

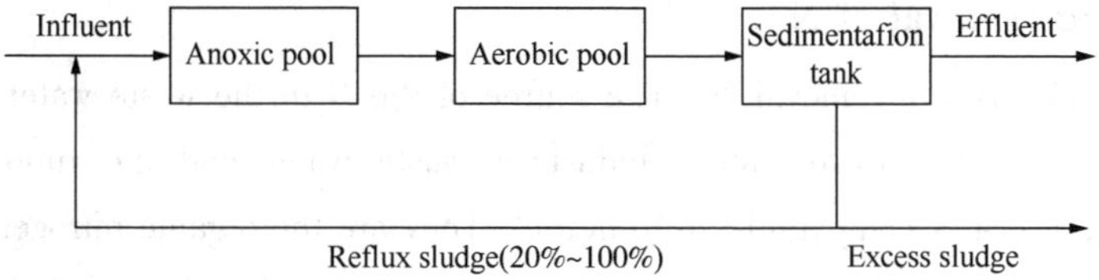

Figure 4-30 The process for removal of N by Ludzack and Ettinger

At 1970s, Barnard improved the process by Ludzack and Ettinger (Figure 4-31). That is the A/O process that it is widely used in practice. In the process, the mixed liquid in the oxic tank and the sludge after the sedimentation simultaneously returned to the anaerobic tank. Therefore, lots of nitrate returned to the anaerobic tank under the enough carbon source in the raw waste water, resulting in the appearance of the complete denitrification. If we add the anoxic tank before the anaerobic tank, that is the A/A/O process with the removal of phosphorus (Figure 4-32).

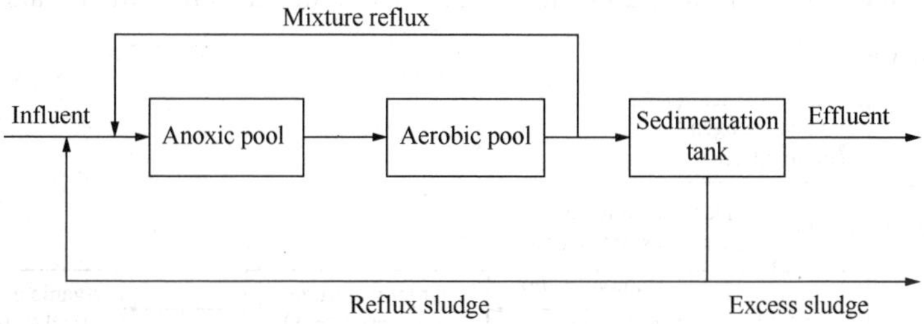

Figure 4-31 The modified process for removal of N by Ludzack and Ettinger

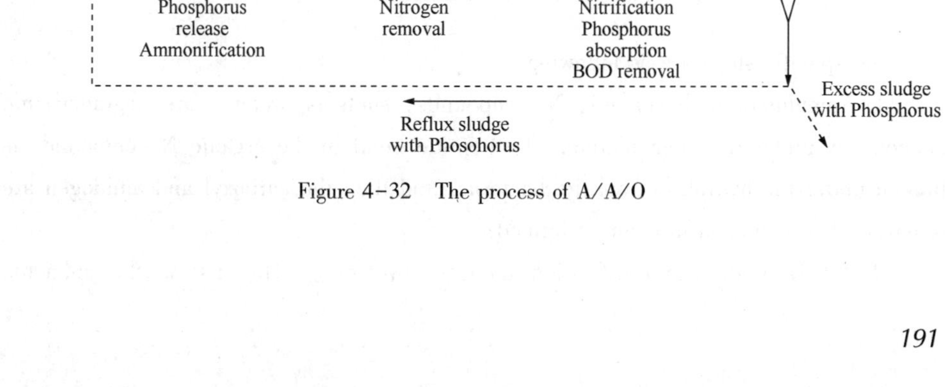

Figure 4-32 The process of A/A/O

4.1.6.2 The removal of N

(1) Mechanism of removal N The source of the N in the waste water is various, such as the domestic waste water, industrial waste water and the runoff from the farms. The forms of N compounds are different. They are the organic nitrogen, ammonia nitrogen, nitrite nitrogen, nitrate nitrogen. The index for the N is as followed. Total nitrogen (TN), total kjeldahl nitrogen (TKN), nitrate nitrogen (NO_3-N), nitrite nitrogen (NO_2-N), ammonia nitrogen (NH_3-N). The nitrate nitrogen (NO_3-N) and nitrite nitrogen (NO_2-N) can be called nitric nitrogen (NO_x-N). The TKN is the sum of the organic nitrogen and ammonia nitrogen. The TN is the sum of the TKN and the inorganic N (NO_x-N+NH_3-N).

The denitrification can be described as followed (Figure 4-33). The organic N is oxidized into NH_3-N under the bacteria in the aerobic tank, and the NH_3-N is transformed into NO_x-N under the nitric bacteria in the same tank. The mixed liquid goes to the anaerobic tank and then the nitrate N is reduced into nitrogen and it goes to the air from the tank.

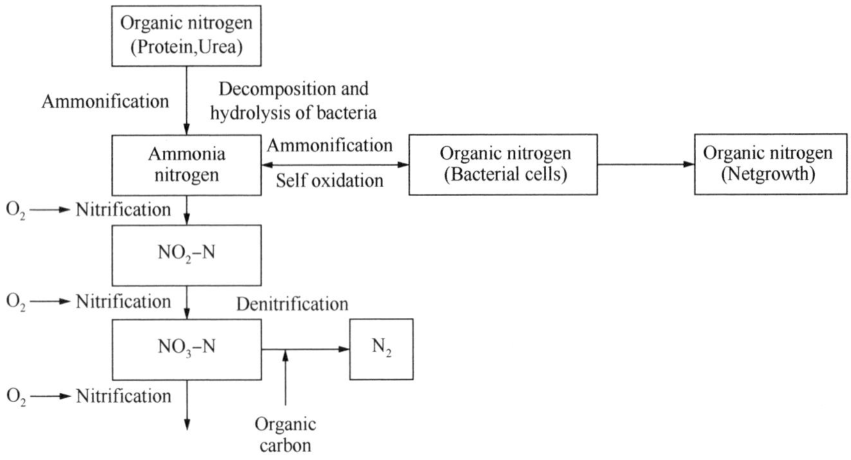

Figure 4-33 The transformation of N

The specific steps are as following.

① Ammoniation. The organic N compounds, such as protein, are degraded into ammonia N under the ammoniation. The peptide bond in the organic N compounds is broken under the hydrolase from the bacteria, and then the carboxyl and amidogen are removed. The ammonia nitrogen is formed.

② Nitrification. The nitrification includes two steps. The first is the ammonia nitrogen is transformed into nitrate nitrogen under thenitrosobacteria. The second is the nitrate nitrogen is transformed into nitrite nitrogen under the nitrobacterium.

$$NH_4^+ + 1.5O_2 \longrightarrow NO_2^- + 2H^+ + H_2O + 243 \sim 352kJ$$

$$NO_2^- + 0.5O_2 \longrightarrow NO_3^- + 64.5 \sim 86.3kJ$$

The nitrosobacteria and the nitrobacterium are the chemoautotrophic bacteria. They can use the energy in the oxidization and compound the organic matter for the cell. In the process the need for the oxygen is huge. The removal of 1 g NH_3–N is to consume 4.33 g O_2, and to build 0.15 g new cell, and to reduce 7.14 g alkalinity ($CaCO_3$), and to consume 0.08 g inorganic carbon. During the process, the pH is about between 7~8.

③ Denitrification. Under the anoxic conditions, the NO_3^- is transformed into the NO_2^- and N_2. Denitrifying bacteria is the facultative heterotrophic bacterium. The reaction is that they can use the kinds of organic matter as the electron donor and the nitric nitrogen (NO_x–N) as the electron acceptor. Therefore, the reaction is the anaerobic respiration, at the same time, the organic matter is consumed and the nitric nitrogen is reduced into N_2. The reduction of 1 g NO_3–N is to consume 2.47 g methyl alcohol (about 3.7 mg COD), and to build 0.45 g new cell, and to produce 3.57 g alkalinity ($CaCO_3$), and to consume 0.08 g inorganic carbon. During the process, the pH is about between 7~8 and the ratio of BOD_5 to TN is more than 4.

(2) Factors affecting the removal of N.

① Temperature. The ratio of the denitrification is affected by the temperature, according to the equation.

$$q_{D,T} = q_{D,20}\theta^{(T-20)}$$

In the equation, $q_{D,T}$—The ratio of the denitrification at temperature T℃, gNO_3^-–N/(gVSS · d);

$q_{D,20}$—The ratio of the denitrification at temperature 20℃, gNO_3^-–N/(gVSS · d);

θ—The coefficient, 1.03~1.05, in general, it is 1.09.

The effect of the temperature on the ratio of the denitrification is related to the form of the denitrification tank (activated sludge or biofilms) and the nitrate load. When the nitrate load is low, the effect of the temperature on the ratio of the denitrification is small. On the other hand, when the nitrate load is high, the effect of the temperature on the ratio of the denitrification is big. In general, the favorite temperature is 20~40℃.

② Dissolved oxygen. The inhibition of the dissolved oxygen onthe denitrification ratio is that the molecule oxygen will compete the electron donor with the nitrate, and the molecule oxygen can decrease the synthesis of the nitrate reductase and its activity. On the other hand, in the denitrifying bacteria, some enzyme only compound under the oxygen. Therefore, the alternation of aerobic and anaerobic is favorable. For the suspended activated sludge system, the dissolved oxygen should keep less than 0.3 mg/L. However, for the biofilm, the content of the dissolved oxygen could be higher. In a word, the best for the denitrification is no oxygen in the system.

③ Alkalinity and pH. The best pH is between 7.0 and 7.5 for the denitrification. But the denitrification could produce the alkalinity, it is favor of the stability of the suitable pH for the system.

④ Carbon source. It is well known that the nitrate is the electron receiver, while the carbon source is the electron donor for the denitrification. Therefore, the carbon source is necessary. The carbon source is different, so the ratio of the denitrification is different. The methyl alcohol is the best carbon source. The domestic waste water, the beer waste water, the volatile organic compound (VOC), molasses, citric acid, acetone can be as the carbon source.

⑤ The ratio of C to N. The consumption of carbon source (BOD_5) is 2.86 g per 1 g nitrate reduction in theory, but for the waste water, the ratio of C to N should be higher.

⑥ Toxic matters. The sensibility of the denitrifying bacteria to the toxic matters is lower than that of the nitrifying bacteria. During the running, the permission content of the toxic matters should be gotten by the experiments.

(3) The designment of the A/O process.

① Sludge age. In general, the sludge age is enough long to keep the nitrifying bacteria breeding and growth.

$$\theta_c \geqslant 3\theta_c^m$$

In the equation, θ_c—The sludge age, d.

θ_c^m—The reciprocal of the biggest specific growth rate of the nitrifying bacteria, d. It is closely related with the water temperature. In winter, the water temperature is 8~10℃, θ_c^m is about 10 d. Therefore, the sludge age is 30 d.

② The volume of the aerobic tank . The volume of the aerobic tank is calculated by the following equation.

$$V=\frac{YQ(S_0-S_e)\theta_c}{X(1+k_d\theta_c)}$$

Or

$$V=\frac{Y'Q(S_0-S_e)\theta_c}{X'}$$

In the equation, V—The volume of the aerobic tank, m^3;

Y—The yield coefficient, kgVSS/kgBOD_5;

Y'—The net yield coefficient, kgVSS/kgBOD_5;

Q—The amount of waste water, m^3/d.

S_0—The content of BOD_5 in the influent, mg/L;

S_e—The content of BOD_5 in the effluent, mg/L;

θ_c—SRT, d;

k_d—Endogenous respiration coefficient, d^{-1};

x— The content of the volatile suspended solids in the mixed liquid, MLVSS, mg/L, in practice, 2,000~4,000mg/L;

X'—The content of the suspended solids in the mixed liquid, MLVSS, mg/L.

Y' is calculated by the following equation.

$$Y'=\frac{Y}{1+k_d\theta_c}$$

The volume of the aerobic tank is also calculated by the following equation.

$$V=\frac{Q\,S_0 Y\,\theta_c}{X}$$

Additionally, the relationship between sludge age and the removal of BOD load is as followed.

$$\frac{1}{\theta_c}=YN'_s-k_d$$

$$N'_s=\frac{Q(S_0-S_e)}{VX}$$

In the equation, N_s—The removal of BOD load, kg/d.

③ The volume of the anaerobic tank. The volume of the anaerobic tank is calculated by the denitrification rate and the content of nitrate.

$$V_{DN}=\frac{1000N}{\gamma_{NR}X}$$

$$N=N_0-N_w-N_e$$

In the equation, V_{DN}—The volume of the anaerobic tank, m^3;

N—The content of the reduced nitrate, $kgNO_3^--N/d$;

γ_{NR}—The denitrification rate, $kgNO_3^--N/(kgMLSS \cdot d)$;

N_0—The content of nitrate in the influent, kg/d;

N_e—The content of nitrate in the effluent, kg/d;

N_w—The content of nitrate in the residual sludge, kg/d, The percentage of N element is 12.4% in the cell.

In practice, it is good that the HRT in the anaerobic tank is less than 2 h. Generally, the HRT is 1.0~1.5 h.

④ The oxygen demand on the nitrification. The oxygen demand is calculated by the following equation.

$$O_2 = Q \times \frac{S_0 - S_e}{1 - e^{-kt}} - 1.42P_x \frac{VSS}{SS} + 4.33Q(N_0 - N_e) - 0.54P_x \frac{VSS}{SS} - 2.6Q\Delta NO_3$$

In the equation, O_2—The oxygen demand for the removal of BOD_5 and nitrate, kgO_2/d;

Q—The amount of the waste water, m^3/d;

S_0—The content of BOD_5 in the influent, mg/L;

S_e—The content of BOD_5 in the effluent, mg/L;

P_x—The amount of the discharged residual sludge, kg/d;

$\frac{VSS}{SS}$— The percentage of the volatile solids in the activated sludge, %;

N_0—The content of ammonia nitrogen in the influent, mg/L;

N_e—The content of ammonia nitrogen in the effluent, mg/L;

ΔNO_3—The content of nitrate which is reduced, mg/L.

⑤ The addition of the carbon source and the alkalinity. The ratio of the BOD content to the TN content is higher than 3~5. If the ratio is less than 3, the external carbon source is necessary. 1 mg/L nitrate is to consume 1.90 mg/L CH_3OH. If the content of DO is higher in the anaerobic tank, it is necessary to add the CH_3OH. In practice, 1 mg/L DO needs 0.67 mg/L CH_3OH.

⑥ The backflow ratio. The backflow ratio is closely related with the denitrification. If the denitrification is higher, the backflow ratio of the mixed liquid is bigger. In practice, the ratio of the mixed liquid is at the range of 3~6, while the ratio of the residual sludge is at the range of 0.7~1.

4.1.6.3 The removal of P

(1) Mechanism of biological method for phosphorus removal Phosphorus is the

limit nutrient for the algae in water bodies. Therefore, controlling the P drainage is favor of the decrease of the eutrophication of the water bodies.

It is well known that the P in the natural water bodies is main from the waste water, such as the faeces, detergent, pesticide, and the industrial waste water with high content of P. The forms in the waste water are organic P, inorganic P and polyphosphates. Therefore, the removal of P from the waste water is necessary.

Generally, some bacteria and algae can store huge amount P, more than the demand of the cell, and they can release the P under the anaerobic conditions. According to the excessive ingestion under the aerobic conditions and the release of the P under the anaerobic conditions, the strategy for the removal of P was proposed by the control of the aerobic conditions and the anaerobic conditions. That is the biological removal of P.

Under the aerobic conditions, the accumulation of P in the bacteria can be described in the equation.

$$C_2H_4O_2+0.16NH_4^++1.2O_2+0.2PO_4^{3-}\longrightarrow$$
$$0.16C_5H_7NO_2+1.2CO_2+0.2(HPO_3)+0.44OH^-+1.44H_2O$$

Under the anaerobic conditions, the release of P in the bacteria can be described in the equation.

$$C_2H_4O_2+0.16NH_4^++0.96NO_3^-+0.2PO_4^{3-}\longrightarrow$$
$$0.16C_5H_7NO_2+1.2CO_2+0.2(HPO_3)+1.4OH^-+0.96H_2O+0.48N_2$$

The component of the organic matter is similar as acetic acid.

Under the anaerobic conditions, the release of P in the bacteria can be briefly described in the equation.

$$C_2H_4O_2+(HPO_3)+H_2O\longrightarrow(C_2H_4O_2)_2+PO_4^{3-}+3H^+$$

The biological removal of P is finished by the phosphorus – accumulating bacteria. Under the aerobic conditions, the P is accumulated excessively by the bacteria and more than their physiology need. In addition, the adsorbed P is stored as the polymerization forms in the bacteria. Therefore, the content of P is very high in the activated sludge. The drainage of the sludge with high content of P is to remove the P from the waste water. There are some factors affecting the biological removal of P as followed.

① The alternation of anaerobic/aerobic conditions. Although the phosphorus – accumulating bacteria only adsorb the P under aerobic conditions, the anaerobic conditions are favor of the selection of the bacteria. The result is that a part of the bacteria in the activated sludge is the phosphorus–accumulating bacteria. Under the anaerobic conditions, the bacteria adsorb the organic matter and it is stored in the bacteria, while the other heterotrophic aerobic bacteria do not uptake the organic matter. And then under the aerobic conditions, the biodegradable organic matter is finished, while the phosphorus–accumulating bacteria can uptake polyhydroxybutyrate (PHB) for the growth. The removal of P is showing in Figure 4–34.

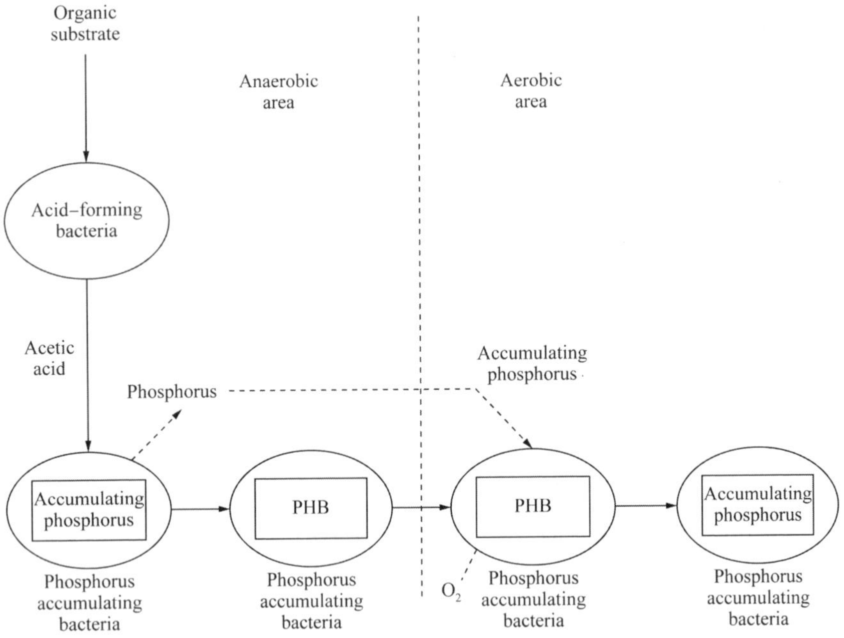

Figure 4–34 The removal of P

② Nitrate and dissolved oxygen. The backflow sludge to the anaerobic tank has nitrate and dissolved oxygen, and this inhibit the release of P under the anaerobic conditions. The denitrifying bacteria and the common heterotrophic aerobic bacteria consume the biodegradable organic matter under the nitrate and dissolved oxygen, resulting in the biodegradable organic matter decrease. It is showing that this is inhibiting the release of P for phosphorus – accumulating bacteria under anaerobic conditions.

③ Sludge age. It is well known the biological removal of P is finished by the drainage of residual sludge. In practice, shorter sludge age is better for the removal of P.

④ Temperature and pH. In practice, the range of temperature from 10℃ to 30℃ is favor of the removal of P, while the suitable pH should be at 6~8.

⑤ The ration of C to N. In practice, higher C/N (BOD_5/TP), better removal of P. The ratio is higher than 30 for the biological removal of P.

(2) Mechanism of chemical method for phosphorus removal The phosphorus in the waste water reacts with the chemical matters to form the insoluble phosphate precipitate, and then it is removed by the solid-liquid separation. The calcium salt, aluminum salt and ferric salt can be used as the chemical matters. For example, lime [$Ca(OH)_2$], aluminum sulfate [$Al_2(SO_4)_3 \cdot 18H_2O$], sodium aluminate ($NaAlO_2$), ferric trichloride ($FeCl_3$), ferric sulfate [$Fe_2(SO_4)_3$], ferrous sulfate ($FeSO_4$), ferrous chloride($FeCl_2$).

① Calcium salt. The P was removed by the formation of calcium phosphate depending on the reaction between the $Ca(OH)_2$ or CaO and the P. The main reaction is as followed.

$$5Ca^{2+}+7OH^-+3H_2PO_4^- \longrightarrow Ca_5OH(PO_4)_3+6H_2O$$

Another reaction is as followed.

$$Ca^{2+}+CO_3^{2-} \longrightarrow CaCO_3$$

The value of pH should be higher than 10 when the Ca^{2+} is used to remove P, but the exorbitant pH can destroy the breeding and activity of the microorganisms. Therefore, the reaction should be placed before or after the aeration tank.

② Aluminum salt and ferric salt. The reaction between Al^{3+}, Fe^{3+} and P is similar, because the characteristic of Al^{3+}, Fe^{3+} is the same. The main reaction is as followed. In the equation, the Me^{3+} presents the Al^{3+}, Fe^{3+}.

$$Me^{3+}+H_2PO_4^- \longrightarrow MePO_4+2H^+$$

Another equation is as followed.

$$Me^{3+}+3HCO_3^- \longrightarrow Me(OH)_3+3CO_2$$

③ Ferrous salt. Since the price of the ferrous salt is cheaper than ferric salt, it is often used for the precipitant. There are two ways for the removal of P by the ferrous salt. The first, the ferrous ions are oxidized into ferric ions under the oxygen. In practice, the ferrous ions are added into the aeration tank. The equations are as followed.

$$Fe^{2+}+0.25O_2+H^+ \longrightarrow Fe^{3+}+0.5H_2O$$

$$Fe^{3+}+H_2PO_4^- \longrightarrow FePO_4+2H^+$$

$$Fe^{3+}+3HCO_3^- \longrightarrow Fe(OH)_3+3CO_2$$

In general, the dose of the ferrous salt is 20~40 g Fe/m^3, and the consumption is

0.14 O_2/gFe^{2+}. The consumption of oxygen is not higher, compared with the organic matter oxidization.

The second is the co-precipitation between the ferrous ions and calcium ions. The method is efficient to remove P from waste water by the co-precipitation between the ferrous ions and calcium ions. But till now, no research can describe the precipitate.

The chemical removal P process is divided into 4 steps, such as precipitation reaction, condensation reaction, flocculation reaction, solid-liquid separation. The reaction speed is fast for the precipitation reaction and condensation reaction. They can be placed into one unit.

4.1.6.4 The process of removal N and P

The process of removal N and P consists of the three parts, anaerobic tank, anoxic tank and aerobic tank. The different combination of the three tanks is the different process.

(1) Anoxic-oxic process (A_N/O) The process consists the anoxic tank and aerobic tank and sedimentation tank (Figure 4-32). The function of the process is to remove organic matter, at the same time, it is to get good removal N. The process is also called preposition denitrification. The typical characteristic is the anoxic tank is before the aerobic tank. The advantage of the designment is the carbon source is used by the denitrifying bacteria in the anoxic tank, at the same time, the nitrate in the backflow from the aerobic tank is reduced to N_2. And then, the mixed liquid comes into the sedimentation tank for the separation of solid and liquid. A part of solid returned to the anoxic tank to replenish the activated sludge, and the mixed liquid with high content of nitrite returned to the anoxic tank too. The advantages of the process are as followed.

① No adding external carbon source and reducing the organic load for the aerobic tank. The running cost is low.

② The process is simple and short. Only one backflow system for activated sludge backflow and one backflow system for mixed liquid. The invest on the construction is low.

The disadvantage of the A_1/O process is that the removal of N is not high. The limit comes from the mixed liquid backflow. The relationship between the removal of N (η) and the backflow ratio (r) is as followed.

$$r=\frac{\eta}{1-\eta}$$

It is showing that η is up to 90%, when the r is 900%. In practice, the r is controlled at the range of 70%~80%. In addition, there is no function to remove P from the A_1/O process. Another disadvantage is that the denitrification is observed in the sedimentation tank under the improper running, resulting in the sludge floating.

(2) Anaerobic-aerobic process (A_P/O) The process consists the anaerobic tank and aerobic tank and sedimentation tank (Figure 4-35). The function of the process is to remove organic matter, at the same time, it is to get good removal P. The process is as followed. The municipal and backflow sludge come into the anaerobic tank, they are mixed under the propeller agitator. The phosphorus-accumulating bacteria release lots of P from its body and uptake a part of the organic matters. And then the mixed liquid goes to the aerobic tank. The organic matters are degraded under the metabolism of bacteria and the P is excessively adsorbed by the phosphorus-accumulating bacteria. After the reaction in the aerobic tank, the mixed liquid goes to the sedimentation tank for the separation. Therefore, the P is removed by the drainage of residual sludge. However, for waste water with low temperature and low organic matter, it is suitable to use the bypass unit to remove P. Under the good running of aerobic tank, P is over 2.5% in the residual sludge and the removal of BOD_5 in the whole process is similar as the traditional activated sludge. HRT is short for the process, 3~6 h. For the anaerobic tank, the HRT is 1~2 h, while the HRT is 2~4 h in the aerobic tank. The ratio of HRT in the anaerobic tank and the aerobic tank is (1 : 2) ~ (1 : 3). The removal of P is 70%~80% and the P content is lower than 1.0 mg/L in the effluent.

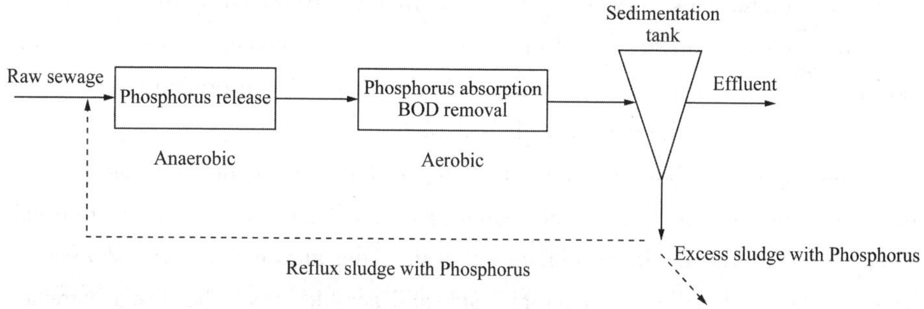

Figure 4-35 Anaerobic-aerobic process

The characteristics are as followed.

① The anaerobic tank is in front of the aerobic tank. It is favor of inhibiting the growth of filamentous bacteria. SVI is less than 100 in the mixed liquid. The sludge is easily settled in the sedimentation tank and it is favor of reducing the organic matter load.

② P is over 2.5% in the residual sludge. The compost of the residual sludge is good.

③ The process is simple.

The process is suitable for the waste water with low ratio of TP to BOD. If the ratio of TP to BOD is high, low BOD load results in the low residual sludge. The treatment on the waste water is unsatisfactory. In addition, low influent amount is resulting in long retention time of waste water in the sedimentation tank. This improves the release of P from the phosphorus-accumulating bacteria under the anaerobic conditions.

(3) Anaerobic - anoxic - aerobic process (A/A/O) The process consists the anaerobic tank and anoxic tank andaerobic tank and sedimentation tank (Figure 4-32). The function of the process is to remove organic matter, at the same time, it is to get good removal N and P.

① The basic process. The anaerobic tank is added into the anoxic-aerobic process, so the mixed liquid is returned to the anaerobic tank for the removal N by denitrification.

For the anaerobic tank, the function is P release from the phosphorus-accumulating bacteria and soluble organic matter is utilized by the bacteria. Another function is to remove ammonia nitrogen by the synthesis of cells.

For the anoxic tank, the nitrate is reduced to N_2 by the denitrifying bacteria using the carbon source in the waste water.

For the aerobic tank, organic matter is utilized and degraded by the bacteria. After the ammoniation and nitrification of the organic matter, the content of ammonia nitrogen is down obviously, but the content of nitrate is up. Under the absorption of phosphorus-accumulating bacteria, the content of P is down.

According to the above analysis, the process has the function of removal organic matter and N and P. The removal of organic matter and the nitrification are finished in the aerobic tank. The denitrification is finished in the anoxic tank. The P removal is finished in the combination of anaerobic tank and aerobic tank. The design parameters are described in Table 4-2.

Table 4-2 the main designment parameters for the technologies to remove N and P

Items	A_N/O	A_P/O	A/A/O
BOD load /[$kgBOD_5$/(kgMLSS · d)]	0.05~0.10	0.4~0.7	0.1~0.2
Total N load /[kgTN/(kgMLSS · d)]	≤0.05	—	—

续表

Items		A_N/O	A_P/O	A/A/O
MLSS/(g/L)		2.5~4.5	2.0~4.0	2.5~4.5
SRT/d		11~23	3.5~7	10~20
O_2 demand/($kgO_2/kgBOD_5$)		1.1~2.0	0.7~1.1	1.1~1.8
Y/($kgVSS/kgBOD_5$)		0.3~0.6	0.4~0.8	0.3~0.6
HRT/h	total	8~16	3~8	7~14
	Anoxic	0.5~3.0	—	0.5~3.0
	anaerobic	—	1~2	1~2
R of sludge/%		50~100	40~100	20~100
R of mixed liquid/%		100~400	—	≥200
η/%	BOD_5	90~95	80~90	85~95
	TN	60~85	—	55~80
	TP	—	75~85	50~75

② The characteristics of the A/A/O process:

• The specific bacteria live in the different tank with anaerobic, anoxic and aerobic environment conditions. The process can remove organic matter and N and P.

• The process is simple and the total HRT is shorter, compared with the other processes with the removal of N and P.

• It is favor of inhibiting the growth of filamentous bacteria. SVI is less than 100 in the mixed liquid. The sludge is easily settled in the sedimentation tank.

• The content of P is high in the residual sludge, and it is over 2.5%.

• The denitrification is affected by the backflow ratio, while the removal of P is affected by the content of DO and the nitrate in the backflow of activated sludge.

③ The influence of environment conditions on the A/A/O:

• Biodegradable organic matter. Phosphorus-accumulating bacteria uptake the biodegradable organic matter for the metabolism and release P from the body. While the content of the organic matter is low, it is to worsen the release of P by the phosphorus-accumulating bacteria. In practice, the ratio of dissolved P to soluble BOD_5 should be less than 0.06 at the anaerobic stage.

If the ratio C to N is high, especially, high soluble BOD_5, the denitrification rate of nitrate is high. HRT at anoxic stage is about 0.5~1.0 h. If the ratio of C to N is low, HRT at anoxic stage is about 2~3h. In practice, the ratio of COD to TKN should be more than 8, the removal of N is up to 80%.

- SRT. SRT in A/A/O is longer due to the generation, compared with the other activated sludge. In theory, SRT in A/A/O is about 25 days. However, the removal of P from the waste water is depending on the drainage of residual sludge. Therefore, the drainage period should be 5 ~ 8 days. In practice, SRT in A/A/O is about 15 ~ 20 days.
- DO. Low content of DO inhibits the breeding of the nitrifying bacteria, and the worst is the nitrifying bacteria are eliminated due to the low DO. To get the high denitrification, the ammonia nitrogen should be completely transformed into nitrate in the aerobic tank. In addition, the consumption on DO is big due to the organic matter degradation and the P adsorption by the phosphorus-accumulating bacteria. Therefore, the content of DO should meet the demand of the three reactions. However, the DO is not too high in the aerobic tank, because the DO comes to the anaerobic tank and the anoxic tank with the backflow of activated sludge and mixed liquid. This influences the release of P from the phosphorus - accumulating bacteria and the denitrification of nitrate. In practice, the content of DO should be 1.5~2 mg/L.

For the denitrification of nitrate in the anoxic tank, the denitrifying bacteria used the chemical oxygen in the nitrate to transform the nitrate to nitrogen. If the content of DO in the anoxic tank is high, this inhibits the denitrification. It is noticeably, it is bad to control the content of DO in the backflow of mixed liquid from aerobic tank to anaerobic tank by decreasing aeration amount in aerobic tank. A part of organic matter will not be oxidized by the bacteria in the aerobic tank, and then the residual organic matter comes to the sedimentation tank, resulting in the denitrification appears. The N_2 affects the separation of solid and liquid in the sedimentation tank. In practice, the content of DO should be less than 0.2 mg/L in anaerobic tank, and the DO is less than 0.5 mg/L in anoxic tank.

- The volume of nitrification area and denitrification area. In practice, the ratio of the volume of denitrification area to the volume of aeration tank is 0.2~0.3. However, the biggest ratio is up to 0.5 because the water quality is bad or the removal of N is high.
- The sludge load (N_s). In general, the Ns is less than 0. 18 $kgBOD_5/(kgMLSS \cdot d)$ in the aerobic tank. On the contrary, the amount of heterotrophic bacterium is higher than the nitrifying bacteria and inhibits the nitrification. The N_s is less than 0.10 $kgBOD_5/(kgMLSS \cdot d)$ in the anaerobic tank. Therefore, the range of Ns is narrow in A/A/O process.

● TKN/MLSS. The higher N load affects the nitrification. In general, the TKN/MLSS is less than 0.05 kgTKN/(kgMLSS · d).

● Sludge return ratio and mixed liquid return ratio. The denitrification efficiency can be increased by increasing the mixed liquid return ratio. However, the higher mixed liquid return ratio is not better, because lots of DO and nitrate come into the anoxic tank and destroys the denitrification conditions. In addition, the consumption on power is bigger. In practice, the range of the mixed liquid return ratio is 100% ~ 600% depending on the removal of N.

The sludge return ratio is 25% ~ 100%. Too higher, more DO goes to the anaerobic tank, affecting the anaerobic conditions (DO < 0.2 mg/L), resulting in the P release worse. Too lower, it is difficult to keep the normal content of activated sludge (2,500 ~ 3,500 mg/L), affecting biochemistry rate.

● Water temperature. The enzyme of the bacteria is sensitive to water temperature. The best temperature is 30 ~ 35℃ for nitrifying bacteria. The nitrification rate and oxidizing rate on organic matter decrease obviously if the water temperature is less than 10℃. For the denitrification, the best water temperature is 20 ~ 38℃. The denitrifying bacteria growth rate decreases obviously if the water temperature is less than 15℃. The denitrifying bacteria growth stops if the water temperature is less than 3℃.

For the phosphorus-accumulating bacteria, the effect of water temperature is little because there are three kinds of phosphorus-accumulating bacteria suitable for high, middle and low temperature.

● Alkalinity. The nitrification and denitrification produce and consume the alkalinity, respectively, resulting in the change of pH. Higher nitrification rate is depending on the suitable range of pH in 7.8 ~ 8.4. If pH is lower than 6 or higher than 9, the nitrification stops. If the pH range is not in 6.5 ~ 7.5, the denitrification rate decreases.

④ The problems of the A/A/O process and the modification. It is impossible to get the good removal on N and P at the same time. If the denitrification is good, the removal of P is bad. Vice versa, the removal of P is good, while the denitrification is bad. In practice, the requirement on the environmental conditions is different. In the same process, it is impossible to meet the different requirement by the removal of N and P. Sludge age plays the most important role in the removal of N and P. SRT is about 25 d for the removal of N, while it is about 5 ~ 8 d for the removal of P. In addition, to get the better removal of N, the nitrification in aerobic tank should be complete. Therefore, the sludge return ratio should be higher. But for the removal of P, it is not good, because the higher return brings more nitrate into the anaerobic tank, affecting the release of P under anaerobic conditions. On the contrary, if the nitrification is not good, it is favor of the release of P under anaerobic conditions.

According to the above analysis, the modification is proposed as followed (Figure 4-36).

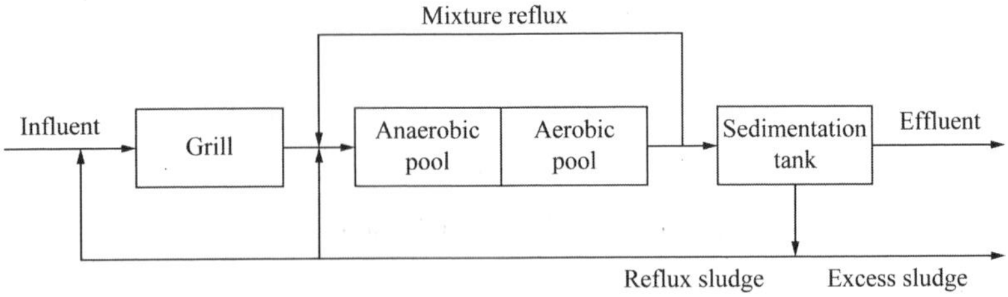

Figure 4-36 The modified process

The return of the sludge is divided into parts and comes into the anaerobic tank and the anoxic tank, respectively, reducing the backflow sludge to the anaerobic tank and reducing the content of nitrate and DO. In practice, the sludge return ratio is about 10% to the anaerobic tank under the total sludge return ratio of 60% ~ 100%.

(4) the designment of the A/AO process The calculation of the aeration tanks is similar as the traditional activated sludge. The other design parameters are as followed.

Table 4-3 The design parameters

Item	Sludge load/ [$kgBOD_5/(kgMLSS \cdot d)$]	The ration of anaerobic stage to aerobic stage	MLSS/(g/L)	SRT/d
A_N/O	BOD-SS<0.18 TKN-SS<0.18	1 : (3~4)	2~3	>20
A_P/O	BOD-SS>0.10	1 : (2~3)	2~3	2~5
A/A/O	0.10<BOD-SS<0.18 General, 0.14	1 : (3~4)	2~3	15~20

4.2 For Industrial waste water

The use of various chemicals in the treatment of domestic waste waters is not a new concept. Documented chemical treatment reports date back to 1740, with an extensive use of chemicals to aid in the sedimentation process being used in Europe since the late

1800's.

By 1934, waste water treatment plants were present in 3 Canadian cities and 34 U.S. cities, where some form of chemical precipitation process was being used. During World War Ⅱ, and for several years thereafter, there was a decline in chemical precipitation processes due to limited availability of these chemicals, and the varying costs sometimes led to poor quality effluent being emptied into the receiving lake or river.

In municipal as well as industrial waste water treatment, a wide range of materials may have to be removed by the treatment process. To cope with this variety, certain properties are modified at one time, and some materials are eliminated as a group:

① Acidity and alkalinityshould be neutralized.

② Oils and greases must be removed.

③ Heavy metals (e.g. chromium, lead, and mercury) and some ions such as cyanide are toxic, and they should be removed almost completely.

④ Soluble organics, including products of plant and animal life are harmful because biomass growth exhausts the dissolved oxygen in the water. This lowers the residual oxygen content in the receiving lake or river. These materials, however, can be decomposed with the aid of appropriate micro-organisms.

⑤ Colloidal particles make water translucent or turbid. Suspended solids then form to settle and form sludges. This process removes color, turbidity and odor.

Increased requirements for the removal of soluble organic compounds led to secondary treatment as discussed earlier. Biological treatment gained favour since chemicals were in short supply.

After a decade of research in the area of physical and chemical processes by waste water treatment, experts widely agree that a resurgence in the use of chemicals to remove nutrients such as phosphorus and nitrogen is absolutely essential in order to sustain healthy aquatic environments in lakes and rivers.

In the early 1970's, effluent limits became stricter, forcing the chemical process industries to upgrade their waste water treatment systems or to create new systems. Today, the limits on pollutants are much tighter as there are new restrictions on over one hundred specific priority pollutants which are mainly organic chemicals and heavy metals.

4.2.1 Anaerobic biological treatment

Anaerobic biological treatment is a series of processes in which microorganisms break down biodegradable material in the absence of oxygen. After undergoing such a treatment, water can be safely released back into the environment. The biological agents used in the process are anaerobic microorganisms that consume or break down biodegradable materials in sludge, or the solid portion of waste water following its filtration from polluted water. Anaerobic biological treatment is used for industrial or domestic purposes to manage waste or to release energy.

4.2.1.1 The principle

Anaerobic biological treatment is also known as anaerobic digestion due to the action of the microorganisms. That is, they are essentially "digesting" the polluted parts of the water. An excellent way to decrease the amount of organic matter leftover in things such as sewage and leftover food, anaerobic digestion is typically apart of any biological waste water treatment system.

Usually, the anaerobic process takes place in sealed tanks, located either above or below the ground. The digestion process begins with bacterial hydrolysis of the input materials to break down insoluble organic polymers, such ascarbonhydrates, and make them available for other bacteria. Acidogenic bacteria then convert the sugars and amino acids into carbon dioxide, hydrogen, ammonia and organic acids. Acetogenic bacteria then convert these resulting organic acids into acetic acid, along with additional ammonia, hydrogen, and carbon dioxide. Finally, methanogens convert these products to methane and carbon dioxide. The methanogenic archaea populations play an indispensable role in anaerobic waste water treatments (see Figure 4-37).

There are four key biological and chemical stages of anaerobic biological treatment: hydrolysis, fermentation, acetogenesis, methanogenesis.

(1) Hydrolysis The reaction of the organic matter with the water is called hydrolysis. In most cases, biomass is made up of large organic polymers. For the bacteria in anaerobic digesters to access the energy potential of the material, these chains must first be broken down into their smaller constituent parts. These constituent parts, or monomers, such as sugars, are readily available to other bacteria. The process of breaking these chains and dissolving the smaller molecules into solution is called hydrolysis. Therefore, hydrolysis of these high-molecular-weight polymeric components is the necessary first step in anaerobic digestion. Through hydrolysis the complex organic molecules are broken down into simple sugars, amino acids, and fatty acids.

Acetate and hydrogen produced in the first stages can be used directly by methanogens. Other molecules, such as volatile fatty acids (VFAs) with a chain length

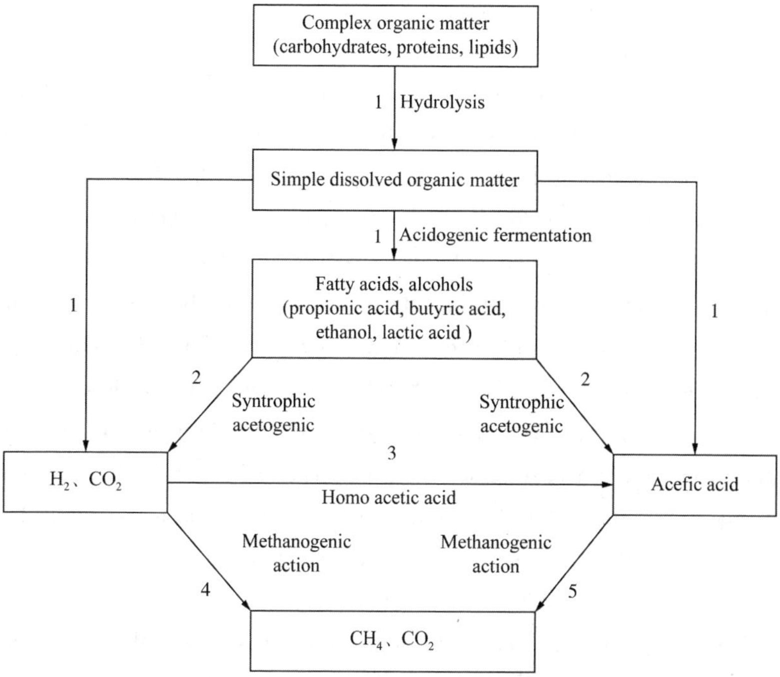

Figure 4-37 The anaerobic biological metabolism

1—Fermentative bacteria; 2—Hydrogen-producing acetogenic bacteria; 3—Homoacetogenic bacteria; 4—Methanogenic bacteria (H_2, CO_2) ; 5—Methanogenic bacteria (acetic acid)

greater than that of acetate must first be catabolised into compounds that can be directly used by methanogens.

In practice, the process of the hydrolysis is slow. It is called the limit step of the reaction velocity for the anaerobic biological treatment on the waste water with big molecule organic matter. The key is the effective contacting of the organic matter with the extracellular enzyme. Therefore, the digestion on the big molecule organic matter is slower than the small organic matter.

There are many factors affecting the hydrolysis speed and efficiency.

① Water temperature.

② Retention time for the organic matter in the reactors.

③ The component of the organic matter (the ratio of lignin, carbohydrate, protein and fat).

④ The size of the organic matter granular.

⑤ pH.

⑥ The content of ammonia nitrogen.

⑦ The content of the hydrolysis products.

Lots of experiments show that it is impossible to make difference between the hydrolysis and the fermentation, even though under the strict controlling conditions, the exception is the technology with the dosing of hydrolytic enzyme. The main reason is that the microorganism is the same for the hydrolysis and the acidogenesis.

(2) Fermentation The fermentation is the organic matter not only the electron donor but also electron receiver. During the process, the small molecule is transformed into the end-product (VFA) in the fermentation bacteria. The end-products are VFA, alcohol, lactic acid, H_2, CO_2, NH_3, H_2S, et al. The new cells are formed.

In the fermentation, there are lots of kinds of fermentation bacteria. The fermentation bacteria are a mixed bacteria group. The main is the obligate anaerobe and facultative anaerobe. The effect of the water temperature on the fermentation bacteria is very big. For the middle temperature, the fermentation bacteria include *bacteroides*, *clostridium*, *butyrivibrio*, *eubacterium*, *bifidbcterium*, et al.

The bio-chemical reaction is affected by the two factors. One is the component and the content of the substrate. The other is the kinds of the metabolite and the continuous bio-chemical reaction. Bigger content of substrate, quicker of the bio-chemical reaction. The accumulation of the metabolite will affect the bio-chemical reaction.

It is well known that the aim of the fermentation is to provide the suitable substrate for themethanogenesis. The rank of the best end - product of the fermentation is organic compound with one carbon, acetic acid, ethyl alcohol and butyrate. The worst end-product of the fermentation is lactic acid and propionic acid. The transformation from propionic acid to acetic acid is very slow. Therefore, the accumulation of propionic acid will result in the failure of the methanogenesis. It is possible for the lactic acid transformed into propionic acid.

(3) Acetogenesis During the stage, the end-product of the fermentation is transformed into the acetic acid, H_2, CO_2 which is the substrate of the methanogenus under the syntrophic acetogenic bacteria. At the same time, the new cells are formed.

Under the anaerobic conditions, the bacteria producing the acetic acid are divided two kinds. One is heterotrophs which uptake the organic matter to produce the acetic acid and H_2. The other is mixer nutrient bacteria which uptake the organic matter to produce the acetic acid. They also uptake H_2 and CO_2 to produce the acetic acid.

The hydrogen partial pressure is kept at low level because the hydrogen produced by the syntrophic acetogenic bacteria is used by the methanogenus. The syntrophic acetogenic bacteria can provide acetic acid and H_2 for the methanogenus and improve the growth of the methanogenus. On the contrary, the methanogenus consumed the hydrogen to decrease the hydrogen partial pressure. It is favor of the growth of the syntrophic acetogenic bacteria. The production and the consumption on hydrogen by different species bacteria is called interspecies hydrogen transfer.

In these years, lots of syntrophic acetogenic bacteria is found, such as *syntrophomonas*, *syntrophobacter*, *clostridium*, *pelobacter*, et al.

(4) Methanogenesis During the stage, the methanogenus transformed the acetic acid, methanoic acid, methyl alcohol, methylamine and H_2/CO_2 into CH_4, CO_2 and new cells. In the anaerobic biological treatment, the methanogenesis is final, and the methanogenus is the most important bacteria.

The methanogenus is a kind of strictly anaerobic archaebacteria which can uptake H_2, CO_2 and simple organic matter. In anaerobic reactor, the amount of methanogenus is up to $10^6 \sim 10^8$ count/L. In UASB, it is up to 10^9 count/L.

The production of CH_4 is described in the followed equations.

$$CH_3COO^- + H_2O \longrightarrow CH_4 + HCO_3^-$$

$$HCO_3^- + H^+ + 4H_2 \longrightarrow CH_4 + 3H_2O$$

$$4CHOOH \longrightarrow CH_4 + 3CO_2 + 2H_2O$$

$$4CH_3OH \longrightarrow 3CH_4 + CO_2 + 2H_2O$$

$$CH_3COOH + 4H_2 \longrightarrow 2CH_4 + 2H_2O$$

The anaerobic biological treatment of the organic matter is according to the above four steps in theory, but in the reactor, the reactions maybe simultaneous, not step by step. After the four steps, the organic matter is transformed into the CH_4 with high caloricity. Therefore, the technology is a simple method to recycle bioenergy.

4.2.1.2 The development of the anaerobic biological treatment

At 1895, Donald designed the first anaerobic digestion tank in the world. At 1896, a continuous anaerobic digestor was founded in UK. In the 1940s, the first continuously mixed anaerobic digestor was founded in Australia. At 1956, Schroefer et al invented anaerobic contact process. At 1967, Young and McCarty invented anaerobic filter. At 1974, Lettinga invented upflow anaerobic sludge bed (UASB). And then, attached growth type anaerobic biological treatment is developed, such as anaerobic filter, anaerobic expanded bed, anaerobic fluiding bed, anaerobic rotating disc, anaerobic baffled reactor, and so on.

From 1980s, new and highly efficient anaerobic processes were developed. For example, upflow anaerobic bed-filter (UBF), expanded granular sludge bed (EGSB), internal circulation reactor (ICR).

Some sketch maps are as followed(Figure 4-38, Figure 4-39).

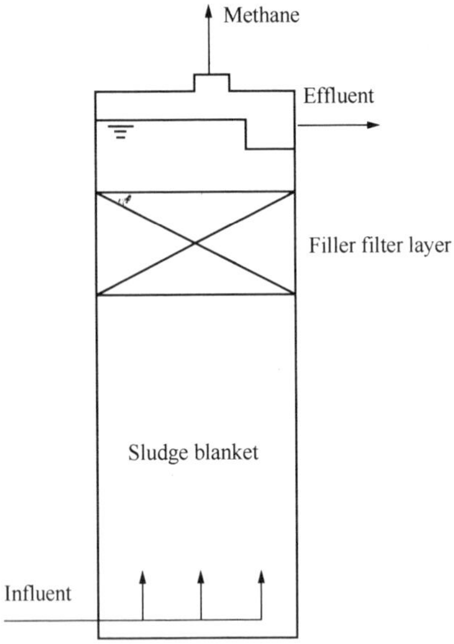

Figure 4-38 Sketch map of UBF

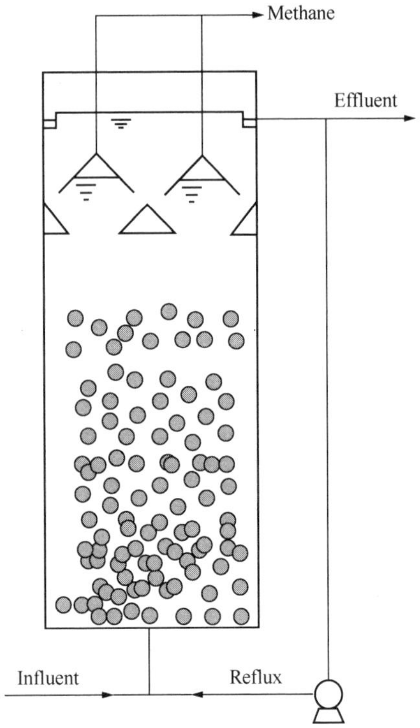

Figure 4-39 Sketch map of EGSB

The design parameters of the main anaerobic biological reactors are in the Table 4-4.

Table 4-4 some anaerobic biological technologies

Anaerobic biological technologies	HRT	Process object	Design load/ [kg/(m^3 · d)]	Application	Running temperature
Digestion tank	half year ~ one year (sludge)	Domestic waste Water and sludge	—	Practice	Normal
Common digester	20 ~ 30 d	Sludge	1.0 ~ 1.5(VSS)	Practice	Middle, high
High-rate digester	7 ~ 10 d	Sludge	3.0 ~ 3.5(VSS)	Practice	Middle, high
Anaerobic contact process	0.5 ~ 6 d	Organic waste water	2 ~ 6(COD)	Practice	Middle
Anaerobic filter	0.9 ~ 8 d	Organic waste water	3 ~ 10(COD)	Pilot-running	Middle

续表

Anaerobic biological technologies	HRT	Process object	Design load/ [kg/(m^3 · d)]	Application	Running temperature
Upflow anaerobic sludge bed	6~20 h	Organic waste water	6~15(COD)	Pilot-running	Middle
Anaerobic expanded bed	6~24 h	Organic waste water	4(COD)	Lab-scale	Normal
Anaerobic fluiding bed	0.5~4 h	Organic waste water	9~13(COD)	Lab-scale	Normal
Anaerobic rotating disc	8~18 h	Organic waste water	8~33[kgCOD/(m^2 · d)]	Lab-scale	Normal
Anaerobic baffled reactor	6~ 26 h	Organic waste water	8~36(COD)	Lab-scale	Normal

4.2.1.3 The affecting factors on the anaerobic biological treatment

(1) Water temperature Water temperature is the important factor affecting the anaerobic biological treatment.

① Water temperature affects the activity of the enzyme.

② Water temperature affects the production amount and the component of the methane.

③ Water temperature affects the characteristic and the component of the sludge.

④ Keeping the running temperature increases the running cost.

The range of the water temperature for the anaerobic biological reaction is very wide (5~83℃). The range of the methanogenesis is 4~100℃. The bacteria are divided into low temperature bacteria, middle temperature bacteria and high temperature bacteria. Therefore, the anaerobic biological reaction is divided into low temperature reaction, middle temperature reaction and high temperature reaction.

In each temperature range, the growth rate of the bacteria gradually increases with the increase of the temperature. Now, the temperature is the best growth temperature for the bacteria. But the temperature is over the bets temperature, the growth rate of the bacteria sharply decreases.

(2) pH and alkalinity pH is another important factor affecting anaerobic biological treatment. For different species of the bacteria, the suitable pH is different. For the acid-forming bacteria, the pH is 3.5~8.0, while the best pH is 6.5~7.5. The metabolism rate and the growth rate are affected by pH value. For the methanogenus, the suitable pH value is 6.5~7.8.

There is a ratio between the value of pH, the content of hydrocarbonate and CO_2 in the anaerobic reactors. Generally, the alkalinity should be in 2,000~4,000 mg/L in the reactor with reasonable running, while the normal range is 1,000~5,000 mg/L.

(3) Redox potential The redox potential (ORP) affects the ratio of the obligate anaerobe and facultative anaerobe. For the aerobic bacteria, the suitable redox potential is 300 ~ 400 mV. The bacteria can grow when the ORP is more than 100 mV. The facultative anaerobe is in aerobic respiration when the ORP is more than 100 mV. When the ORP is less than 100 mV, the facultative anaerobe is in fermentation. The anaerobic bacteria only grow under 100 mV. The acid-forming bacteria is in normal physiological activity at the range of −100 ~ +100mV. For the methanogenus, the suitable ORP is less than −300mV.

(4) Nutrient The enough nutrient is the guarantee of the excellent growth of the bacteria in the anaerobic biological treatment. The nutrient depends on the component of the cell (Table 4-5).

Table 4-5 The chemical component of the methanogenus (g/kg dry cell)

Element	Content	Element	Content
N	65	Fe	1.8
P	15	Ni	0.10
K	10	Ge	0.075
S	10	Cu	0.060
Ca	4	Zn	0.060
Mg	3	Mn	0.020

The smallest content of the element is estimated by the equation.

$$e = 1.14 \times \mathrm{COD} \times Y \times \rho_{\mathrm{cell}}$$

in the equation, e—The smallest content of the element, mg/L;

COD—The content of organic matter in the influent, mg/L;

Y—The productivity of the cell, gVSS/gCOD;

ρ_{cell}—The content in the cell, mg/g dry cell.

For the waste water without acidification, Y is 0.15, and COD : N : P = 350 : 5 : 1. For the waste water with complete acidification, Y is 0.03, and COD : N : P = 100 : 5 : 1.

(5) Inhibiting matters There are toxic matters in the industrial waste water to inhibit the activity of the bacteria, resulting in the failure of the reactor running.

The inhibition on the methanogenesis is referred to the special content of the chemical matter resulting in the activity of the methanogenus down. It is described as IC_{50}.

The toxicity of the heavy metals on the methanogenesis is showed in the Table 4-6.

Table 4-6 IC_{50} of the anaerobic sludge for different heavy metals

Heavy metals	IC_{50}/(mg/L)	Substrate	Inoculum
Cr^{3+}	350	Acetic acid	Digestion sludge
	>240	Acetic acid	Flocculent sludge
Cr^{4+}	≈30	Acetic acid	Flocculent sludge
Cu^{2+}	15	Acetic acid	Digestion sludge
	75	Acetic acid, butyric acid, Propionic acid	Digestion sludge
Ni^{2+}	200	Acetic acid	Digestion sludge
Zn^{2+}	≈300	—	Flocculent sludge
	≈90	Acetic acid	Flocculent sludge
Pb^{2+}	≈300	Acetic acid	Flocculent sludge
Cd^{2+}	80	Acetic acid	Digestion sludge

The toxicity of the organic matter on the methanogenesis is showed in the Table 4-7.

Table 4-7 IC_{50} of the anaerobic sludge for different organic matter mg/L

Organic matter	IC_{50} for the bacteria without domestication	IC_{50} for the bacteria with domestication	Organic matter	IC_{50} for the bacteria without domestication	IC_{50} for the bacteria with domestication
Chloroform	0.5	45	Cyanide	1.0	25.0
Chloropropane	8.0	—	Benzene	≈40	—
Pentachlorophenol	≈1.0	>5.0	Ethylbenzene	340	—
Methanal	100	400			

(6) Organic matter load The organic matter load presents the relationship between the substrate and bacteria. The organic matter affects the sludge growth, sludge activity and organic matter degradation. Increasing the organic matter load is favor of the growth of bacteria and the degradation of organic matter, at the same time, it is beneficial to reducing the volume of the reactor.

In practice, for the treatment on the common industrial organic waste water, the organic matter load is 0.1~0.5 $kgBOD_5$/(kgMLVSS · d) for the aerobic process, while the load is 0.5~1.0 $kgBOD_5$/(kgMLVSS · d) for the anaerobic process. The content of the sludge in the anaerobic process is higher than that in the aerobic process.

4.2.1.4 The advantages of the anaerobic biological treatment

① The application is wide, compared with aerobic processes. The anaerobic biological process is suitable for the sludge and the waste water with different content. The content of COD is up to thousands mg/L, even up to ten thousand mg/L.

② The running cost and energy consumption is low. The economy is good. In addition, the need on the nutrient is low, compared with the aerobic processes. For the aerobic processes, BOD : N : P = 100 : 5 : 1, while the ratio is (300~500) : 5 : 1 for the anaerobic biological process. The production rate of residual sludge is low in anaerobic biological process, 0.03~0.15 VSS/kgCOD, while the rate is 0.4~0.6 VSS/kgCOD for the aerobic processes.

③ It is producing lots of energy. The end-product of the anaerobic biological treatment is CH_4 and CO_2. The caloricity of CH_4 is up to 39300 kJ/Nm^3.

④ The organic matter volume load is high, compared with the aerobic processes. In practice, the organic matter volume load is 0.5~1.0 $kgBOD_5/(m^3 \cdot d)$ for the aerobic processes, while the load is 5~10 $kgBOD_5/(m^3 \cdot d)$ for the anaerobic processes, even up to 50 $kgBOD_5/(m^3 \cdot d)$.

4.2.1.5 The disadvantages of the anaerobic biological treatment

① The content of COD in the effluent is high, compared with the aerobic process.

② The reaction rate is slow. For the initial start, the time is long, up to 8~10 weeks.

③ The anaerobic bacteria are sensitive to the toxic matters.

④ The odor is common in the anaerobic biological treatment. It is easy to pollute the environment around the plants.

In developing countries, government-funded programs are available to help power single homes and farms with the biogas produced by on-site anaerobic digestion. The United Nations also offers funding to anaerobic digestion programs in the developing world, provided proof of reduced gas emissions is available.

In the United States, anaerobic biological treatment is usually part of a municipal waste water treatment plant. Still, anaerobic digestion is also used by single-family homes in areas not connected to a municipal system and by businesses with on-site waste water treatment facilities.

4.2.2 Upflow anaerobic sludge blanket reactor

Upflow anaerobic sludge blanket (UASB) technology, normally referred to as

UASB reactor (Figure 4-40), is a form of anaerobic digester that is widely used in the treatment of waste water.

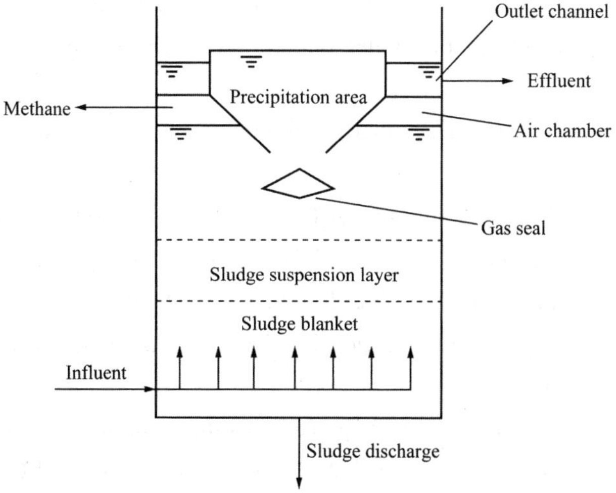

Figure 4-40 Sketch map of UASB

The UASB reactor is amethanogenic (methane-producing) digester that evolved from the anaerobic digester. A similar but variant technology to UASB is the expanded granular sludge bed (EGSB) digester.

UASB uses an anaerobic process whilst forming a blanket of granular sludge which suspends in the tank. Waste water flows upwards through the blanket and is processed (degraded) by the anaerobic microorganisms. The upward flow combined with the settling action of gravity suspends the blanket with the aid of flocculants. The blanket begins to reach maturity at around 3 months. Small sludge granules begin to form whose surface area is covered in aggregations of bacteria. In the absence of any support matrix, the flow conditions create a selective environment in which only those microorganisms, capable of attaching to each other, survive and proliferate. Eventually the aggregates form into dense compact biofilms referred to as "granules". A picture of anaerobic sludge granules can be found here.

Biogas with a high concentration of methane is produced as a by-product, and this may be captured and used as an energy source, to generate electricity for export and to cover its own running power. The technology needs constant monitoring when put into use to ensure that the sludge blanket is maintained, and not washed out (thereby losing the effect). The heat produced as a by-product of electricity generation can be reused to heat the digestion tanks.

The blanketing of the sludge enables a dual solid and hydraulic (liquid) retention time in the digesters. Solids requiring a high degree of digestion can remain in the reactors for periods up to 90 days. Sugars dissolved in the liquid waste stream can be converted into gas quickly in the liquid phase which can exit the system in less than a day.

UASB reactors are typically suited to dilute waste water streams (3% TSS with particle size >0.75 mm).

Conventional treatment settles sludge which is then digested, and then aerates the remaining liquids which use bacteria to oxidize the potential digester fuel, and uses up energy to drive the compressors. The result is that on a standard western treatment works the energy produced from settled sludge digestion is all used by the aeration process, with little power export.

With UASB the aeration the whole process of settlement and digestion occurs in one or more large tank(s). Only the UASB liquids, with reduced BOD, needs to be aerated.

This leads to a halving of the aeration energy and a doubling of the power generated from digestion, leading over all to a tripling of power generated.

UASB has many features in the aspect of treating waste water.

UASB solves air, liquid and solid problems during different kinds of reactors. The treatment effect is reliable in treating industrial or domestic waste water and it currently has been extensively applied for food, beer and beverage, paper making, chemical, municipal and other industries.

4.2.3 The other typical technologies

4.2.3.1 The oil waste water and petrochemical industry waste water

(1) The oil waste water The oil waste water mainly comes from the petroleum, petrochemical industry, steelworks, mechanical manufacture, et al.

The oil is divided into floating oil, dispersed oil, emulsified oil and soluble oil.

① Floating oil. The size of the floating oil is big, up to 100 μm, floating on the water.

② Dispersed oil. The size of the oil particle is about 10~100 μm, suspended in the water. But the dispersed oil is not stable. After a long time, it is transformed into the floating oil.

③ Emulsified oil. The size of the oil is less than 10 μm, generally, 0.1~2 μm.

④ Soluble oil. The size of the soluble oil is less than the emulsified oil.

(2) The petrochemical industry waste water

The characteristics of the waste water are as followed.

① The amount of the waste water is big.

② The component of the waste water is complex.

③ The content of the organic matter is high.

④ The waste water consists many types of heavy metals.

(3) The typical process The petrochemical industry waste water is attributed to the organic matter and it is mixed with the other waste water in the same plant and then the mixed liquid can be treated together. The typical process is as following (Figure 4-41).

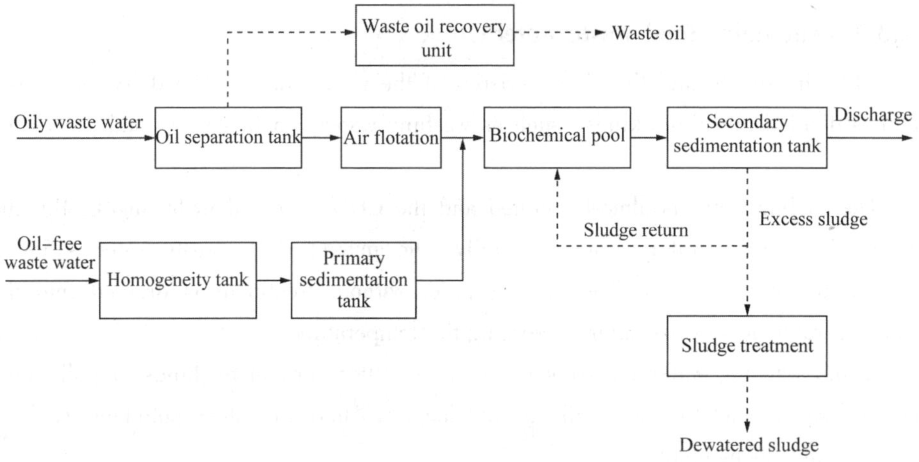

Figure 4-41 The process of the oil waste water and petrochemical industry waste water

4.2.3.2 The papermaking waste water

The main pollutants are as followed.

① Suspended solids. The solids consist settling suspended solids and non-settling suspended solids. The main suspended solids are fiber and fine fiber.

② Biodegradable organic matter. They mainly come from the raw materials, such as hemicellulose, methyl alcohol, acetic acid, formic acid and carbohydrate.

③ Non-biodegradable organic matter. They are mainly lignin and macromolecule carbohydrate.

④ Toxic matters. The toxic matters are abietic acid, unsaturated fatty acid, sulfuretted hydrogen, dimethyl sulfide, many kinds of chlorinated organic compounds, dioxin, and so on.

⑤ Acid and alkali. The acid and alkali change the pH in the waste water, up to 1.2 or 12.

⑥ Color. The color main comes from the residual lignin.

The typical process is as following (Figure 4-42).

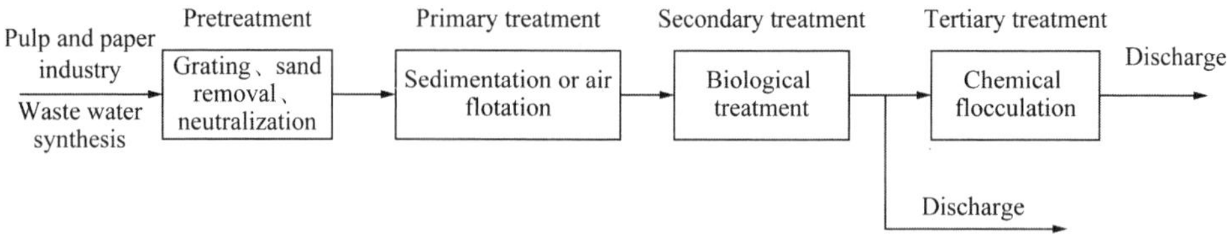

Figure 4-42 The process of the papermaking waste water

4.2.3.3 The dairy food waste water

(1) The source and the characteristics of the waste water The dairy food waste water is divided into three kinds, such as washing water, cooling water and processing water.

The cooling water is almost polluted and the COD is less than 50 mg/L. But the amount of the waste water is big. Generally, the amount of the waste water is 5 ~ 20 times of the milk amount. The cooling water without treatment is drained into the receiving water or recycled after decreasing the temperature.

In the washing water and processing water, there are many kinds of pollutants. Such as casein, lactoprotein, milk fat and lactose. There are other pollutants such as detergent and disinfectants.

(2) The technology system of the waste water The technology depends on the property of the waste water. For the cooling water, it is recycled after cooling process.

There are two typical processes as followed.

① The main equipment is the two-stage high-rate biological filter (Figure 4-43). The sand, oil and suspended solids are removed in the grit chamber, oil separator, respectively. There is a sedimentation tank to remove the biofilm in the middle of the two-stage filters. The second sedimentation tank is to improve the quality of the effluent. The measure of alternating water can be taken to avoid of uneven organic matter load for the two biological filters.

② The main equipment is stabilization tank (Figure 4-44). The waste water successively comes into the anaerobic tank, facultative pond and aerobic tank. The pollutants are removed before the fishpond and a part of the pollutants are transformed into the food of the fish.

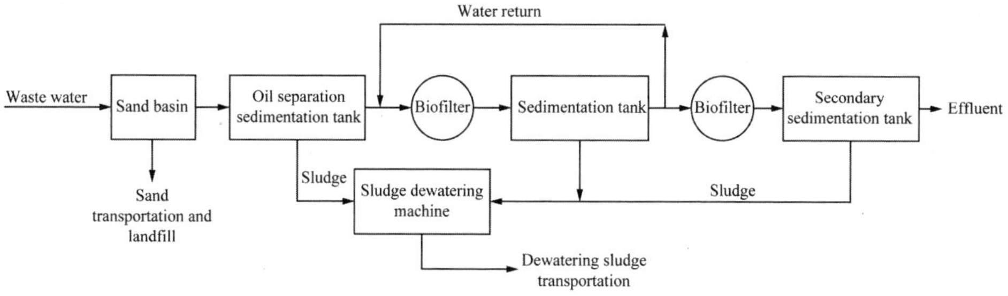

Figure 4-43 The typical process with two-stage high-rate biological filter

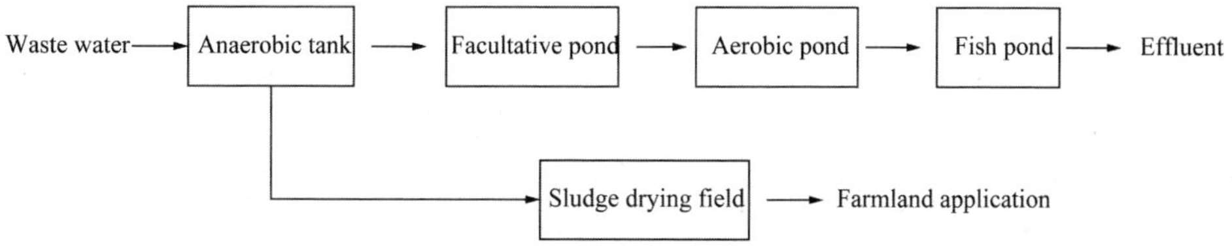

Figure 4-44 The process of the dairy food waste water

4.2.3.4 The beer waste water

(1) The quality and quantity of the beer waste water The content of the beer waste water is very high. Although it is not toxic, it is easily corrupt, resulting in the consumption on dissolved oxygen in the receiving water.

The main water quality indexes are showed in the Table 4-8.

Table 4-8 the main water quality indexes in the beer waste water

pH	Temperature/℃	COD/(mg/L)	BOD_5/(mg/L)	Alkalinity/(mg/L)	SS/(mg/L)	TN/(mg/L)	TP/(mg/L)
5~6	16~30	1,000~2,500	700~1,500	400~450	300~600	25~85	5~7

(2) The technology system of the beer waste water In practice, the main pollutants are the organic matter in the beer waste water and the biodegradability is good. The biological process is chosen for the treatment of the waste water.

Two-stage aerobic biological treatment is used widely due to the higher content of organic matter. And then the COD and BOD_5 content are less than 100 mg/L and 20 mg/L, respectively.

In practice, the anaerobic biological treatment is used as the first stage treatment, and the removal of COD is up to 85%. The COD is about 300 ~ 500 mg/L in the effluent. And then the aerobic biological treatment is necessary to meet the demand of the drainage standard.

The typical process is anaerobic and aerobic process (Figure 4-45)

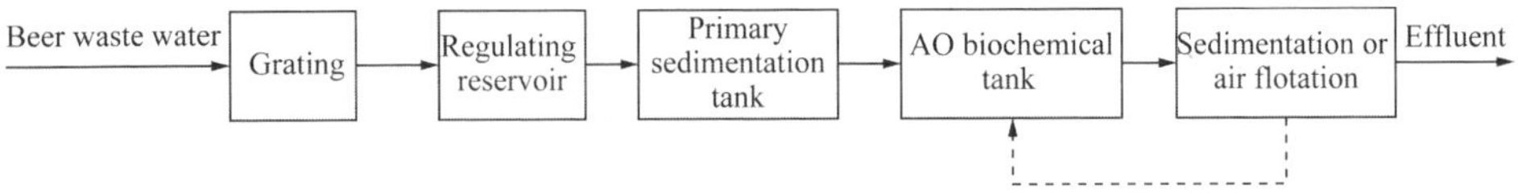

Figure 4-45 The process of the beer waste water

The advantages of the process are as followed.

① Most of the organic matter is removed in the anaerobic tank and the power is saved obviously due to the aeration.

② The COD volume of the new type of anaerobic tank is about 6 ~ 8 kg COD/(m^3 · d). the volume of the anaerobic tank is smaller than that of the aerobic tank for the treatment of the same quantity of the beer waste water.

③ The sludge age is long in the anaerobic tank and the residual sludge is little.

④ The organic pollutants can be transformed into the methane.

4.2.3.5 The textile printing and dyeing waste water

There are many kinds waste water in the textile printing and dyeing. For the clean waste water, such as cooling water, it is recycled after the cooling treatment. For the processing waste water, it is treated before the discharge, because the waste water has the high content of organic matter and color.

The textile printing and dyeing waste water is divided into four kinds. ① Dyeing waste water from wool spinning industry. ② Dyeing waste water from cotton industry. ③ Dyeing waste water from silk textile industry. ④ Dyeing waste water from jute spinning industry.

The typical process is as followed.

① Submerged biofilm reactor—coagulation and sedimentation tank.

The process is showed in the Figure 4-46.

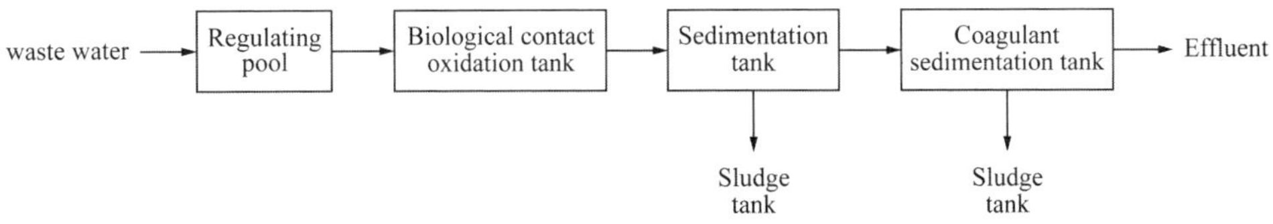

Figure 4-46 Biological—chemical process

The submerged biofilm reactor has the characteristics of the activated sludge and the biofilm. The activity of the bacteria is strong and the ratio of *F/M* is big. In addition, the organic matter load is high and the retention time is short. There is no sludge backflow. The problem is the removal of the nonbiodegradable organic matter and color is low. In the coagulation and sedimentation tank, aluminum chloride and sodium hypochlorite are used to remove the color. The dose of the aluminum chloride is 0.1% ~ 0.3% in practice. Sometimes, it is to get the good removal of color under the dose of 0.5%. The problem of the process is the high amount of sludge and the high content of water. It is necessary to add slaked lime and ferric trichloride for the sludge dewatering.

② Slag filter—chemical coagulation process

The slag filter—chemical coagulation process is suitable for different textile printing and dyeing waste water. The efficiency is excellent.

The aluminum chloride can be used as the coagulant. The process is showed in the following Figure 4-47.

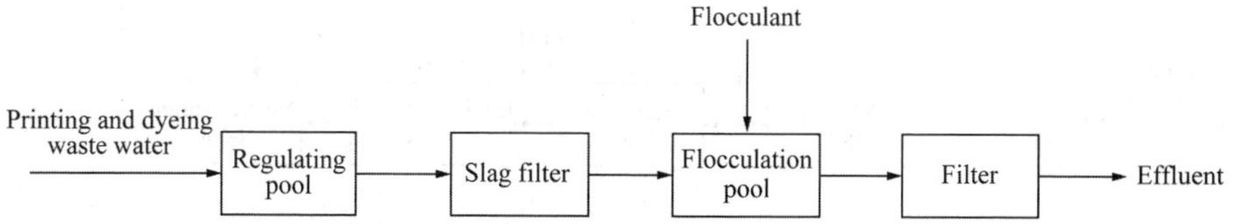

Figure 4-47 Slag filter—chemical coagulation process

③ Anaerobic—aerobic—biological carbon process

The process is showed in the following Figure 4-48.

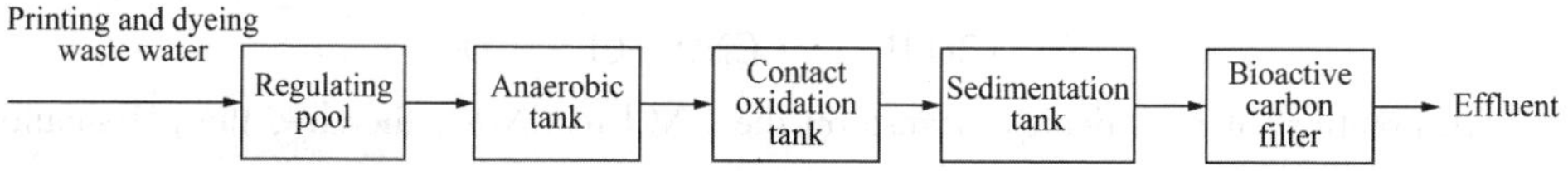

Figure 4-48 Anaerobic—aerobic—biological carbon process

Anaerobic process is not the traditional anaerobic digestion in the above process. There is only hydrolysis and acidification. The hydraulic retention time is 8 ~ 10 h. The aim of the anaerobic tank is to improve the biodegradability.

The COD_{Cr} is 670 ~ 825 mg/L, and the BOD_5 is 120 ~ 192 mg/L and the color is 55 ~ 85 times in the influent, while the removal of the COD_{Cr}, BOD_5, and color is 80% ~ 90%, 94% ~ 96% and 70% ~ 80%, respectively.

3.2.3.6 The toxic industrial waste water

(1) Cyanide waste water There are free cyanide and complex ion, such as copper, cadmium, silver, zinc during the process of the cyanide electroplating. The heavy metals come into the waste water after cyanide breaking. Therefore, the treatment of the cyanide waste water includes the cyanide treatment and the heavy metals treatment.

There is rich experience to treat the cyanide waste water in China, such as alkaline chlorination process, electrolytic oxidation, activated carbon adsorption, ion exchange, ozone and ferrous sulfate method. Nowadays, the alkaline chlorination process is used widely.

Take alkaline chlorination process as an example (Figure 4-49).

NaClO, chlorinated lime or chlorine is used to destroy the cyanide. The principle is the oxidation of ClO^-. The equation is as followed.

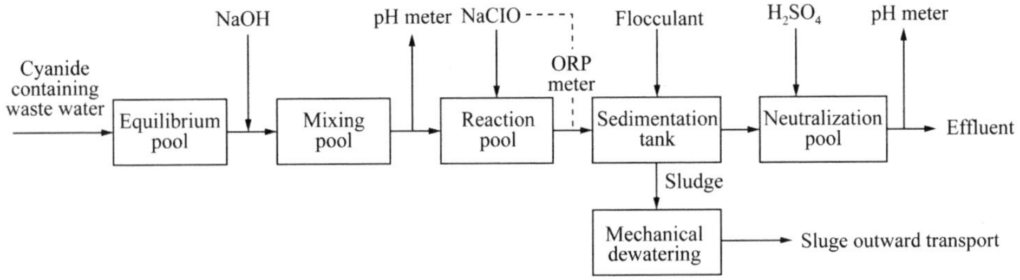

Figure 4-49 Alkaline chlorination process

$$CN^- + ClO^- + H_2O \longrightarrow CNCl + 2OH^-$$

$$CNCl + 2O\ H^- \longrightarrow CNO^- + Cl^- + H_2O$$

In practice, it is difficult to transform the CNCl to CNO^-, thereby, the pH should be more than 11.

The waste water comes into the pipeline mixer after the pH of the waste water is regulated in the equilibrium pool, and then the alkali liquor is added into the pipeline mixer. The pH is stable about 10~12. The waste water comes into the reaction pool. The NAClO is added into the reaction pool to break the cyanide. And then the waste water goes to sedimentation tank for the separation of solid and liquid. It is necessary to neutralize the pH before the discharge.

(2) Chromate waste water The chromate waste water mainly comes from chroming process, passivating process and aluminum anodizing process. Generally, the content of the Cr^{6+} is about 20~150 mg/L in the washing water with chromate, while the content of

the Cr^{6+} is about 200~300 mg/L in the passivating process. In addition, the waste water consists the Cr^{3+}, Fe^{3+}, Ni^{2+}, Zn^{2+} and sulfuric acid, nitric acid, et al.

The treatment methods include chemical method, ion exchange, electrolysis, activated carbon adsorption, evaporation, et al.

The chemical reduction is used widely (Figure 4-50).

The main mechanism is to reduce the Cr^{6+} to Cr^{3+} by ferrous sulfate, sulfite and sulfur dioxide. And then the Cr^{3+} reacted with the OH^- to form the precipitate. The equation is as followed.

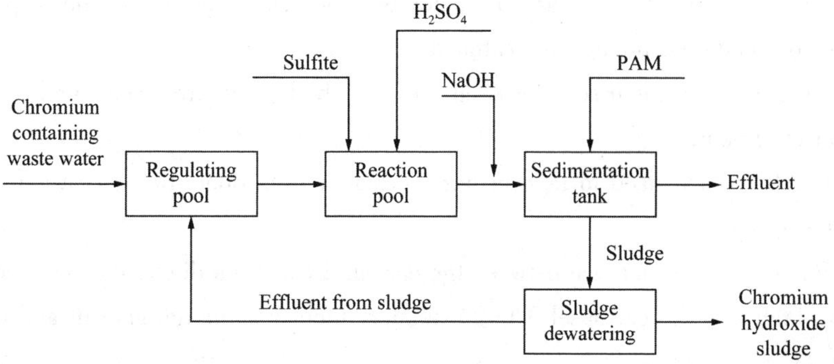

Figure 4-50 The chemical reduction process

$$2H_2Cr_2O_4+3Na_2SO_3+3H_2SO_4 \longrightarrow Cr_2(SO_4)_3+3Na_2SO_4+5H_2O$$

$$Cr_2(SO_4)_3+6NaOH \longrightarrow 2Cr(OH)_3\downarrow+3Na_3SO_4$$

The retention time in the reaction tank is about 10~30 min, the real reductant is 1.3~1.5 times of the theory. pH is about 2~3, and pH is 7~9 after the regulation.

In practice, the investment and the running cost of the technology is low.

In addition, ferriter technique is also used widely (Figure 4-51).

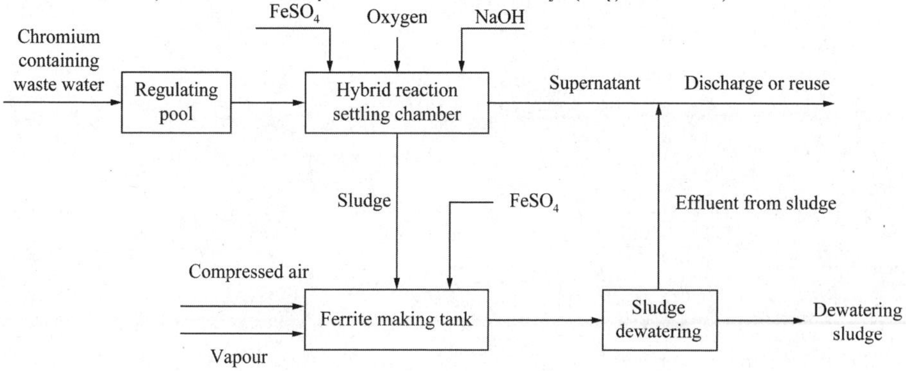

Figure 4-51 The ferriter technique process

The main mechanism is to reduce the Cr^{6+} to Cr^{3+} by ferrous sulfate. And then the Cr^{3+} and the other heavy metals (M^{n+}) reacted with the OH^- to form the precipitate. It is described as $M \cdot M(OH)_n \cdot Fe(OH)_3$.

The process is as followed (Figure 4-51).

Questions

1. What is the difference between MBR and MBBR process?

2. What is the difference between AO and AAO process?

3. Please draw the diagram of anaerobic biological treatment and explain the principles of anaerobic biological treatment.

4. Making a comparison between aerobic biological treatment and anaerobic biological treatment.

5. Is AB method used to recover organic matter from domestic waste water? Please clarify the reason.

6. What are the difference between the carrousel oxidation ditch, the carrousel 2,000 oxidation ditch and the carrousel 3,000 oxidation ditch. For the designment and running, is there any distinction?

7. SBR is used widely in the treatment on waste water, please clarify the modifications and development on SBR.

8. For the oil waste water, how many are there types of oil? How to remove different type of the oils.

9. There are many types of anaerobic biological process. Please give the list and describe the characteristics, designment and running for one type of anaerobic biological process.

Chapter 5 Typical technology on rural waste water

Emphasis: Characteristics of rural waste water. Collection of rural waste water. Selection and designment and running of typical technologies on rural waste water.

Nut: The collection of rural waste water. The Selection on typical technologies. The running and management of treatment units.

5.1 Production and collection of rural domestic waste water

5.1.1 Production of rural domestic waste water

Rural domestic waste water includes toilet feces, kitchen waste water, washing waste water and bathing waste water. The waste water mainly includes domestic waste and human excrement. China has approximately 600,000 administrative villages and 2.5 million natural villages, producing more than 9 billion tons of domestic waste water each year. Thus, the effective treatment of domestic waste water has become a critical component of rural environmental improvement. Accordingly, our country has actively promoted rural waste water treatment in recent years. As of 2015, the central government had allocated a total of 31.5 billion yuan of special funds for rural environmental protection (rural energy conservation and emission reduction funds), supported 78,000 organic villages across the country to carry out comprehensive environmental remediation, and built 248,000 sets of domestic waste water treatment facilities. By 2020, it is expected that an additional 130,000 organic villages would have completed comprehensive environmental improvement, thereby raising the accumulated number to more than one-third of the total number of organic villages in the country. Moreover, according to the statistics of the Ministry of Ecology and Environment, during the 13th Five-Year Plan period, 150,000 administrative villages have completed the comprehensive improvement of the rural environment. In 2019, the number of waste water treatment plants in our country's constructed towns and townships was 11,486 and 1,830 respectively; the daily waste water treatment capacity of our country's constructed towns and townships reached 24.77 million cubic meters per day and 1.09 million cubic meters per day, respectively, as shown in Figure 5-1.

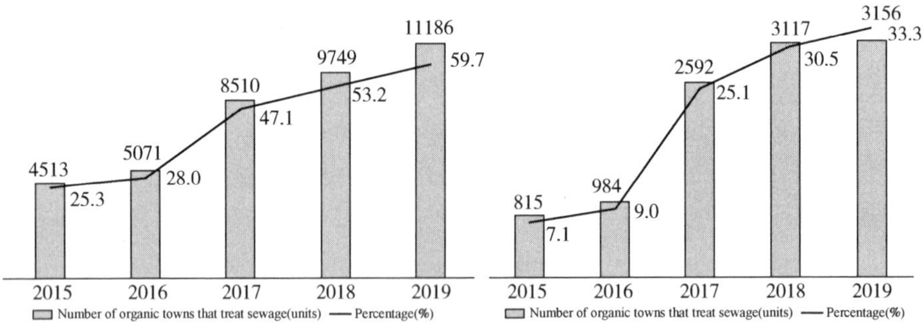

Figure 5-1 2015~2020 China's construction of towns and township waste water treatment plants and daily treatment capacity (unit: unit, ten thousand cubic meters/day).

Note: 2015~2019 data comes from the Ministry of Housing and Urban-Rural Development, no relevant official data for 2020 are forward-looking estimates.

5.1.2 Collection of rural domestic waste water

In comparison with urban domestic waste water, rural domestic waste water discharge is scattered and has myriad destinations, including various methods such as courtyard sprinkling, septic tank effluent return to the field, takeover discharge, decentralized facility treatment and other methods, which make it difficult to build a collection pipe network. Therefore, for the collection of rural domestic waste water, it is not appropriate nor sufficient to adopt the traditional urban waste water collection and treatment mode. Various collection and processing modes must be adopted based on the characteristics of the countryside, as well as topography and landforms, in accordance with the specific conditions of the particular locality. At present, the rural domestic waste water treatment in our country is primarily comprised of three waste water treatment modes, namely household waste water treatment, centralized waste water treatment in villages, and centralized treatment in the urban waste water pipeline network.

5.1.2.1 Household waste water treatment

For single-family and multi-family villagers whose households are scattered over long distances. Set up single-family and multi-family decentralized waste water treatment facilities nearby. The suitable waste water volume of decentralized waste water treatment facilities is generally not more than 2 m^3/d, serving a population of about 6~20 people, which corresponds to 1~5 households. On the basis of single-family collection system, waste water pipes or ditches of 1~5 households have been introduced into small waste water treatment facilities. This treatment mode has the characteristics of flexible layout, cost-saving pipe network laying, and simple construction. It is suitable for villages where farmers are scattered, with complex terrain conditions, unplanned construction layouts, and collective collection of the waste water is nearly impossible. It is usually used in the central and western regions of our country where the villages are scattered.

5.1.2.2 Village centralized waste water treatment

This treatment mode is the main mode of rural waste water treatment. By laying a pipe network throughout the village, the waste water produced by all farmers is collected through the pipe network and connected to the nearby waste water treatment facility. Unified treatment is accomplished through the construction of unified waste water treatment facilities or waste water treatment stations, which typically employ a combination of traditional biological treatment and ecological treatment. This treatment mode has the advantages of stable operation, high treatment efficiency, and small footprint. It is suitable for areas far from cities and towns that have densely distributed villages, a high population density, a high waste water discharge, and better economic conditions. It is typically appropriate for rural areas in eastern and northern China with densely distributed villages and better economic foundations.

5.1.2.3 Incorporated into the urban waste water pipeline network for centralized treatment

For villages close to the urban municipal waste water pipe network, only the waste water collection pipe network system should be constructed, and their domestic waste water can be directly connected to the existing urban municipal waste water pipe network for unified treatment. That is, after the waste water in the village is collected through the waste water pipes, it is connected to the nearby municipal waste water pipe network, and the urban waste water treatment plant is used for unified treatment. This model has the characteristics of low investment, short construction period, quick results, and convenient unified management. It is suitable for villages with better economic conditions in the suburbs of cities and towns, or villages close to waste water treatment plants and municipal pipe networks. It not only saves investment in waste water treatment facilities in rural areas, but also treats waste water uniformly by urban waste water treatment plants, resulting in good waste water treatment effects and operational and management guarantees.

5.1.3 Water quality characteristics of rural domestic waste water

Due to the particularity of its geographical location, population size, economic development, and residents' production and lifestyles, rural waste water discharges are quite different from those of urban areas. Rural domestic waste water is mainly waste water from toilet flushing and waste water from washing, rice, vegetable and bathing. Waste water mainly contains domestic waste and human excrement. Generally, it does not contain toxic substances, but often contains nutrients such as nitrogen and phosphorus, as well as a large number of bacteria, viruses and parasite eggs. Due to different living habits, lifestyles, economic levels, etc., the water quality and quantity of domestic waste water in rural areas vary greatly. The waste water has the following characteristics and problems.

① The distribution of waste water is relatively scattered, involving a wide range and a high degree of randomness, making prevention and control very difficult. The collection system is imperfect, coupled with extensive discharge, and basically no waste water treatment facilities.

② The concentration of rural domestic waste water is low but the changes are significant.

③ The nature of most rural domestic waste water is the same, the water quality fluctuates greatly, and the biochemical property is strong.

④ Albeit the waste water discharged from the toilet is of poor quality, it can be used as fertilizer in the septic tank.

Nowadays, our country's rural economy is developing rapidly, the living standards of rural residents have been significantly improved, and rural domestic waste water has taken on new characteristics. Rural domestic waste water can be divided into black water and gray water. Gray water refers to the waste water produced in the kitchen and bathroom, and black water mainly refers to the waste water produced by feces and toilet flushing. Rural domestic waste water is seriously polluted, and its composition is as shown in Table 5-1.

Table 5-1 Composition characteristics of rural domestic waste water

Component	Annual load/ (kg/person)	Gray water/ %	Urine/ %	Stool/ %
Nitrogen	4~5	About 3	About 87	About 10
Phosphorus	About 0.75	About 10	About 50	About 40
Potassium	About 1.8	About 34	About 54	About 12
COD	About 30	About 41	About 12	About 47

5.2 Typical technologies

5.2.1 Oxidation lagoon

5.2.1.1 Overview

The oxidation lagoon is called stabilization lagoon. The definition is the suspended growth biological treatment system without sedimentation tank and backflow of sludge. The system stabilizes and degrades the organic matter depending on natural organism. The name of the oxidation lagoon is coming from the construction. Generally, the oxidation lagoon is a waste water pond which there is the causeway and impervious barrier. The waste water comes into the pond from one end and comes out from the other end. The speed of the waste water is very slow. Under the sunshine, the organic matter is degraded by the metabolism of the microorganisms and the clarification of the water plants. In general, it is not necessary to keep the biomass in the oxidation lagoon. The retention time of the waste water is similar as the retention time of sludge, so the retention time of the sludge is very long, up to several days.

The oxidation lagoon is divided into four kinds, which are showed in the Table 5-2.

Table 5-2 The kinds of the oxidation lagoon and their parameters

Item	Aerobic lagoon	Facultative lagoon	Anaerobic lagoon	Aerated lagoon
Water depth/m	0.4~1	1~2.5	>3	3~5
HRT/d	3~20	5~20	1~5	1~3
BOD load/[g/(m^2·d)]	1.5~3.0	5~10	30~40	20~40
BOD removal/%	80~95	60~80	30~70	80~90
BOD degradation	Aerobic	Aerobic, Anaerobic	Anaerobic	Aerobic
Sludgedecomposition	No	Anaerobic	Anaerobic	Aerobic or anaerobic
Photosynthetic reaction	Yes	Yes	—	—
Algae/(mg/L)	>100	10~50	0	0

(1) Aerobic lagoon The water depth is small, and the sunshine goes to the bottom of the lagoon. The oxygen comes mainly from the algae. The waste water in the lagoon is aerobic. The organic matter is degraded by the aerobic microorganisms.

(2) Facultative lagoon The water depth is big, and the sunshine does not go to the bottom of the lagoon. In general, the sunshine can go to the depth of 0.5 m. At the

depth, the Photosynthetic reaction of algae is vigorous. Therefore, the dissolved oxygen is plenty and the waste water is under the aerobic conditions. At the bottom, there is the settling sediments with anaerobic fermentation under the anaerobic conditions. The area between the aerobic area and the anaerobic area is facultative. The clarification of the organic matter depends on the aerobic and anaerobic bacteria.

(3) Anaerobic lagoon The water depth is big, and the BOD load is high. The waste water is under the anaerobic conditions. The anaerobic reaction includes hydrolyzation, acetogenesis, methanogenesis.

(4) Aerated lagoon The oxygen comes from the surface aerator. The waste water in the lagoon is aerobic. The HRT is short. The growth and the photosynthetic reaction of algae is inhibited due to the mixing.

In addition, there are advanced lagoon and ecological lagoon according to the function. The advanced lagoon is only to treat the secondary effluent to meet the demand of the receiving water body or the reclaimed water. The lagoon is to further remove the residual organic matter, bacteria, N and P. As for the ecological lagoon, it is to clarify the waste water by the water plants or the aquatic animals, such as reed, duckweed, fish, et al. The most important is the bacteria and algae in the lagoon. A long food chain is built by the different consumers.

5.2.1.2 The oxidation lagoon systems

The component of the ecological system in the oxidation lagoon is biological parts and abiological parts. The biological parts include bacteria, algae, protozoa, metazoan, hydrophyte, and aquatic animals. The abiological parts include the sunshine, the wind, the temperature, the BOD loading, pH, DO, CO_2, N and P. Facultative lagoon is the typical lagoon (Figure 5-2).

The aerobic, anaerobic and facultative heterotrophic bacteria are the main consumer of the organic matter in oxidation lagoon. These bacteria use the organic matter as the carbon source, and uptake the energy produced during the degradation. The aerobic bacteria live in the aerobic lagoon and the aerobic area in the facultative lagoon. On the contrary, the anaerobic bacteria live in the anaerobic lagoon and the anaerobic area in the facultative lagoon. For the aerobic lagoon, the oxygen comes from the atmospheric reoxygenation and the photosynthetic reaction of the algae. It also comes from the surface aerator. When the ecological system is in balance in the oxidation lagoon, the biomass of the bacteria is balanced. When the multi-stage oxidation lagoon is used, the biomass of the bacteria decreases with the increase of the stage.

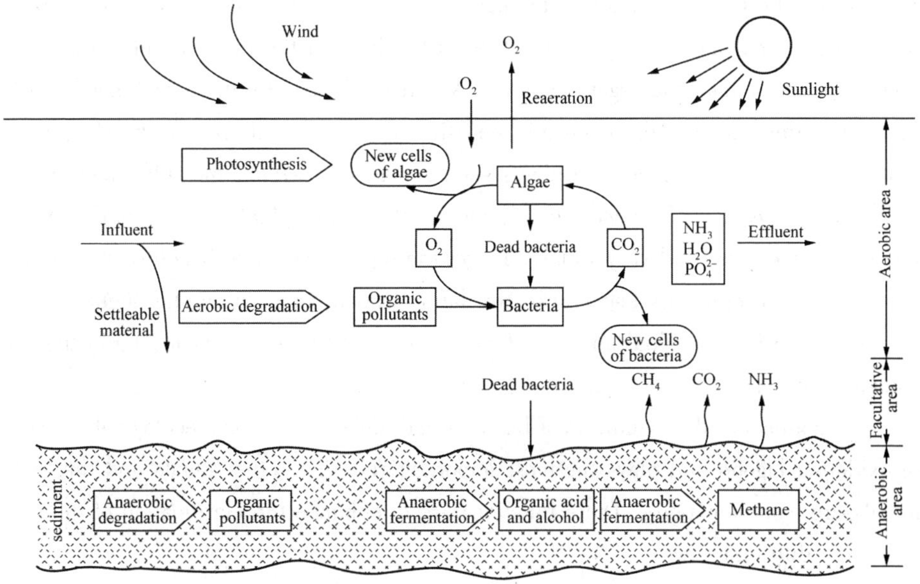

Figure 5-2 The purification principle of the oxidation lagoon

Algae plays the important role in the oxidation lagoon. It can form the algae and bacteria symbiosis. Algae is autotrophic bacteria and can produce the oxygen by the photosynthetic reaction. At the same time, the algae build the new cell by the inorganic carbon, N and P for self-breeding. In the oxidation lagoon, the algae are main *green algae*, *green-blue algae*, *Euglena*, *and Chlamydomonas.* The heterotrophic bacteria use the dissolved oxygen in the water to degrade the organic matter. The products are CO_2, NH_4^+, NO_3^-, and H_2O. The products are the food of the algae. During the biological reaction, the bacteria and the algae improve each other and form the symbiosis. The result is that the organic matter decreases, and the algae and inert matter increase. Those matters are discharged with the effluent.

The degradation on the organic matter by the bacteria is as followed.

$$C_6H_{12}O_6+6O_2 \longrightarrow 6CO_2+6H_2O+\text{energy}$$

The photosynthetic reaction is showed as the following equation.

$$NH_4^++5CO_2+2.5H_2O \longrightarrow C_5H_9O_{2.5}N+5O_2$$

There are zooplankton which eat the bacteria and algae in the oxidation lagoon, such as cladoceran, metazoan with shell. These zooplankton can swell the bacteria and algae and suspended organic matter, at the same time, they produce the emplastic to promote the flocculation of the small suspended solids and to clarify the water. The plankton are the final consumer of the algae and bacteria in the oxidation lagoon, but in the aquatic animal lagoon they are the food of the fish. In the plant lagoon, the aquatic vascular plants uptake the organic matter, N and P, especially the heavy metals by the roots. After the harvest, the economic benefits have been achieved. Generally, in the plant lagoon, the floating aquatic plants such as *water hyacinth*, the submerged aquatic plants such as *potamogeton malaianus*, emergent aquatic plant such as *reeds*, *Scirpus tabernaemontani* are used very widely. To get the bigger economic benefits, it is to keep gooses, ducks or omnivory fishes in the oxidation lagoon. In addition, the higher animals can uptake the food residual and the plankton, control the algae bloom, build the good ecological systems.

It is noticeably the symbiosis of bacteria and algae is the basic ecological system in the oxidation lagoon and plays the most important role. The other plants and the higher animals are complementary. There functions are to enhance the clarification of the waste water from different ways.

5.2.1.3 The principles

The clarification mechanism of the oxidation lagoon is similar as the natural water bodies. The mechanisms are as followed.

① The dilution. The waste water is diluted in the oxidation lagoon under the wind and flowing conditions.

② The sedimentation. The suspended solids are flocculated by the secreta of the bacteria and settled under the gravity.

③ The metabolism of the microorganisms. The aerobic and facultative heterotrophic bacteria are the main consumer of the organic matter in oxidation and facultative lagoon. At the bottom of the facultative lagoon and the anaerobic lagoon, the anaerobic bacteria degrade the organic matter. The end products are CH_4, CO_2 and sulfur alcohol.

④ The plankton. The plankton can swell the free bacteria, the algae, the colloidal organic matter and fine particles. In addition, the plankton can produce the secreta with the flocculation.

⑤ The aquatic vascular plants. The plants can uptake N and P nutrients. Their roots can adsorb the heavy metals. In addition, the roots and the stems can produce the media for the bacteria and microorganisms.

The clarification is different according to different types of the lagoon.

① The aerobic lagoon. The function is similar as the aerobic area in the Figure 5-2.

For the aerobic lagoon, the BOD loading is small and the DO is more than 1 mg/L. In addition, the water depth of the lagoon is less than 0.5 m, so the sunshine can go to the bottom. The main clarification on the waste water depends on the algae and bacteria symbiosis. The DO comes from the photosynthetic reaction of the algae and the atmospheric reoxygenation due to the wind, resulting in the water in aerobic state. The aerobic heterotrophic bacteria degrade the organic matter by the metabolism. The end products, CO_2, are the carbon source of the algae. In the aerobic lagoon, the microorganisms are very rich and the biomass are very big. The BOD loading is low in the aerobic lagoon. It is suitable for the municipal waste water treatment. It can be as the advanced treatment for the secondary effluent. The inflowing water quality is as followed. The content of BOD_5 is less than 30 mg/L, and the COD is less than 120 mg/L and the SS is at the range of 30~60 mg/L. Therefore, the aerobic lagoon is suitable for the rural waste water treatment.

② The facultative lagoon. The function is showed in the Figure 5-2.

There is aerobic process and anaerobic process in the facultative lagoon. It is called facultative lagoon. Therefore, there is aerobic layer and anaerobic layer in the facultative lagoon. The metabolism and the microorganisms in aerobic layer are similar as the aerobic lagoon, but the content of the algae is less than the aerobic lagoon. The function of the anaerobic layer is same as the anaerobic lagoon. The liquid end products are H_2O, amino acid and organic acid. The gaseous end products are CO_2, CH_4, which get away from the water or are utilized by the bacteria when passing the aerobic layer or are used by the algae. For the facultative layer, the DO is very low. Generally, it is aerobic in the day and it is anaerobic at night. There are facultative microorganisms in the layer, which can use the free dissolved oxygen or can use the chemical oxygen in NO_3^- or CO_3^{2-}. In the day, the function of the facultative layer is similar as the aerobic layer, while at night, the function is similar as the anaerobic layer. The facultative lagoon is not only suitable for the domestic waste water and municipal waste water, but also the waste water coming from the paper making, the coal chemical industry, the petrochemical industry. As for the organic industrial waste water with high content, it is used after the anaerobic lagoon as the secondary stage lagoon.

③ The anaerobic lagoon. The function is showed in the Figure 5-3.

The degradation of the organic matter depends on the anaerobic microorganisms. Therefore, the function is limited by the anaerobic fermentation. The structure of the anaerobic lagoon lies on the anaerobic reaction. In the anaerobic lagoon, there are hydrolyzation,

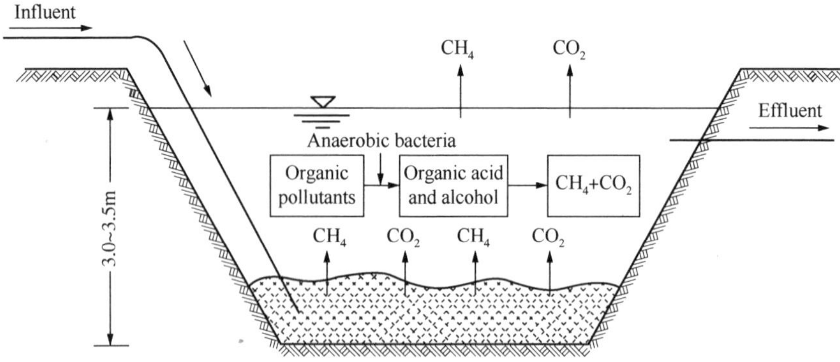

Figure 5-3 The purification of the anaerobic lagoon

acetogenesis, methanogenesis microorganisms. Generally, the anaerobic lagoon is suitable for the organic industrial waste water with high content, such as food industry, meat processing industry, stockyards. However, the content of the organic matter in the effluent of the anaerobic lagoon is higher, so it is necessary to further treat it by the facultative lagoon and aerobic lagoon.

5.2.1.4 The aerated lagoon

The aerated lagoon is the artificially enhanced oxidation lagoon with the mechanical aerator for the aeration and mixing of the waste water in the lagoon. The aerated lagoon is suitable for those area, which the occupied area is limited, is not enough to build the lagoon with the natural clarification. Another explanation is that the facultative lagoon over the BOD loading is rebuilt to the aerated lagoon. The designment aim is to meet the standard of the secondary treatment. The growth of the algae stops due to the mixing of the waste water.

The aerated lagoon is divided into aerobic aeration lagoon and facultative aeration lagoon, depending on the numbers, the density of the aerators and the intensity of the aeration. The surface mechanical aerator is used widely. Sometimes the blast aeration is used. When the power of the aerator is enough big to promote the whole sediments suspension and provide the enough oxygen, the lagoon is the aerobic lagoon (Figure 5-4). When the power of the aerator is to promote the part of the sediment suspension and the other parts of the sediments still at the bottom and provide the oxygen that is not enough for the demand, it is facultative aeration (Figure 5-5).

The aerated lagoon is attributed to the oxidation lagoon, but it is different from the other lagoons with the natural clarification. The aerated lagoon is a process between

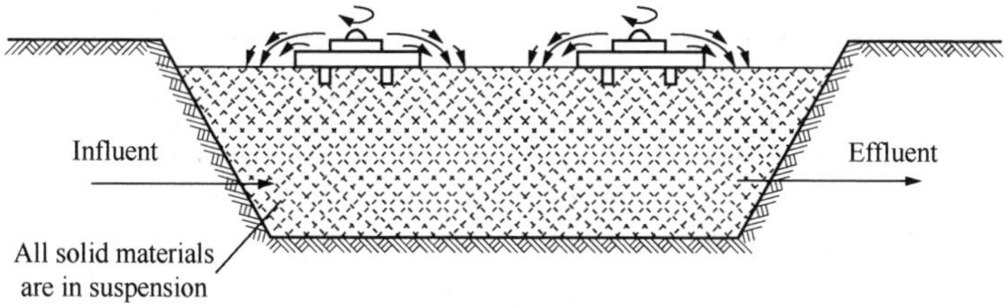

Figure 5-4 The aerobic aeration lagoon

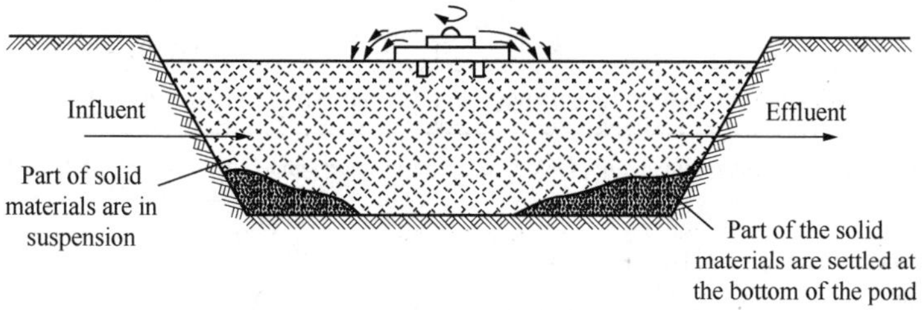

Figure 5-5 The facultative aeration lagoon

extended aeration activated sludge and oxidation lagoon. It is similar as the activated sludge without the backflow of sludge. The clarification efficiency of the aerated lagoon is higher than the other oxidation lagoons due to the artificial enhancement. The retention of waste water is short in the aerated lagoon and the effective volume and the occupied area is small, but the problems are the consumption on energy is high and the running cost is high. The cost is lower than that of the activated sludge.

5.2.1.5 The designment of oxidation lagoon

The biological reaction in the oxidation lagoon is very complex, such as the metabolism of the microorganisms, the growth of the algae and the adsorption and usage of the aquatic plants and aquatic animals. In addition, the reaction is related with the local climate, so it is difficult to build a model for the calculation. Nowadays, the calculation depends on the experience.

The designing parameters of the oxidation lagoon are showed in the Table 5-3.

Table 5-3 The designing parameters of the oxidation lagoon

Parameters	Types of the lagoon			
	Aerobic	Facultative	Anaerobic	Aerated
Water depth/m	0.4~1	1~2.5	3~5	3~5

续表

Parameters		Types of the lagoon			
		Aerobic	Facultative	Anaerobic	Aerated
HRT/d	Ⅰ	20~30	20~30	3~7	1~3
	Ⅱ	10~20	15~20	2~5	—
	Ⅲ	3~10	5~15	1~3	—
BOD_5 loading/ [g/(m^2 · d)]	Ⅰ	1~2	3~5	20(28~66①)	10~20
	Ⅱ	1.5~2.5	5~7	30(40~100①)	20~30
	Ⅲ	2~3	7~10	40(66~200①)	20~40
BOD_5 removal/%		80~95	60~80	30~70	80~90

Annotation: Ⅰ, The average temperature<8℃; Ⅱ, The average temperature with 8~16℃; Ⅲ, The average temperature>16℃.

① BOD_5 volume loading, gBOD_5/(m^3 · d).

① Aerobic lagoon and facultative lagoon. For the aerobic lagoon and the facultative lagoon, the efficiency depends on the area of the lagoon, especially the aerobic lagoon. The area of the lagoon is gotten by the following equation.

$$F=\frac{S_0Q}{q}$$

In the equation, F—The surface area, m^2;

Q—The average amount of waste water, m^3/d;

S_0—The content of BOD_5 in the influent, kg/m^3;

q—BOD_5 volume loading, kg/(m^2 · d).

② Anaerobic lagoon. For the anaerobic lagoon, the effective volume of the anaerobic lagoon is gotten by the following equation.

$$V=\frac{S_0Q}{1000\,q_v}$$

In the equation, V—The effective volume, m^3;

Q—The average amount of waste water, m^3/d;

S_0—The content of BOD_5 in the influent, kg/m^3;

q_v—BOD_5 volume loading, kg/(m^2 · d).

③ Aerated lagoon. The main parameters of the aerated lagoon are BOD_5 volume loading, HRT, specific power of aeration. When the HRT is less than 24 h, it is controlled by the aeration. When the HRT is more than 24 h, it is controlled by the mixing. In practice, the HRT and the specific power of aeration is limited. On the contrary, it is over the limit, the consumption on energy is higher than that of the secondary treatment in waste water treatment plant, resulting in the loss of the competition. The specific power of aeration is 5~6 W/m^3 in the aerobic aerated lagoon, while the value is 1~2 W/m^3 in the facultative aerated lagoon. In practice, the removal of the pollutants is followed. BOD_5 is 60%~90% (BOD_5 in the influent is 200~500 mg/L). COD is 70%~90% and TSS is 70%~90% (TSS in the influent is 200~500 mg/L).

The comparison of the design parameters between aerated lagoon and aerobic lagoon is showed in the Table 5-4.

Table 5-4 The comparison of the design parameters

Parameters	High speed aerobic lagoon	Aerobic lagoon	Advanced lagoon	Aerated lagoon
Flowing	Moderate mixing	Moderate mixing	Moderate mixing	Complete mixing
Area/×$10^4 m^2$	0.25~1.0	<4	1~4	1~4
Running	Series	Series or parallel	Series or parallel	Series or parallel
HRT/d	4~6	10~40	5~20	3~10
Water depth/m	0.3~0.45	1.0~1.5	1.0~1.5	2~6
pH	6.5~10.5	6.5~10.5	6.5~8.0	6.5~8.0
Temperature range/℃	5~30	0~30	0~30	0~30
Optimal temperature/℃	20	20	20	20
BOD_5 loading/[g/(m^2·d)]	1.0~30.0	4.0~12.0	<1.5	—
BOD_5 removal/%	80~95	80~95	60~80	80~95
End products	Algae, CO_2, cell	Algae, CO_2, cell	Algae, CO_2, CH_4, cell	CO_2, cell
Algae/(mg/L)	100~260	40~100	5~10	—
Effluent TSS/(mg/L)	150~300	80~140	10~30	80~250

In addition, the design parameters of the aerated lagoon and the anaerobic lagoon and the facultative lagoon are showed in the Table 5-5.

Table 5-5 The design parameters of the different lagoon

Parameters	Anaerobic lagoon	Facultative lagoon	Aerated lagoon
Flowing	—	—	Complete mixing
Area/×$10^4 m^2$	0.2~1	1~4	1~4

续表

Parameters	Anaerobic lagoon	Facultative lagoon	Aerated lagoon
Running	Series	Series or parallel	Series or parallel
HRT/d	20~50	7~30	3~10
Water depth/m	2.5~5	1~1	2~6
pH	6.8~7.2	6.5~9.0	6.5~8.0
Temperature range/℃	6~50	0~50	0~30
Optimal temperature/℃	30	20	20
BOD_5 loading/[g/(m^2 · d)]	20.0~50.0	1.5~8.0	—
BOD_5 removal/%	50~85	80~95	80~95
End products	CO_2, CH_4, cell	Algae, CO_2, CH_4, cell	CO_2, cell
Algae/(mg/L)	0~5	20~80	—
Effluent TSS/(mg/L)	80~160	40~100	80~250

5.2.1.6 The characteristics of oxidation lagoon

Nowadays, there are 10,000 oxidation lagoons in over 40 countries of the whole world.

(1) The advantages:

① The construction investment is small and it is only 1/3~1/2 of the traditional waste water treatment plant.

② The easy running and management. The people for management and maintenance are few. The wind power is important for the atmospheric reoxygenation. The reasonable designment is killing of the consumption on power. The running cost is only 1/5~1/3 of the traditional waste water treatment plant.

③ The realization of the waste water recycling. After the treatment of the oxidation lagoon, the effluent quality meets the standard of the farm irrigation. The sediments can be as the organic fertilizers for the farm.

④ Landscaping.

⑤ The residual sludge is very small, and it is only 1/10 of the activated sludge.

⑥ The strong resistance on the hydraulic shocking.

(2) The disadvantages:

① The occupied area is huge.

② The efficiency is not stable due to the season, temperature and sunshine.

③ The ground water is easily polluted, once the impermeability is not good.

④ It is odor and lots of flies and mosquitos.

The comparison is showed in the Table 5 - 6 between the oxidation lagoon and traditional activated sludge.

Table 5-6 The comparison between the oxidation lagoon and traditional activated sludge

A/A/O	Oxic ditch	Oxidation lagoon
Complex process Lots of structures Troublesome running	Simple process Few structures Simple running	Simple process. Few structures. Running stable. Simple management. No backflow of sludge. No discharge of residual sludge for many years.
High construction investment	Construction investment is saving 15%~20%	Low construction investment. Occupied area is huge.
High running cost	High running cost	Low running cost. The effluent quality meets the standard of the farm irrigation.
It is difficult to remove N and P simultaneously. It is not suitable for the waste water with low BOD content.	N and P removal simultaneously. It is suitable for the waste water with low BOD content.	N and P removal simultaneously. The strong resistance on the hydraulic shocking. The realization of the waste water recycling.

5.2.2 The wetlands

The waste water is sprayed on the marsh, in which the soils are saturated with water and there are many reeds, cattails, et al. The waste water is flowing along a certain direction. Under the systematic function of the plants and the soil, the water is clarified.

Natural wetland is using the depression, reed ponds with the artificial construction. There is guiding dam making the waste water flowing along a certain direction. The water depth is generally 0.3~0.8 m and it is suitable for the growth of the emergent aquatic plant. The Figure 5-6 is the natural wetlands treatment systems.

Constructed wetland is a waste water treatment process to simulate the natural wetland. The clarification of the waste water is finished under the systematic function of the chemical, physical and biological action by the filtration, adsorption, sedimentation, ion exchange and plant adsorbing and metabolism of the microorganisms. The control and the treatment ability of the constructed wetland is higher than that of the natural wetland.

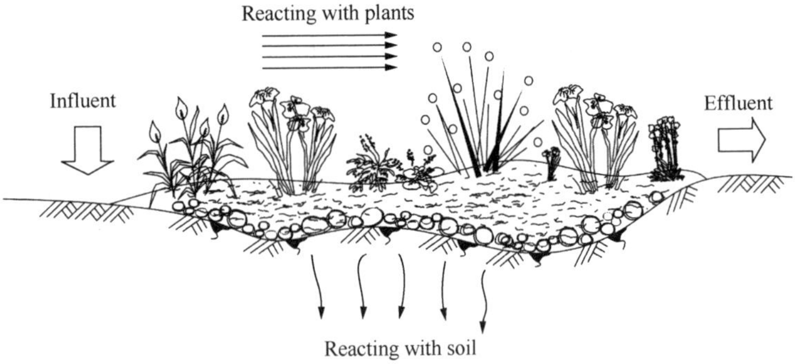

Figure 5-6 The natural wetlands treatment systems

5.2.2.1 The types of the constructed wetland

The constructed wetland is divided into water surface wetland and infiltration wetland according to the flowing. The infiltration wetland is divided into horizontal subsurface flow constructed wetland and vertical subsurface flow constructed wetland.

The structure of the water surface wetland is showed in the Figure 5-7. The tank or the ditch is built and there is impervious barrier at the bottom of the wetland. The soil is filled into the tank or the ditch. The vascular plant species are planted in the soil. The waste water comes into the wetland by the water distributor at one end. The water flows to the other end with the shallow water layer and then the treated water goes to the collection ditch at the other end. There is the free water surface during the flowing. In the wetland, the water depth is very small, in general, 0.1 ~ 0.6 m. The depth of the water is close to the natural wetland. The organic matter is adsorbed by the biofilm on the stem of the vascular plant species under the water table. The filters and the roots are not used sufficiently in the water surface wetland. The use on the project of the water surface wetland is little due to the icy and the low efficiency in winter, although the investment is low.

The infiltration wetland is showed in the Figure 5-8.

The bed is constructed and is filled with the media to provide the good conditions for the reeds. There is impervious barrier at the bottom of the wetland with a slope. The waste water flows in the bed. The clarification of the wetland is two parts. One part is the interception of the biofilm on the stem of the plants, the rich roots and the filters to increase the efficiency. The other part is the waste water flowing under the ground. Therefore, the heat preservation is good and the effect of the climate is little. It is suitable for the high content of organic matter. The construction investment of the infiltration wetland is higher than that of the water surface wetland. In general, the

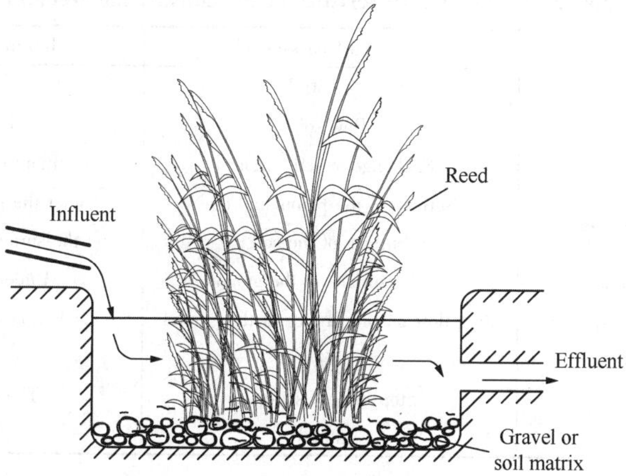

Figure 5-7 The water surface wetland

investment is 4 ~ 8 times of the water surface wetland. The main cost is about the gravels. The waste water flows horizontally on the media with the roots of the advanced plants, and this is called horizontal subsurface flow constructed wetland. The waste water flows vertically passing the media with the roots of the advanced plants.

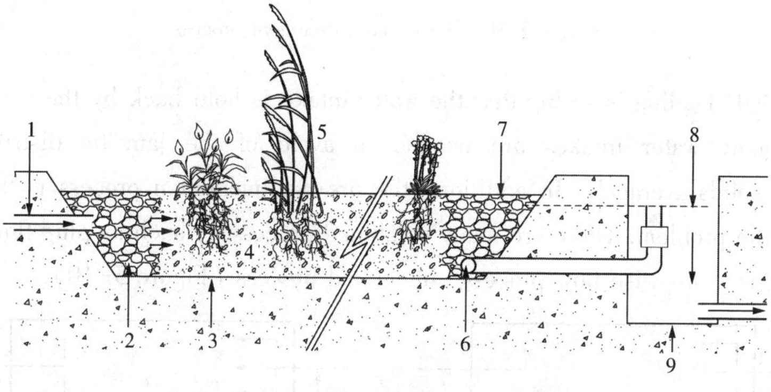

Figure 5-8 The infiltration wetland

1—Influent after mechanical pretreatment;2—Water distribution area filled with rock blocks;
3—Impermeable layer;4—Bed materials(coarse sand, gravels, chippings) ;5—Plants;6—Outlet catch pipe;
7—Water catchment filled with rock blocks;8—Using overflow pipe to maintain a
constant water level in the reed bed;9—Outlet channel

The characteristics of the water surface wetland and the infiltration wetland are showed in the Table 5-7.

Table 5-7 The characteristics of the constructed wetland

Items	Water surface wetland	Infiltration wetland
Water depth/m	2~30	0
Media	Raw soil	Artificial media
Oxygen	Exchange of soil surface	Plant transportation
Effective function area	Surface of the plants and the Surface of the soil	Under the ground, roots and the surface of the media
Water distribution	On the surface of the ground	Under the ground
Water flowing way	Plug flow on the surface of the ground	Underflow under the ground
Climate effect	Big	Small
Running	Stop in cold seasons	The whole year
Water-resisting layer	Natural	Artificial

5.2.2.2 The combination of the constructed wetland

The pretreatment and the aftertreatment processes are needed to keep the stability and increase the lifetime and the treatment ability of the wetlands. The processes are showed in the Figure 5-9.

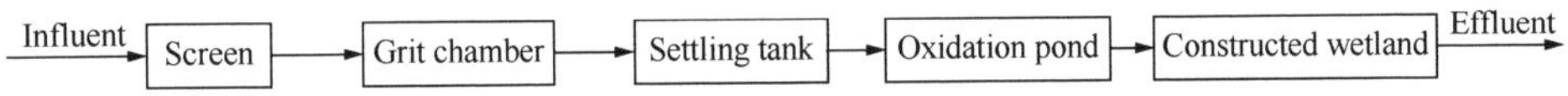

Figure 5-9 The wetland treatment process

The BOD loading is so big that the water intake is hold back by the pollutants. In general, many water intakes are needed to avoid of the jam by distributing the suspended solids averagely. In addition, the pre-sedimentation process is favor of the solving of the problem. There are many combined processes, such as plug flow process, reflow process, stepwise flow process, integrated process (Figure 5-10).

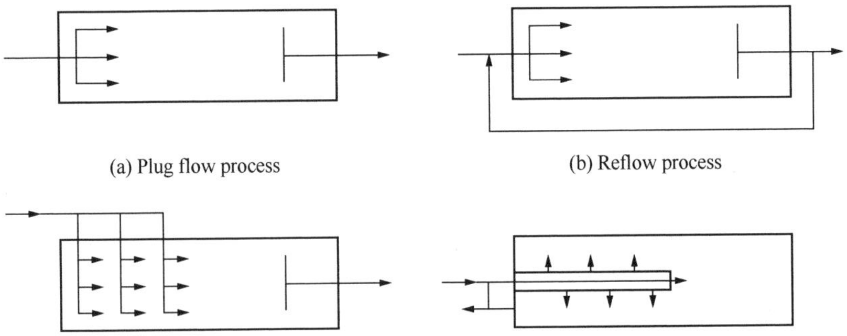

(a) Plug flow process　(b) Reflow process

(c) Stepwise flow process　(d) Integrated process

Figure 5-10 The combined processes of the wetlands

The reflow process is to dilute the organic matter and suspended solids content in the influent and to increase the oxygen content and to reduce the odor in the effluent. The backflow of the effluent is favor of the nitrification and denitrification. The ladder process is to avoid of the jam of the ahead of the bed to make the average growth of the plants. This is favor of the denitrification in the end of the bed. The comprehensive process is to reduce the BOD loading by the backflow of effluent and the distribution into the middle of the bed.

5.2.2.3 The mechanisms

The constructed wetland is a combined system with the soil, the plants and the microorganisms. The microorganisms play the most important role in the degradation of the organic matter. The roots transport the oxygen to the soil around the roots, while it is anaerobic condition far away the roots. The change of the micro-environment enhances the removal of the complex organic matter and non-biodegradable organic matter. The most part of organic matter depends on the microorganisms, but the pollutants, such as heavy metals, sulfur, P, are removed by the adsorption of the soils and the plants.

① SS. The settled pollutants can be quickly removed by the filtration and sedimentation and the suspended solids can be removed by the adsorption of the growth of the microorganisms, the surface of the media in the wetland. The removal efficiency of the SS is excellent and the content of SS in the effluent is less than 10 mg/L. Generally, the pretreatment on the total SS is necessary to avoid of the jam at the water intake, such as sedimentation tank.

② Organic matter. The organic matter includes the particulate organic matter and the dissolved organic matter. For the former, it is quickly removed by the sedimentation and filtration, while for the latter, it is biodegraded by the microorganisms. The aerobic respiration, anaerobic reduction and sulfate reduction contribute 'the removal of BOD. About 50% of BOD in the influent is removed at the ahead of the bed. Generally, BOD is about 10 mg/L in the effluent.

③ Denitrification. The mechanism of removal N is evaporation, ammoniation, nitrification, denitrification and utilization by plants and adsorption by media. The main mechanism is nitrification, denitrification.

④ P removal. The P removal is related with the types of the soils. The dissolved inorganic P is adsorbed by the Al, Fe, Ca salts and settled. Therefore, the removal of P depends on the content of the Al salts in the soils. The P removal of the wetland is adsorption of the soil and media, the adsorption by the plants and the sedimentation. The

P removal is about 30%~40% in the wetland.

⑤ Metals. The mechanism of the metal removal is ion exchange, plant adsorption, chemical sedimentation and oxidation by the microorganism.

5.2.2.4 The advantages and the disadvantages

The advantages of the wetland are as followed.

① The construction investment is low, and the running cost is little. Generally, the construction investment is only 1/3 ~ 1/2 of the traditional treatment processes. If the scale is less than 50 m^3/d, the construction investment is higher. For the other scale, the construction investment is about 800 ~ 2,500 Yuan (RMB)/m^3, and the running cost is about 0.06~0.30 Yuan (RMB)/m^3.

② Easy maintaining. No special people are needed, except of the wetland with backflow.

③ Effective management on the waste water.

④ Resistance on the shocking.

⑤ Economic benefit and social benefit.

The disadvantages of the wetland are as followed.

① Huge occupied area.

② The design parameters are not accurate, and it is shortage of the experience on long time running.

③ It is affected by plant diseases and insect pests.

④ Easy blocking.

⑤ Reeds recession appears for the reed wetland.

5.3 Commonly used integrated treatment device for rural domestic sewage treatment

5.3.1 Vertical flow type AAO integrated device

The AAO process is an anaerobic – anoxic – aerobic waste water treatment process. The waste water is routed through anaerobic zones for phosphorus release, anoxic zones for nitrogen removal, and aerobic zones for nitrification and phosphorus removal, resulting in better nitrogen and phosphorus removal and a shorter total hydraulic retention time (HRT) than other processes. The process equipment has a simple structure, low operating costs in sewage treatment work, and is suitable for popularization and use in most rural areas. The vertical flow type AAO integrated device in Figure 5-11 can treat thousands of tons/day of township domestic sewage. Based on the AAO process, the sewage flows vertically and uses the gravity of the sludge to settle naturally. The aerobic zone uses deep water aeration to increase the utilization rate of oxygen. The reactor integrates the anaerobic zone, anoxic zone, aerobic zone and precipitation zone that are separately set up in the conventional AAO process. After aeration treatment in the aerobic zone, it enters the inclined pipe sedimentation zone through the water hole and guided by the water retaining jacket for solid - liquid separation. The supernatant flows into the disinfection channel from the clear water pipe through the water outlet weir to form up-to-standard discharge. The sludge that sinks to the bottom of the anaerobic zone is discharged from the sludge pipe to achieve the purpose of sludge discharge.

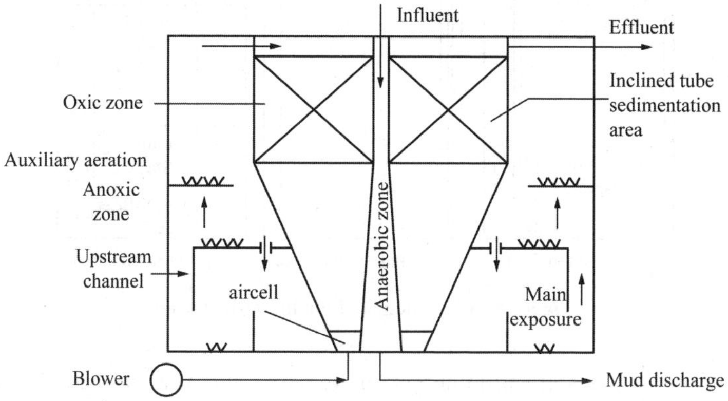

Figure 5-11 Vertical flow type AAO integrated device

5.3.2 Integrated air lift SBR device

Sequencing batch reactor activated sludge process (SBR), also called intermittent activated sludge process, is the pioneer of sewage activated sludge biochemical treatment system. Compared with the continuous flow system, the process has the advantages of high purification efficiency, high impact load resistance, good sludge sedimentation effect, prevention of sludge expansion, and no need to set up sedimentation tanks and sludge return devices. Figure 5-12 shows the integrated air lift SBR device. Compared with traditional SBR, this device uses an air lifter instead of pumps and decanters. The amount of air required for aeration, sewage, and sludge lifting can be supplied by only one micro air pump, which is convenient for maintenance and management. The device integrates three functional units of adjustment tank, reaction tank, and mud storage tank. There is no sedimentation tank. The overall shape of the reactor is a cylindrical structure, and the three functional units are equally divided in a radial shape. Compared with the square reaction tank, reducing the dead angle is more conducive for the uniform mixing of sludge and sewage.

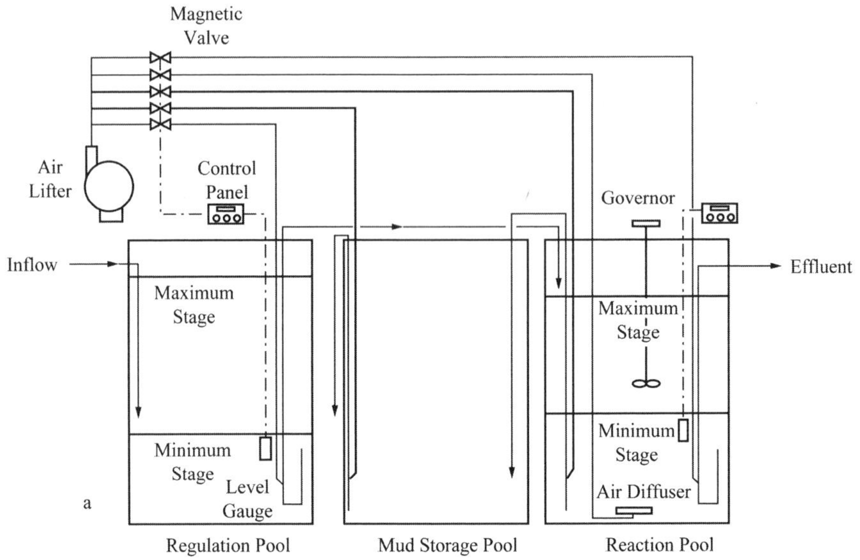

Figure 5-12 Integrated air lift SBR device

The device uses a self-priming pump to pump the raw sewage that has passed through the grille into the adjustment tank through the inlet pipe deep into the bottom of the adjustment tank to evenly distribute the water, which plays a role in adjusting the water quality and quantity. The electric water inlet valve of SBR is controlled by the liquid level gauge installed in the regulating tank and the reaction tank, and the electric drain valve is controlled by the liquid level gauge installed in the reaction tank. If the device encounters accidents such as low water level in the regulating tank, the reaction tank enters the idle process, and a small amount of aeration is performed regularly to maintain the sludge activity. Each process of the device is regulated by an automatic control system. When the mass concentration of DO stabilizes at about 3 mg/L, aeration for 3 h, precipitation for 1.5 h, the COD, NH_4^+-N, and TN indicators of the reactor effluent can meet the requirements of GB 18918—2002 Class Ⅰ level B, and the effluent TP can reach Class Ⅱ emission requirements. It has the characteristics of small investment, small land occupation, short construction period, simple operation and stable water quality. Under the premise of considering the actual economic and engineering durability conditions in the rural areas, as well as the operating level and maintenance capabilities of the personnel. The integrated reactor is very suitable for the actual situation, in line with the new rural construction project and energy saving and emission reduction. It also provides a reference for sewage treatment in small towns, suburban living areas, resorts, and villa areas that are relatively scattered and difficult to be integrated into the urban pipeline network.

5.3.3 Integrated biological filter

Biological Aerated Filter (BAF) is a fixed-bed biofilm technology for sewage. This process combines biological oxidation and SS retention. Air is supplied to the film-coated granular suspended filler through compressed aeration, and periodic operation is realized through backwashing and regeneration. Figure 5 - 13 shows the integrated biological filter. The device is cylindrical in shape, with three-stage filters arranged in series from the inside to the outside in the form of concentric circles. The Ⅰ tank (anoxic tank) does not aerate during normal operation. The bottom of the filter tank is equipped with a membrane pore gas flushing pipe. Between the upper and lower filter screens/plates is 7 ~ 11 mm ceramsite filler, and the filler height is 3.0 m. A single-hole membrane aeration system is installed at the bottom of the Ⅱ tank (aerobic tank) for aeration. The filter material adopts broken floating ceramsite with a particle size of 4 ~ 7 mm, and the thickness of the filter material is 3.5 m. The Ⅲ tank (anoxic tank) is not aerated during operation, the filter material is broken-shaped floating ceramsite with a particle size of 4 ~ 7 mm, and the thickness of the filter layer is 3.1 m. The density of suspended ceramsite infiltration particles is (0.95 + 0.5) g/cm, and the lift pump is a port sewage pump with aeration blower.

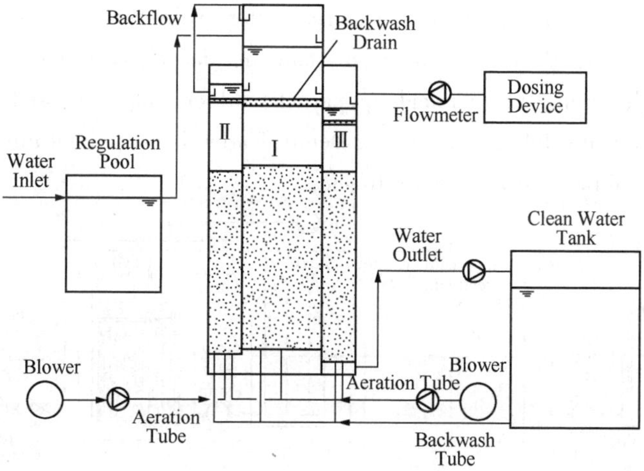

Figure 5-13 Integrated biological filter

Sewage enters the first-stage downflow pretreatment denitrification expanded bed biological filter (Ⅰ tank) from top to bottom in the integrated biological filter. In the filter tank, the suspended filter media intercepts most of the suspended matter. Denitrifying bacteria use carbon source in raw water to denitrify and remove nitrogen and most organic pollutants; The first-stage effluent enters the second-stage up-flow fluidized-bed biological aerated filter (Ⅱ tank) through the lower pipe from bottom to top, which mainly conducts nitrification reaction; The second-stage effluent enters the third-stage downflow biological aerated filter (Ⅲ tank) for nitrogen and phosphorus removal from top to bottom, which is mainly used as the best place for intensified nitrogen and phosphorus removal as well as SS removal. Compared with the existing process, the integrated biological filter has the advantages of high denitrification and phosphorus removal efficiency, high organic load, small footprint, and high degree of automation.

5.3.4 Integrated MBR device

The integrated MBR process refers to a new type of sewage treatment technology that combines membrane separation and biofilm treatment. It increases the concentration of activated sludge through the principle of membrane separation, thereby strengthening the function of the bioreactor and eliminating the secondary settling tank, ensuring stable operation even when sludge expansion occurs. Figure 5-14 shows the integrated MBR device. The device intercepts large debris through the grille to ensure the removal of the influent particles before the pump, and adjusts the water quality and volume through the adjustment pool. The hydrolysis acidification tank mainly uses the pre-reaction mechanism of anaerobic biochemical treatment, and the microorganisms participating in the reaction are mainly facultative bacteria. After hydrolysis and acidification, the organic pollutants in the macromolecular state are decomposed into small molecular substances with strong biochemical properties, which improves and enhances the biodegradability and solubility of sewage. Denitrification is carried out through the anoxic tank to convert nitrate nitrogen into nitrogen.

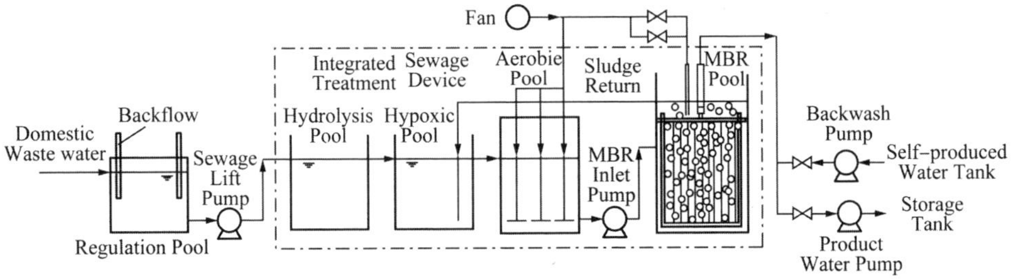

Figure 5-14 Integrated MBR device

The sewage passes through the automatic grille to remove the larger particles of impurities and then enters the adjustment tank and mixes evenly. The sewage is lifted by the sewage lift pump to the hydrolysis tank for hydrolysis and acidification. The effluent flows into the anoxic tank, and then enters the aerobic tank to degrade most of the organic matter, and then is pumped into the MBR tank. The organic matter is further degraded in the MBR, and bacteria, SS and other substances are intercepted through the micropores of the MBR, and the water is discharged to the storage tank for the greening of the plant area. This device is used to treat sewage when the influent pH is 6~9, the average mass concentrations of COD, BOD_5, NH_4^+-N, and SS are 367.5 mg/L, 200 mg/L and 27.1 mg/L, 143.6 mg/L, respectively. The average removal rates of COD, BOD_5, NH_4^+-N, and SS were 87.6%, 97%, 87.2%, and 99.2%, respectively, reaching the first-level emission standards in DB 37/676—2007.

5.4 Long-term operation and management mechanism of rural waste water treatment technology

The management mechanism of rural domestic waste water treatment facilities mainly includes three mechanisms: territorial (village and town) self-management, third-party operation management, and integrated management of construction and operation.

5.4.1 Territory (village and town) self-operation and maintenance management mechanism

This mechanism is mostly used in villages and towns that have low economic development levels, have just started sewage treatment, or have relatively scattered facilities. Because villages and towns do not pay enough attention to the operation and maintenance of sewage treatment facilities, and villagers lack professional knowledge of sewage treatment technology and facilities, maintenance and repair becomes difficult when failures occur, and as such are easily abandoned. The investigation of the operation of sewage treatment facilities in the Taihu Lake Basin revealed that in this mode, abnormal operation of facilities is more common, so regular follow-up inspections are required during facility maintenance, and relevant technical training and professional guidance should also be strengthened.

5.4.2 Third-party operation management mechanism

5.4.2.1 Government outsourcing management

Government outsourcing management mechanism has become common. Generally, the government invests in the construction of rural domestic sewage treatment facilities, and entrusts a third party (enterprise or institution with professional capabilities) to perform operation, maintenance and management; The local government or village collectively owns the property rights of the facilities, supervises and manages the operation of the facilities, and pays fees to third parties based on the performance of waste water treatment.

5.4.2.2 Facility rental service mechanism

The facility leasing service mechanism is a new type of market-oriented operation mechanism explored in Chongqing and other regions. Villages and towns entrust a third-party company in the form of leased facilities to treat sewage up to standards and pay related treatment fees. The property rights of sewage treatment facilities belong to a third party, and the government or villages and towns shall be the owners, with sewage treatment fees determined by the treatment effect. The facility can also be removed based on the current situation, and the form of cooperation is more flexible. The use of third-party operation management has significantly improved the professional level of local rural domestic sewage treatment, which is conducive for the long-term operation of facilities.

5.4.3 Construction and operation integrated operation management mechanism

The construction and operation integrated management mechanism is to integrate the construction of facilities with the later operation and management. The local government in charge of the project allocates project funds to the enterprise in installments based on the operational performance, which will encourage the enterprise to ensure the effective operation of sewage treatment facilities. For example, since 2015, Changshu City, Suzhou, Jiangsu Province has adopted an integrated model of "construction-operation-management". Sign a concession agreement between the rural domestic waste water treatment facility project and the enterprise, and authorize the contracting company to undertake the investment (financing), construction and maintenance of the project. Within the concession period stipulated in the agreement, they are permitted to construct and operate specific facilities, recover investment and make profits. The government has the power to supervise and regulate the construction and operation of infrastructure. After the expiration of the concession period, the contracting party's enterprise transfers the facility to the government department for free or paid.

Questions

1. There are many kinds of wetlands, please present the advantages and disadvantages.
2. Please clarify the mechanism on waste water for wetland.
3. The characteristics of rural waste water and the source of rural waste water.
4. Why is it difficult to collect rural waste water?
5. What are the kinds of the oxidation lagoon, please present the parameters.
6. The clarification mechanism of aerobic lagoon.
7. Please make a comparison between oxidation lagoon and wetland.
8. Wetland can be the advanced treatment for domestic waste water, please present the reason.
9. There are many kinds of integrated treatment devices for rural domestic sewage. Presenting some kinds and showing the characteristics.
10. In your opinion, when people select the integrated treatment device, what characteristics should be focused?

Chapter 6 Control and use on sludge

Emphasis: Characteristics of residual sludge. Sludge treatment methods. Sludge dewatering and thickening. Comprehensive utilization and ultimate disposal of the sludge.

Nut: Sludge amount calculation. Sludge dewatering and thickening. Sludge utilization and ultimate disposal.

6.1 Overview

The sludge comes from the solids which were separated by the screen, the grit chamber, the settling tank, the floating tank, et al. and which were the end-product by the bacteria during the biological treatment. Generally, the quantity is about 0.3% ~ 0.5% to the water amount (water content is about 97%). The sludge must be controlled and disposed properly because there are lots of toxic matters in the sludge, such as, parasitic ovum, pathogenic microorganism, synthetic organics, and heavy metals.

The treatment rate and degree of waste water increase year by year, and the amount of sludge increases sharply, due to the fast urbanization and the improved environmental quality standard. Nowadays, the amount of dry sludge is 5.32 million tons and the increase rate of each year is up to 10%.

6.1.1 The principle for the sludge treatment

The aims of the sludge treatment are as followed.

① To be sure the running of the waste water treatment plant and to be sure the efficiency.

② To be sure the proper disposal of the toxic and harmful matters in the sludge.

③ To be sure the stabilization of the organic matter in the sludge.

④ To be sure the recycle of the useful matters in the sludge.

According to the above aims of the sludge treatment, the principle of the sludge treatment is followed.

(1) Reduction After the separation in the second sedimentation tank, the water content of the residual sludge is very big, up to 99.99%. Therefore, the volume of the sludge is very big. It is not beneficial of storage and transport. For the reduction, the key is the thickening. The water content is reduced from 99.2% to 97.5% after the thickening, but the sludge is still liquid. And then the dewatering is the important step to reduce the water content after the thickening. After the dewatering, the water content is about 60% ~ 80%, but the sludge is solid now. It is easy for storage and transport. After the treatment, the water content in the dry sludge is lower than 10%. 100 kg wet sludge is transformed into 5 kg dry sludge.

(2) Stabilization For the storage and transport of sludge, the stabilization is necessary because the content of organic matter in the sludge is about 60% ~ 70% that is easy corrupt and produce stench. The widely used methods are aerobic or anaerobic digestion. They can transform the organic matter to stable end-product. Another method is adding the chemical matters to kill the bacteria or inhibit the activity of the bacteria. For example, the addition of lime to increase pH is favor of the inhibition on the bacteria. It is noticeably that the sludge is not suitable for the long storage because the activity of the bacteria is restored due to the pH decrease.

(3) Non-harm There are lots of parasitic ovum, pathogenic microorganism, synthetic organics, and heavy metals in the sludge, especially in the primary sludge. The content of pathogenic microorganism is up to 10^6 count/g in the faeces from the people with infection. If the above toxic matters do not be removed from the sludge, it has the chance going into the environment. Therefore, it is necessary to remove the toxic matters.

(4) Reclamation In these years, the single treatment on the sludge has been transformed into the reclamation. The sludge has been widely used for the agriculture, building materials and fuels. The disposal technologies include compost, building materials, methane, adsorbent, the degradable plastic, the fodders for fish or fowl, and so on.

6.1.2 The basic method for the sludge treatment

(1) Thickening Try the best to separating water from the sludge using the gravity or floating thickening.

(2) Stabilization Transforming the organic matter to inert matter using

digestion. Inhibiting the activity of the bacteria to avoid the stench.

(3) Conditioning Improving the water separation in the succeed process using heating or chemical matters.

(4) Dewatering To remove water and to decrease the volume of the sludge by using the vacuum, pressure and drying.

6.1.3 The basic process for the sludge treatment

① Raw sludge→Thickening→Digestion→Natural drying→Ultimate disposal;

② Raw sludge→Thickening→Digestion→Mechanical dewatering→Ultimate disposal;

③ Raw sludge→Thickening→Digestion→Ultimate disposal;

④ Raw sludge→Thickening→Natural digestion→Compost→Ultimate disposal;

⑤ Raw sludge → Thickening → Mechanical dewatering → Drying or ignition → Ultimate disposal.

6.2 Sludge characteristics and calculation

6.2.1 The component and the types of the sludge

The sludge consists of solid phase and liquid phase. The solid phase includes the organic matter and inorganic matter. The organic matter consists the toxic matters, organic chemical compound and microorganisms. The inorganic matter includes the toxic organic matters, nutrient, inorganic minerals. The liquid phase includes the water and soluble organic and inorganic matter. The water is referred to the pore water, capillary water, absorbed water and bound water. The characteristics, the component and the source are described in the Table 6-1.

Table 6-1 Characteristics of the different kinds of sludge

Principle	Types	Characteristics
The waste water source	Domestic waste water sludge	Organic matter is higher, and heavy metals are lower
	Industrial waste water sludge	Depending on the water quality

续表

Principle	Types	Characteristics
Sludge component	Organic sludge	Small particles, small density, strong moisture holding capacity, difficult dewatering
	Inorganic sludge	Lots of metal compound, big density, easy settling
	Hydrophilia sludge	Difficult thickening and dewatering
	Hydrophobic sludge	Good thickening and dewatering
	Raw or fresh sludge	No any treatment
Treatment phase	Thickening	Anaerobic digestion and oxic treatment on the sludge
	Digestion	
	Dewatering	
	Drying	
Sludge source	Screenings	Suspended matters by the screening
	Primary sludge	From the primary settling tank
	Residual sludge	From the second settling tank
	Humus sludge	From the biofilm tank
	Chemical sludge	From the chemical sedimentation

6.2.2 The indexes of the sludge

The indexes of the sludge are as followed.

(1) Water content and solid content The definition of the water content is that the percentage of the weight of the water in the sludge to the total sludge (p, %). In general, the water content for the biological treatment process is high, nearly to 1. On the contrary, the other is the solid content which the percentage of the weight of the solid in the sludge to the total sludge (%). Therefore, the sum of the water content and the solid content is 1.

The volume, the weight and the solid content can be described in the following equation.

$$\frac{V_1}{V_2}=\frac{C_2}{C_1}\approx\frac{W_1}{W_2}=\frac{100-P_2}{100-P_1}$$

In the equation, V_1, W_1, C_1—the volume, the weight and the solid content at the water content of P_1;

V_2, W_2, C_2—the volume, the weight and the solid content at the water content of P_2.

In practice, the equation is suitable for the sludge with the water content of 65%.

(2) Dewatering property In practice, the dewatering property is described using the resistivity (r) in the treatment with the filtration to separate water from sludge.

It is well known that the water in the sludge is forced to pass the filter under the force due to the pressure difference at the both sides of the filter. Therefore, the solids are hold back by the filer. This is called dewatering.

The force comes from the following four ways.

① Static pressure due to the thickness of the sludge. For example, natural drying.

② Negative pressure. For example, vacuum dewatering.

③ Adding pressure. For example, filter pressing.

④ Centrifugal force. For example, centrifugal dewatering.

The dewatering is described by the Carman equation.

$$\frac{t}{V}=\frac{\mu \omega r}{2PA^2}V+\frac{\mu R_f}{PA}$$

In the equation, V—filtrate volume, m^3;

t—filtration time, s;

P—filtration pressure, kg/m^3;

A—filtration area, m^2;

μ—macroviscosity of filtrate, $kg \cdot s/m^2$;

ω—The weight of the dry solid retained on the filter for unit filtrate, kg/m^3;

r—resistance, m/kg. The resistance of the weight of the unit dry filter cake on the unit filtration area;

R_f—impedance, $1/m^2$.

According to Carman equation, the relationship between t/V and V is line. Therefore, the slope and intercept are as followed.

$$b=\frac{\mu \omega r}{2PA^2}$$

$$a=\frac{PR_f}{PA}$$

The r is that:

$$r=\frac{2PA^2}{P}\cdot\frac{b}{A}$$

(3) Characteristics property The characteristics property includes organic matter, inorganic matter, the extent of digestion, nutrient, toxic matters, caloricity. The Table 6-2 shows the main characteristics.

Table 6-2 The main characteristics

Item	Primary sludge	Residual sludge	Anaerobic digestion sludge
Dry solid/%	3~8	0.5~1	5~10
Volatile solid/%	60~90	60~80	30~60
Particle density/(g/cm^3)	1.3~1.5	1.2~1.4	1.3~1.6
Relative density	1.02~1.03	1.0~1.005	1.03~1.04
BOD_5/VS	0.5~1.1		
COD/VS	1.2~1.6	2.0~3.0	6.5~7.5
Alkalinity/(mg/L)	500~1,500	200~500	2,500~3,500
pH	5.0~8.0	6.5~8.0	6.5~7.5

(1) Volatile solid It is referred the organic matter in the sludge. The index is measured the loss weight after ignition at 600℃.

(2) The extent of digestion The content of the degradable organic matter.

(3) Relative density of the wet sludge and the dry sludge Relative density of the wet sludge is referred to the ratio of the weight of wet sludge to the weight of the same volume water. It is described in the following equation.

$$p=\frac{p+(100-p)}{\left(p+\frac{100-p}{p_s}\right)\times 1}=\frac{100p_s}{p\,p_s+(100-p)}$$

In the equation, p——relative density of the wet sludge;

p_s——relative density of the dry sludge.

(4) Nutrient The sludge from the municipal waste water treatment plant includes lots of nutrient. Take the following sludge as an example (Table 6-3).

Table 6-3 The nutrient of the sludge in municipal waste water treatment plant in China

Sludge	TN/%	$P(P_2O_5)$/%	$K(K_2O)$/%	Organic matter/%
Primary sludge	2~3	1~3	0.1~0.5	50~60
Activated sludge	3.3~7.7	0.78~4.3	0.22~0.44	60~70
Digestion sludge	1.6~3.4	0.6~0.8		25~30

(5) Toxic matter There is lots of parasitic ovum, pathogenic microorganism,

virus in the sludge. The measurement on the sludge, such as digestion, should be taken before the disposal. The content and the kinds of heavy metals depend on the industrial waste water ratio and the characteristics.

(6) Caloricity The sludge has high caloricity. It is similar as lignite after drying.

6.2.3 The calculation on the sludge amount

The sludge amount in the municipal waste water treatment plant can be estimated by the experience in the Table 6-4.

Table 6-4 The sludge amount in the municipal waste water

Types of sludge	sludge amount/(L/m^3)	Water content/%	Density/(kg/L)
Grits	0.03	60	1.5
Primary sludge	14~16	95~97.5	1.015~1.02
Secondary sludge	10~21	99.2~99.6	1.005~1.008
Biofilm sludge	7~19	96~98	1.02

The primary sludge and the secondary sludge can be estimated by the following equations.

(1) The primary sludge The equation is as followed according to the suspended solids, the amount of waste water, the removal and the content of the water.

$$Q_p = \frac{100\ C_0 l Q}{10^3 (100-p) C}$$

In the equation, Q_p—The amount of primary sludge, m^3/d;

Q—The amount of waste water, m^3/d;

l—Removal, %;

C_0—The content of suspended solids, g/L;

C—The density of the sludge, 1,000 kg/m^3.

The volume of the primary sedimentation tank is described by the following equation for the municipal waste water.

$$V = \frac{SN}{1,000}$$

In the equation, S—the amount of sludge for each one a day, 0.3~0.8 L/(d · cap);

N—the design people, cap.

(2) The residual sludge The residual sludge is estimated by the following

equation.

$$Q_s = \frac{\Delta X}{f X_r}$$

In the equation, Q_s—The amount of residual sludge, m^3/d;

ΔX—The dry weight of the volatile sludge each day, m^3/d;

f—MLVSS/MLSS, in general, 0.75;

X_r—The content of the backflow sludge, g/L.

ΔX is estimated by the following equation.

$$\Delta X = Y(S_a - S_e)Q - K_d V X_v$$

In the equation, $(S_a - S_e)Q$—The degradable amount of organic matter, m^3/d;

VX_v—The amount of the volatile suspended solids, kg;

K_d—Self oxidation rate of the bacteria, d^{-1}.

6.3 Sludge thickening

The thickening is to reduce the volume of the sludge because the water content is high, primary sludge 95%~97%, residual sludge over 99%. The water in the sludge is divided into four parts, pore water (70%), capillary water (20%), surface adsorbed water and bound water (10%). It is showed in the Figure 6-1.

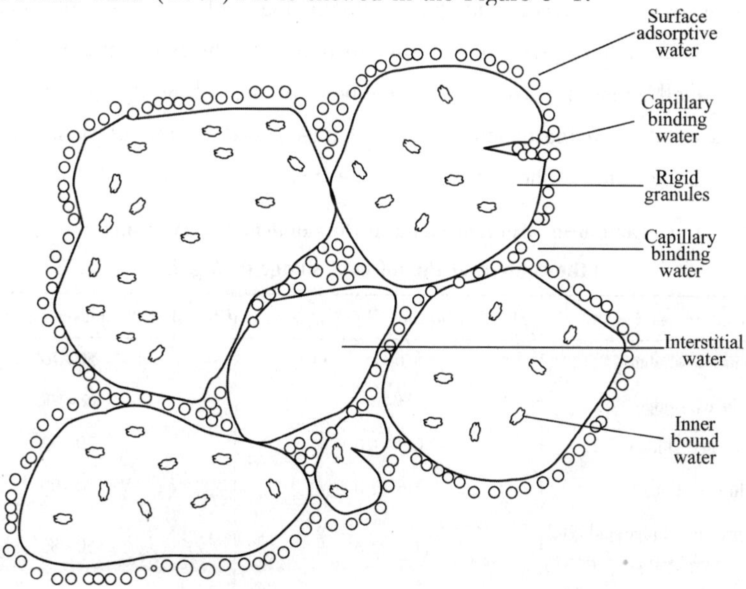

Figure 6-1 The schematic diagram of the water in the sludge

The methods to reduce the water content are as followed. Thickening is used to reduce the pore water. Natural dewatering and chemical dewatering are to remove the capillary water. The ignition is to remove adsorbed water and bound water. The efficiency of the different methods is showed in the Table 6-5.

Table 6-5 The efficiency of the different dewatering

Dewatering		Equipment	Water content/%	Shape
Thickening		Gravity, air flotation, centrifugation	95~97	pasty
Natural drying		Natural drying field, drying field	70~80	Mud cake
Chemical dewatering	Vacuum	Vacuum drum	60~80	Mud cake
	Filter pressing	Plate-and-frame filter press	45~80	Mud cake
	Roller belt	Roller belt filter	78~86	Mud cake
	Centrifugal method	Centrifugal machine	80~85	Mud cake
Drying		Drying equipment	10~40	Granular
Ignition		Ignition equipment	0~10	ash

6.3.1 The gravity thickening

The gravity thickening is attributed to compressed settling. Under the gravity of the sludge, the water is removed. The gravity thickening includes the continuous gravity thickening and the batch gravity thickening. In practice, the batch gravity thickening is suitable for small amount waste water treatment plant, while the continuous gravity thickening is suitable for big or middle amount waste water treatment plant. The running of the gravity thickening is showed in the Table 6-6.

Table 6-6 Running data in the municipal waste water (the content of the influent sludge 0~6 g/L)

Types of the sludge	Sludge solid flux/[kg/(m^2 · h)]	Content of the thickening sludge/(g/L)
Domestic sludge	1~2	50~70
Primary sludge	4~6	80~100
Activated sludge	0.5~1.0	20~30
Humus sludge	1.6~2.0	70~90
Mix of primary sludge and activated sludge*	1.2~2.0	50~80
Mix of primary sludge and humus sludge	2.0~2.4	70~90

6.3.2 The flotation thickening

For the density of the sludge is closed to 1, it is suitable for the flotation thickening. The method is suitable for the hydrophobic sludge which the density is closed to 1. In practice, the pressurized dissolved air floatation is widely used. The coagulant is used to improve the efficiency of the thickening, such as, aluminum salt, ion salt, active silica, polyacrylamide. In addition, for the normal running of the flotation thickening, enough ratio of air to solids is necessary. The value is described by the equation.

$$\frac{A_a}{S}=\frac{S_a R(fP-1)}{C_0}$$

In the equation, A_a—Effective air amount, mg/h;

S—The total solid amount in the influent sludge, mg/h. In practice, 0.005~0.06. In general, 0.03~0.04. By the experiment, $S=Q_0C_0$;

S_d—Solubility of the air in the water under the atmospheric pressure of 0.1 MPa;

P—The pressure of the dissolved air vessel, 2~4 kgf/cm^2;

f—Air saturation factor, 50%~80%;

C_0—The content of the influent sludge, mg/L;

R—The backflow ratio. The ratio is between the pressurized mixed matter with air and water and the amount of the influent sludge, 1.0~3.0.

The surface area is described by the equation.

$$A=\frac{Q_0(R+1)}{q}$$

In the equation, A—The surface area of the flotation thickening tank, m^2;

q—the surface hydraulic load of the flotation thickening tank, m^3/(m^2·h);

Q_0—The amount of the influent sludge, m^3/h.

The hydraulic load and the solid load are showed in the Table 6-7.

Table 6-7 The hydraulic load and the solid load of the flotation thickening tank

Types of sludge	Content of influent sludge/%	Hydraulic load/ [m^3/(m^2·h)]	Solid load/ [kg/(m^2·h)]	Content of solid/%
Activated sludge	<0.5	1.0~3.6	1.04~3.12	3~6
Residual sludge	<0.5	1.0~3.6	2.08~4.17	3~6

续表

Types of sludge	Content of influent sludge/%	Hydraulic load/ [$m^3/(m^2 \cdot h)$]	Solid load/ [$kg/(m^2 \cdot h)$]	Content of solid/%
Residual sludge from high-purity O_2 aeration	<0.5	1.0~3.6	2.50~6.25	3~6
Mixed sludge of primary sludge and residual sludge	1~3	1.0~3.6	4.17~8.34	3~6
Primary sludge	2~4	1.0~3.6	<10.8	3~6

6.4 Sludge digestion

Sludge digestion is divided into anaerobic digestion and aerobic digestion.

6.4.1 Anaerobic digestion

Anaerobic digestionis divided into mesophilic digestion (30~36℃) and thermophilic digestion (50~55℃) according to the characteristic and the suitable temperature for anaerobic bacteria, especially methanogens. The thermophilic digestion is favor of the production of methane, decreasing the digestion time, killing the parasitic ovum. The higher energy consumption is problematic, compared with mesophilic digestion.

Anaerobic digestion is divided into lower load and higher load. The digestion tank with lower load is a closed tank without heating and mixer. In general, for the lower load digestion, the load is at the range of 0.5~1.6 kgVSS/($m^3 \cdot d$). The digestion speed is slow and the digestion time is long. The retention time is about 30~60 d. The load of the higher load digestion with heating equipment and mixer is at the range of 1.6~6.4 kg VSS/($m^3 \cdot d$). The difference between the lower load digestion and the higher load digestion is the continuous influence and effluence. For the former, it is intermittent, while the latter, it is continuous. The shortest retention time is about 10~15 d. The whole tank is in mixing. In practice, the two-stage digestion was used for the higher load digestion. In the designment, for the first stage, there is the heating and mixer and gas-collecting hood, while there is no the heating and mixer for the second stage. The heat for the second stage comes from the first stage. Therefore, the temperature is about 20~26℃, and the production of methane is about 20%. Generally, for the designment of the two-stage digestion tank, the ratio of the first stage tank and the second stage tank is 1 : 1 or 2 : 1 or 3 : 2. The 2 : 1 is used widely.

6.4.2 Aerobic digestion

The microorganism uses the cellular plasm or the self-oxidation to get energy. This is called aerobic digestion. The aerobic digestion is similar as the activated sludge. For the aeration tank, the aeration time is long, up to 10 ~ 20 d, and the bacteria in the sludge is in the endogenous respiration phase. It is to reduce the organic matter in the sludge by aerobic metabolism of the organic matter and the endogenous metabolism of the bacteria, resulting in the reduction of sludge. Generally, the removal of the sludge is up to 80%, while this part is mainly the biodegradable organic matter.

The cellular plasm is reacted under the aerobic conditions according to the equation.

$$C_5H_7NO_2+7O_2 \longrightarrow 5CO_2+3H_2O+H^++NO_3^-$$

By the calculation, the removal of 1 kg cellular plasm consumes 2 kg O_2. In addition, ammonia nitrogen is oxidized into NO_3^-, resulting in the decrease of the alkalinity. It is necessary to regulate pH by the enough alkalinity.

According to the above analysis, the consumption on the energy is higher and the running management is strict and the volume of the tank is big for the aerobic digestion. Therefore, in practice, we only choose the aerobic digestion when the sludge amount is small.

6.5 Sludge conditioning

In general, the resistivity of the sludge is at the range of $(0.1 \sim 0.4)\times10^9$ S^2/g, and the mechanical dewatering is suitable and economic. However, the primary sludge, the activated sludge, the humus sludge and the digested sludge, the resistivity is big (Table 6-8). The mechanical dewatering is not suitable the above sludge. Therefore, the sludge conditioning is necessary. The conditioning is referred to destroy the bound of the sludge and reduce the appetency of the solid and water. The aim is to improve the property for dewatering. There is the chemical conditioning, heat conditioning, freezing solution conditioning, elutriation conditioning and radiation conditioning.

Table 6-8 The resistivity in different sludge

Sludge types	Resistivity	
	S^2/g	m/kg
Primary sludge	$(4.7 \sim 6.2)\times10^9$	$(46.1 \sim 60.8)\times10^{12}$

续表

Sludge types	Resistivity	
	S^2/g	m/kg
Digestion sludge	$(12.6 \sim 14.2) \times 10^9$	$(123.6 \sim 139.3) \times 10^{12}$
Activated sludge	$(16.8 \sim 28.8) \times 10^9$	$(164.8 \sim 282.5) \times 10^{12}$
Humus sludge	$(6.1 \sim 8.3) \times 10^9$	$(59.8 \sim 81.4) \times 10^{12}$

6.5.1 Chemical conditioning

The chemical conditioning is to add the coagulant, coagulant aids into the sludge to make the particles flocculation and to decrease the resistivity and improve the dewatering property. The widely used coagulant is inorganic coagulant and organic polymeric coagulant. The function of the coagulant aids is to regulate pH in the sludge and to provide the skeleton and change the structure of the sludge particle and to improve the coagulation efficiency. In practice, the dose of the coagulant depends on the types and characteristic of the sludge, the degree of the digestion, the content of the solid. Therefore, it is necessary to do experiment in the lab to decide the dose.

6.5.2 Heat conditioning

The function of the heat conditioning is the stabilization of the sludge. In addition, it is to degrade the organic matter, and to destroy the stability of colloid, and to promote the bound water and adsorbed water release. After the treatment of heat, the resistivity is reduced to 1×10^8 S^2/g. Therefore, it is to improve the dewatering property and to kill the parasitic ovum, pathogenic bacteriums and virus. The water content of the mud cake is only 30% ~ 45% after the mechanical dewatering. In practice, the consumption on the energy is big, so the use of heat conditioning is little.

6.5.3 Elutriation conditioning

Elutriation conditioning is suitable for the digestion of sludge. The main reason is that the alkalinity is over 2,000 mg/L in the sludge. It is necessary to neutralize the alkalinity before the chemical conditioning. Otherwise, the dose of the coagulant increased sharply. The aim of elutriation conditioning is to wash the sludge to decrease the alkalinity in the sludge by using the effluent from the waste water plant, potable water, river water.

6.5.4 Freezing solution conditioning

The method is to freeze the water in the sludge, and then heating to dissolve the frozen sludge. During the freezing and dissolving process, the sharp change of the temperature makes the sludge destabilization and make the fine particles to coarse particles. In addition, the cell is destroyed by the sharp temperature and the water in the cell is lost. Therefore, the settling property and the dewatering property are improved.

6.6 Sludge dewatering

After the thickening and the digestion, the water content is still about 95%~97%, and the volume is so big. It is necessary to conduct drying or dewatering for the recycle and the disposal. Sludge dewatering is to remove capillary water and adsorbed water and then the water content is reduced to 60%~80%. Now, the volume is only the 10%~20% of the raw volume before dewatering. Generally, the drying and dewatering is mainly natural drying and mechanical dewatering. The dewatering mechanical is used widely, such as plate-and-frame filter press, belt filter press and centrifugal machine.

6.6.1 The natural drying of sludge

The natural drying of sludge is a simple and economic method and it is widely used. Generally, the sludge drying bed includes the natural filter bed and the artificial filter bed. The former has the good properties of soil and the ground water level is low. The filter for the latter is artificial.

The volume of the sludge is reduced due to the water natural evaporation and penetration on the drying bed, resulting in the mobility of sludge disappearance. The water content is reduced to 60%~70%. But the problem of the method is the demand for lots of labors and bed, the method is still used in small number of small waste water treatment plant.

6.6.2 The mechanical dewatering of sludge

Mechanical dewatering has been the main measurement for the sludge dewatering. It includes vacuum filtration, pressure filtration and centrifugal filtration. The principle followed the Carmon equation. The main equipment for the dewatering is plate-and-frame filter press, belt filter press, centrifugal machine and vacuum filter.

(1) Plate-and-frame filter press Firstly, the plate-and-frame filter press is used for chemical engineering dewatering, but it is used widely in sludge dewatering. The main reasons are as followed. The force due to the pressure is big. The

solid content is high. The filtrate is clean. The consumption on coagulant due to conditioning is low. The recycle rate of the solids is high.

The principle is as followed (Figure 6-2). The filter plate and the filter frame are placed at interval. The two sides of the filter plate are covered by the filter fabric, and then there is the filter room between the filter plate and the filter frame. There is a small hole in the middle of the filter plate and the filter frame above the equipment, and then all the holes is an aisle due to the pressure. The sludge with the pressure of 0.2 ~ 0.4 MPa comes into the aisle and is divided into the filter room. After the filtration, the filter passes the filter fabric under the pressure and comes out the equipment.

The structure of the plate-and-frame filter press is showed in the Figure 6-3.

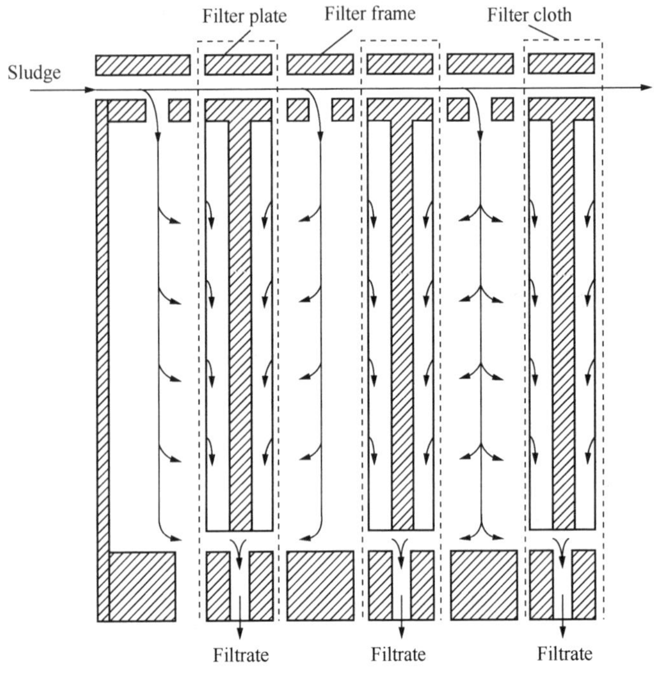

Figure 6-2 The schematic diagram of the plate-and-frame filter press

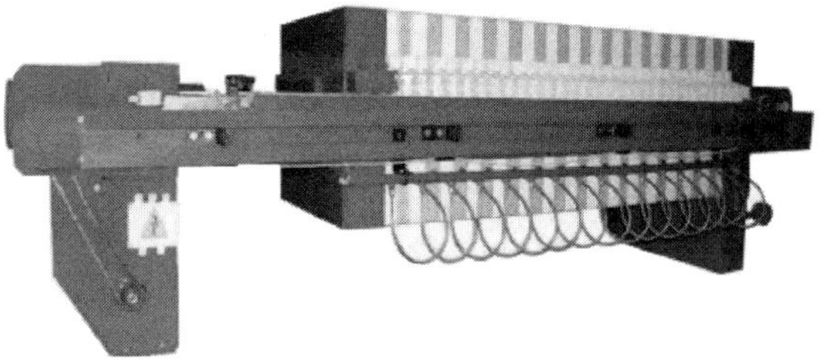

Figure 6-3 The plate-and-frame filter press

(2) Belt filter press The successful develop of the belt filter press is the belt. It comes from the organic polymer. The structure of the belt filter press is showed in the following (Figure 6-4). The advantage of the belt filer press is as followed.

① Continuously running.

② Running and management is simple.

③ The affiliated equipment is little.

Figure 6-4 The belt filter press

Therefore, the investment, the demand on labors, the consumption on the energy and the management fee are low. It is widely used in sludge dewatering. In China, for most newly built waste water treatment plant, the belt filter press is chosen and used.

(3) Centrifugal filter The principle of the centrifugal filter is similar as the centrifugal thickening. The solids and the water are separated under the rotary movement. The separation speed of the particles is higher 1,000 times than that in the sedimentation tank. Therefore, the very small particles can be separated in short time. In addition, the centrifugal filter has some advantages, compared with the other technologies.

① Solid content is high.

② Turbidity of the centrifugal liquid is low.

③ Treatment amount of sludge is big.

④ Investment on the construction is low.

⑤ Occupied area is small.

⑥ Working environment is clean.

⑦ Running is simple.

⑧ No addition or low addition coagulant for conditioning.

The problem of the technology is the higher consumption on energy.

(4) Vacuum filter The advantages of the vacuum filter are as followed.

① Continuously running.

② Running stable.

③ Automatic control.

The problems are as followed.

① Many affiliated equipment.

② Complex processes.

③ Higher running cost.

6.7 Sludge drying and incineration

The heat drying is to remove most capillary water, adsorbed water and bound water. The water content is reduced from 60% ~ 80% to 10% ~ 30%. While the incineration is to remove all the adsorbed water and internal water in the sludge and organic matter, and then the water content is reduced to 0 and changed into ash.

Generally, the investment on the equipment and the running cost are higher. Nowadays, the incineration is only used for the industrial sludge and municipal waste. The following conditions should be considered to use incineration.

① The higher content of the toxic matter and no use for the agriculture.

② The land shortage in the big cities.

③ The higher caloricity of the sludge.

④ Producing power by mixing with the municipal waste.

6.8 Comprehensive utilization and ultimate disposal of the sludge

Ultimate disposal of sludge is the last process, including agricultural fertilizer, building materials, landfill, land use, compost and thermolysis.

6.8.1 Agricultural fertilizer

There are lots of nutrients for plants and organic humus changing the properties of the soil, that the sludge is used as agricultural fertilizer is the best way for the ultimate disposal. However, the parasitic ovum, virus, pathogenic bacterium and heavy metals should be removed or stabilization or compost before the use of fertilizer. For the sludge without digestion, it is corrupt due to the high content of the organic matter. In addition, the water content of the sludge is high, it is difficult to be fertilization.

6.8.2 Building materials

The sludge can be used as the brick and the fibreboard.

The dry sludge can be used to make brick. Generally, the sludge is mixed with the clay to regulate the component of the sludge similar as the brick.

The ash can be used to make brick directly. The suitable ratio for the brick should be 100 : 50 : (25~20) for the ash : clay : silica sand.

The coarse protein (organic matter) and globulin (enzyme) are dissolved in the water, diluted acid, diluted alkali and the water with neutral salt, so the sludge can be used to make the fibreboard. Under the alkali conditions, the properties of the protein change after the heating, pressure. This is called denaturation of protein. The sludge is formed into sludge resin, and then the resin with the dumped fiber, which is decolored and degreased, can be made into the fibreboard

6.8.3 Landfill

Landfill is referred to microbial decomposition. The organic matter in the sludge can be decomposed by the microorganisms in the nature and then sludge volume decreases to meet the demand of sludge reduction and stabilization. Landfill is a mature technology with the following advantages, without highly dewatering, lower investment, bigger capacity, working faster. However, there are some disadvantages. The hazardous matters in the sludge can pollute the groundwater and surrounding soil by penetrating into the soil. It is necessary to prepare bigger ground. The requirements are strict about mechanical properties of soil.

6.8.4 Land use

Sludge can be used as fertilizers or soil improvement material for garden, planting, forestry or agriculture. This is called land use of sludge.

The biggest advantage of land use of sludge is to transform organic matter in the sludge into nutrient of soil. In addition, the rich nutrient of N, P, K and the necessary trace element of Ca, Mg, Zn, Cu, Fe can be used as improvement material to improve the soil by upgrading soil composition, increasing soil fertilizer, promoting growth of plants, accelerating chlorophyll formation and avoiding of plant diseases and insect pests. The other advantages are as following, lower investment, lower energy consumption, lower running fee, and so on.

It is noticed that there is danger for land use of sludge. The first is salt. Once there is higher concentration of salt in the soil, it will upgrade soil electrical conductivity and destroy nutrient homeostasis. The second is pathogen. If the sludge has not been treated before land use, there is lots of pathogen and parasite. This will affect adversely soil, water resource and air during the use of sludge. The third is heavy metals. The forth is organic matter, especially the nondegradable organic matter, such as pesticide, polycyclic aromatic hydrocarbon. Therefore, environmental risk management and control during the whole time is necessary for land use of sludge (Figure 6-5).

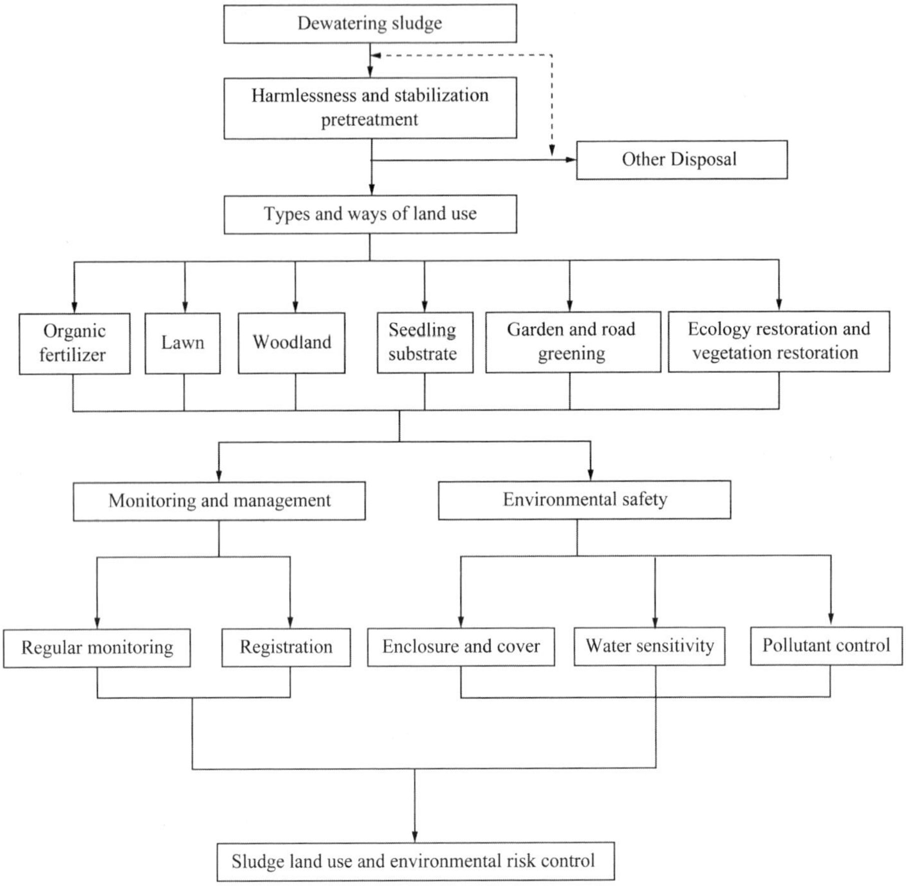

Figure 6-5 Land use and environmental risk control

6.8.5 Compost

Compost is referred to high-temperature fermentation of microorganisms in the sludge. During the compost, the sludge can be mixed with rising agent and conditioning agent at some ratio. The microorganism can oxidatively decompose organic matter in the sludge and transform them into huminoids. Therefore, compost can be attributed to sludge recycling utilization. The advantages of compost are as following: low carbon, conserve energy, low investment, low running fee. Compost can be used as the pretreatment of land use and landfill.

Some notices should be focused during the compost. The first is to shatter sludge into pieces and the particle size of broken sludge is not bigger than 2 cm. The aim is to ensure complete compost without blackheart phenomenon. This is the key for compost. The second is that the sludge should be fully mixed with supplementary material. This is to ensure even fermentation by increasing specific surface area of material and the utilization efficiency.

6.8.6 Thermolysis

Thermolysis is thermal decomposition of sludge under inert gas at atmospheric pressure, transforming organic matter in the sludge into hydrocarbon. Pyrolytic product includes biochar, tar, pyrolytic air and water. It is well known that tar and pyrolytic air can be used as fuel.

Thermolysis can transform organic matter in the sludge from macromolecular matter to small molecule substance under high temperature and anaerobic conditions. The solid form is changed to liquid form and air form. That is to say, the sludge is changed to solid product and air product. Compared with other sludge disposal, thermolysis has following advantages. Non-harm effect is better. Organic matter in the sludge can be decomposed into small molecule substance and carbon. Nitrogen and Sulphur pollutants can be transformed into small molecule oxidizing material. Heavy metals can react with silicon (Si), oxygen (O), aluminum (Al) etc. to form inorganic matter with the characteristics of high stability, complex structure, uninsolution. Therefore, the heavy metals can be fixed in the pyrolytic products. In a word, thermolysis can delete completely organic pollutants and heavy metals in the sludge. Occupy area is small. The main reason is high reaction efficiency and short reaction time under high temperature conditions. Resource efficiency is high. Pyrolytic solid product is a kind of material with high specific surface area and lots of holes. The adsorption capacity of solid product is very good so that the material can be used as adsorbent or soil conditioner. Pyrolytic air product can be used as fuel after purification and its thermal value is similar as gas. Therefore, pyrolytic products can be utilized as resource product to offset the cost of sludge thermolysis.

Questions

1. What is the principle of sludge treatment?

2. What is the basic method for the sludge treatment?

3. In a waste water treatment plant, the water content is about 99.99%, after dewatering, the water content is reduced to 75%, how to calculate the reduction of sludge volume. If the water content is reduced to 55%, what about the sludge volume, please give the reason.

4. What is the resistivity?

5. What are the kinds of water in the sludge.

6. Please clarify the functions of different chemical dewatering.

7. What is sludge digestion? What is the mechanism for sludge digestion.

8. What is sludge conditioning, what is the aim of sludge conditioning.

9. There are many types of sludge utilization, presenting some examples and showing the mechanism.

10. In your opinion, if you want to use compost as sludge utilization for the residual sludge in a domestic waste water treatment plant, what should be noticed?

Chapter 7 Waste water treatment cases

Emphasis: The component of typical cases of waste water treatment. The parameters of treatment cases.

In practice, the component of the municipal waste water is the domestic waste water, industrial waste water and runoff. The domestic waste water is the main part of the municipal waste water. The water quality is as Table 7-1.

Table 7-1 Typical water quality in the domestic waste water

No.	Index	Content/(mg/L)			No.	Index	Content/(mg/L)		
		high	mean	low			high	mean	low
1	Total solid	1,200	720	350	16	Biodegradable COD	750	300	200
2	Dissolved solid	850	500	250	17	Dissolved biodegradable COD	375	150	100
3	Non-volatile solid	525	300	145	18	Suspended biodegradable COD	375	150	100
4	Volatile solid	325	200	105	19	Total nitrogen	85	40	20
5	Suspended solid	350	220	100	20	Organic nitrogen	35	15	8
6	Non-volatile suspended solid	75	55	20	21	Ammonia nitrogen	50	25	12
7	Volatile suspended solid	275	165	80	22	Nitrite	0	0	0
8	Settling matter	20	10	5	23	Nitrate	0	0	0
9	BOD_5	400	200	100	24	Total phosphorus	15	8	4
10	Dissolved BOD_5	200	100	50	25	Organic phosphorus	5	3	1
11	Suspended BOD_5	200	100	50	26	Inorganic phosphorus	10	5	3
12	Total organic carbon	290	160	80	27	Chloride (Cl^-)	200	100	60
13	COD	1,000	400	250	28	Alkalinity ($CaCO_3$)	200	100	50
14	Dissolved COD	400	150	100	29	Oil	150	100	50
15	Suspended COD	600	250	150					

7.1 Traditional activated sludge

The traditional activated sludge is used widely in China, and it is suitable for the big, middle and small waste water treatment plant. The main characteristic is the resistance of the shock load and the efficiency is stable and the designment and running experience is rich.

Take Binhe waste water treatment plant in Shenzhen as an example.

The fluctuation of the water quality in Binhe waste water treatment plant is big. The BOD_5 is 80~450 mg/L, and the mean is 190 mg/L. The SS is 120~240 mg/L, and its mean is 146 mg/L. The COD is 200~2,000 mg/L, and its mean is more than 380 mg/L.

The process is showed in the Figure 7-1.

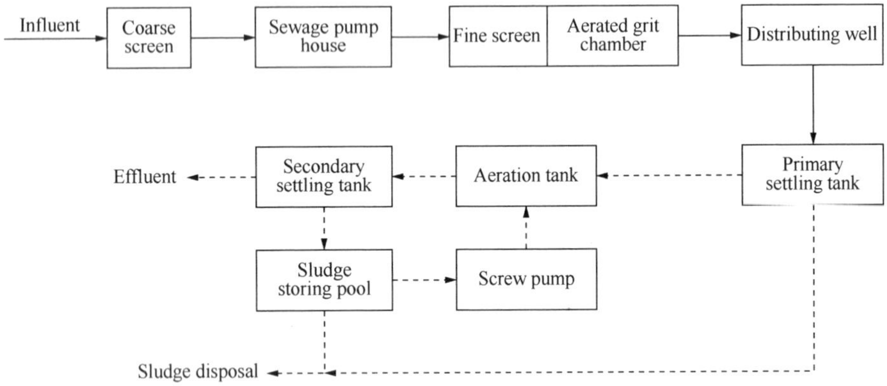

Figure 7-1 The flow chart of Binhe waste water treatment plant

(1) Coarse screen The gap is between the grid is 20 mm.

(2) Aerated grit chamber There is the fine screen before the aeratied grit chamber. The gap is between the grid is 10 mm.

(3) Primary settling tank There are two tanks for the primary sedimentation. They are the circular radical sedimentation basin with the diameter of 25 m. The depth of the water is 3.14 m. The settling time is 1.5 h. The removal of suspended solid is 60% and the removal of BOD_5 is 25%~30%.

(4) Aeration tank There are two tanks as the aeration tank. Each tank has 4 galleries. Total effective volume is 8,350 m^3. The depth is 6 m. The retention time is 8 h. The sludge load is 0.2 $kgBOD_5/(kgMLSS \cdot d)$. The two tanks are parallel running.

(5) Secondary settling tank There are two tanks for the secondary sedimentation. They are the circular radical sedimentation basin with the diameter of 30 m. The depth of the water is

3.97 m. The settling time is 21.5 h.

(6) Backflow sludge pump station There are three screw pumps for the backflow sludge. The backflow rate of the sludge is 85%. No reserve.

(7) Belt Filter Press is used for the sludge dewatering The removal of BOD_5, SS is more than 90% and the removal of COD_{Cr} is more than 80%.

7.2 Anaerobic-aerobic activated sludge process

7.2.1 Anaerobic-aerobic activated sludge process for removal P

The first waste water treatment plant in Dalian city is the earliest demonstration project using the A/O_P process in China. The total amount of the plant is 6×10^4 m^3/d. The first stage is 3×10^4 m^3/d. The raw water comes from the domestic and industrial waste water in the urban built-up area. The water quality is as followed. BOD_5, 200 mg/L, SS, 250 mg/L, COD_{Cr}, 500 mg/L, TN, 30~40 mg/L, TP, 3~5 mg/L. The process is observed in the Figure 7-2.

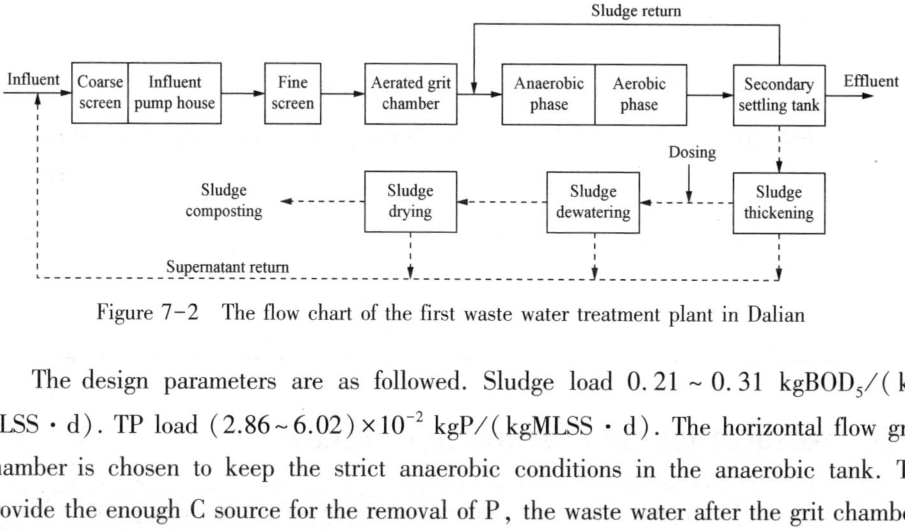

Figure 7-2 The flow chart of the first waste water treatment plant in Dalian

The design parameters are as followed. Sludge load 0.21 ~ 0.31 $kgBOD_5/(kg\ MLSS \cdot d)$. TP load $(2.86\sim6.02)\times10^{-2}$ $kgP/(kgMLSS \cdot d)$. The horizontal flow grit chamber is chosen to keep the strict anaerobic conditions in the anaerobic tank. To provide the enough C source for the removal of P, the waste water after the grit chamber comes into directly the anaerobic stage in the anaerobic-aerobic process. To reduce the P load in the backflow sludge, the P from the overlying water after the thickening and after dewatering was removed and returned to the raw water. The BOD_5 in the effluent is about 10~15 mg/L. The TP in the effluent is less than 0.5 mg/L. The SS in the effluent is about 10 ~ 20 mg/L. The efficiency is better than that in the traditional activated sludge.

7.2.2 Anaerobic/anoxic/aerobic process

The fourth waste water treatment plant was built at 1996. The treatment amount is 10×10^4 m^3/d. The process is anaerobic/Anoxic/aerobic process (Figure 7-3)

The design influent quality:

$BOD_5=120$ mg/L, SS=220 mg/L, $NH_3-N=25$ mg/L, TP=8 mg/L.

The design effluent quality:

$BOD_5<20$ mg/L, SS<20 mg/L, $NH_3-N<15$ mg/L, TP<1 mg/L.

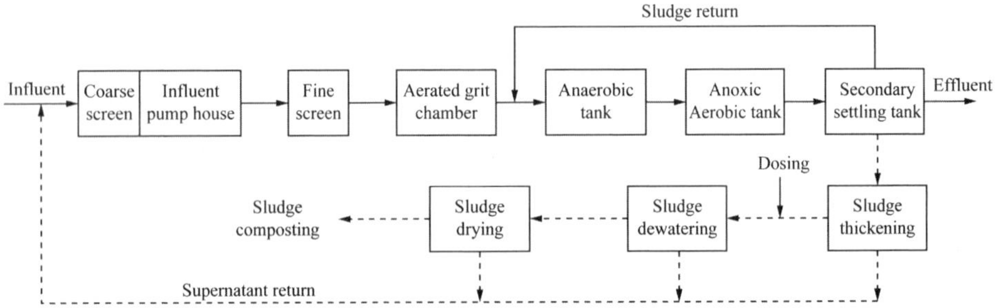

Figure 7-3 The flow chart of the fourth waste water treatment plant in Guilin

The removal of TP is not good according to the practice (Table 7-2).

Table 7-2 The influent and effluent quality in the fourth waste water treatment plant

mg/L

Item	Influent	Effluent	Item	Influent	Effluent
BOD_5	125	4.3	TN	25	6.3
SS	200	4.4	TP	15	8.0

7.3 Adsorption/biodegradation process

The treatment amount of waste water in Zibo waste water treatment plant is 14×10^4 m^3/d. The biggest flow amount is 7,233 m^3/d and the mean is 5,833 m^3/d. The process is as followed (Figure 7-4).

The influent and effluent quality in the waste water treatment plant is showed in the following table.

Table 7-3 The influent and effluent quality in the waste water treatment plant

mg/L

Items	BOD_5	COD	SS	TN	NH_4^+-N	TP
Influent	221~504	526~1,142	274~787	59~117	40~97	9~56
Effluent	10~25	80~120	18~35	40~80	20~60	5~15

The design parameters for aeration are as followed.

(1) A stage aeration tank Sludge load 4.5 $kgBOD_5/(kgMLSS \cdot d)$. Volume load 10 $kgBOD_5/(m^3 \cdot d)$. The retention time 36 min (the mean flowing amount) and 29 min (the biggest flowing amount). MLSS 1.5~2.0 kg/m^3. DO 0.3~0.5 mg/L. Sludge age 4~4.5 d. Sludge backflow ratio 50%~75%. Sludge volume index 60 mL/g. The removal of BOD_5 50% (the adsorption is 85%, and the biodegradation is 15%). O_2/C load 0.4 $kgO_2/BOD_{5\,removal}$. The component of the activated sludge is aerobic bacteria and facultative bacteria.

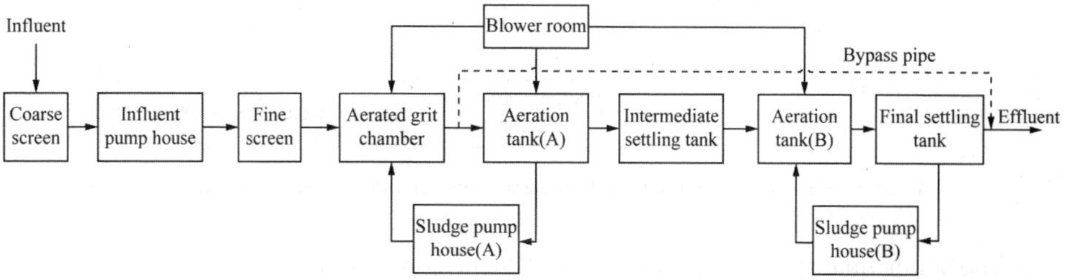

Figure 7-4 The flow chart of the waste water treatment plant in Zibo

(2) B stage aeration tank Sludge load 0.125 $kgBOD_5/(kgMLSS \cdot d)$. Volume load 0.525 $kgBOD_5/(m^3 \cdot d)$. The retention time 6.26 h (the mean flowing amount) and 5.05 h (the biggest flowing amount). MLSS 3.45 kg/m^3. DO 0.7~2.0 mg/L. Sludge age 21 d. Sludge backflow ratio 100%. Sludge volume index 60 mL/g. O_2/C load 0.4 $kgO_2/BOD_{5\,removal}$. O_2/N load 4.4 $kgO_2/NH_{4\,removal}^-$. The component of the activated sludge is strict aerobic bacteria and protozoa.

The practice shows that the efficiency of the AB on the waste water with high content organic matter is excellent. The saving on the energy is more than 20%~30%. Now, the top consumption on the energy is 0.4 kW · h/m^3. The biogas power generation can be used for the aeration in summer and autumn. The energy consumption is reduced to 0.25~0.30 kW · h/m^3. The energy consumption on BOD_5 is 1.6 kW · h/m^3 and 1.0~1.2 kW · h/m^3.

7.4 Oxidation ditch process

7.4.1 A–E oxidation ditch

The design amount is 6×10^4 m^3/d in the Niukouyu waste water treatment plant in Yanshan petrochemical corporation. The waste water is mainly the drainage during the process of making ethylene and the domestic waste water from the district. The main process is Orbal oxidation ditch which is a kind of new technology with the removal N and P (Figure 7–5).

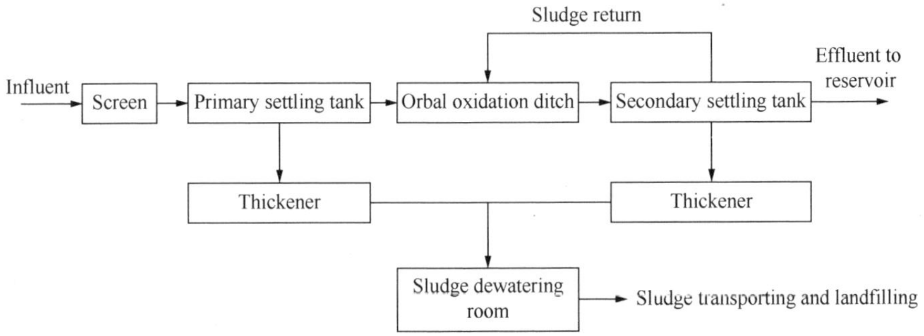

Figure 7–5 The flow chart of the waste water treatment plant in Niukouyu

The water quality in the influent and effluent is as the Table 7–4.

Table 7–4 The water quality in the influent and effluent

Item	Influent	Effluent	Item	Influent	Effluent
COD_{Cr}	300	80	Oil	40	4
BOD_5	180	25	pH	6~9	6~8.5
SS	200	20			

The process included parallel treatments. Each one included one Orbal oxidation ditch and two radial flow sedimentation tanks. There are 24 aeration discs and the outer ditch, the inner ditch and the middle ditch have 8 aeration discs, respectively. The design parameters for each oxidation ditch are as followed.

Design inflow amount 1,250 m^3/h

Hydraulic retention time 14.2 h

SRT 35 d

Effective water depth 4.26 m

MLSS 4,000 mg/L

Sludge load 0.074 $kgBOD_5/(kgMLSS \cdot d)$

Volume distribution 56 : 26 : 18 (outer : middle : inner)

DO 0-1-2 mg/L (outer : middle : inner)

The running parameters are showed in the Table 7-5.

Table 7-5 The running parameters for the Niukouyu waste water treatment plant

Oxidation ditch		Average	Range
Influent flow/(m^3/h)		903	851~937
HRT/h		19	18~21
Effective water depth/m		4.2	
Water temperature/℃		15	13~26
Running aeration discs	Outer ditch	5	
	Middle and inner ditch	3	
The depth of the aeration disc/mm		450	
MLSS/(mg/L)		3,037	2,923~3,245
Backflow ratio/%		61	59~65
SV/%		93	91~94
DO/(mg/L)	Outer ditch	0	0~0.3
	Middle ditch	0.4	0.1~0.9
	Inner ditch	3.5	2.9~3.9

The efficiency is showed in the Table 7-6.

Table 7-6 The efficiency for the Niukouyu waste water treatment plant

Item	Influent/(mg/L)	Oxidation ditch/(mg/L)				Effluent/(mg/L)	Removal in ditch/%	Total removal/%
		Influent	Outer	Middle	Inner			
COD_{Cr}	455	396	37	29	24	28	92	93
BOD_5	—	197	5	4	3	3	>95	—
SS	58	31	—	—	—	12	—	79
NH_3-N	11.82	11.55	—	—	—	—	>99	>99
TKN	—	16.09	1.89	1.09	0.95	0.98	—	—
NO_3^--N	—	1.57	0.72	0.61	0.60	1.39	—	—
TN	—	17.44	1.96	1.26	1.18	1.43	92	—
pH	8.0	—	—	—	—	8.0	—	—

7.4.2 T oxidation ditch

The design amount is 25×10^4 m^3/d in Luofang waste water treatment plant in Shenzhen city. The water quality in the influent and effluent is showed in the Table 7-7.

Table 7-7 The water quality in the influent and effluent

Item	Influent	Effluent	Remark
COD_{Cr}/(mg/L)	250~400	≤60	
BOD_5/(mg/L)	150	≤20	
SS/(mg/L)	150	≤20	
NH_3-N/(mg/L)		≤15	
TP/(mg/L)	4		
PO_4^{3-}-P/(mg/L)		≤0.5	
TN/(mg/L)	30		
pH		6.5~9	
Water temperature/℃	14~28		
Water content/%			≤80

The main process is three-channel oxidation ditch (Figure 7-6). There is the anaerobic tank before the oxidation ditch to improve the P removal. There is the backflow sludge thickening tank before the anaerobic tank to keep the strict anaerobic conditions in the anaerobic tank. HRT is 30 min in the anaerobic tank. The flow in the anaerobic tank is circular plug flow and there is the blender under the water.

The running of the equipment in the oxidation ditch is controlled by the period. Each period is divided into 6 stages, which are showed in the Figure 7-7.

(1) A stage The running time is 1.5 h. The waste water comes into the ditch Ⅰ which is under the anoxic conditions. The submerged blenders are running and the aeration discs are closed. The denitrification is finished in the ditch Ⅰ. A part of mixed liquid in the ditch Ⅰ goes to the ditch Ⅱ and the other part of mixed liquid in the ditch Ⅰ is discharged as the backflow sludge. The submerged blenders and the aeration discs are running. The nitrification is finished in the ditch Ⅱ. The aerobic mixed liquid in the ditch Ⅱ goes to the ditch Ⅲ. The ditch Ⅲ is in the settling and discharge condition. In the ditch Ⅲ, the submerged blenders and the aeration discs are closed.

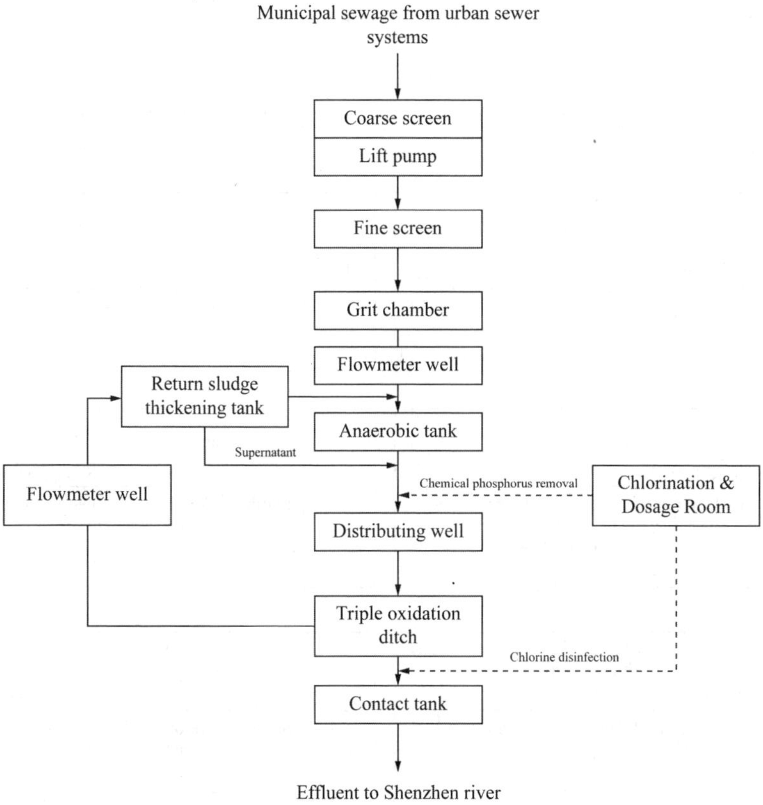

Figure 7-6 The flow chart of Luofang waste water treatment plant in Shenzhen

(2) B stage The running time is 1.5 h. The waste water comes into the ditch Ⅱ which is under the aerobic conditions. The submerged blenders and the aeration discs are running in the ditch Ⅰ. The mixed liquid in the ditch Ⅱ comes into the ditch Ⅰ and the ditch Ⅲ. The ditch Ⅲ is in the settling and discharge condition.

(3) C stage The running time is 1 h. The waste water comes into the ditch Ⅱ which is under the aerobic conditions. The submerged blenders and the aeration discs are running in the ditch Ⅱ. A part of mixed liquid in the ditch Ⅱ goes to the ditch Ⅲ and the other part of mixed liquid in the ditch Ⅱ is discharged as the backflow sludge. In the ditch Ⅰ, the submerged blenders and the aeration discs are closed and the ditch is under the pre-sedimentation conditions. The residual sludge is discharged from the ditch Ⅰ. The ditch Ⅲ is in the settling and discharge condition.

For the D stage, E stage and F stage, the running model is similar as the A stage, B stage and C stage, while the difference is the change of the running of the ditch Ⅰ and the ditch Ⅲ.

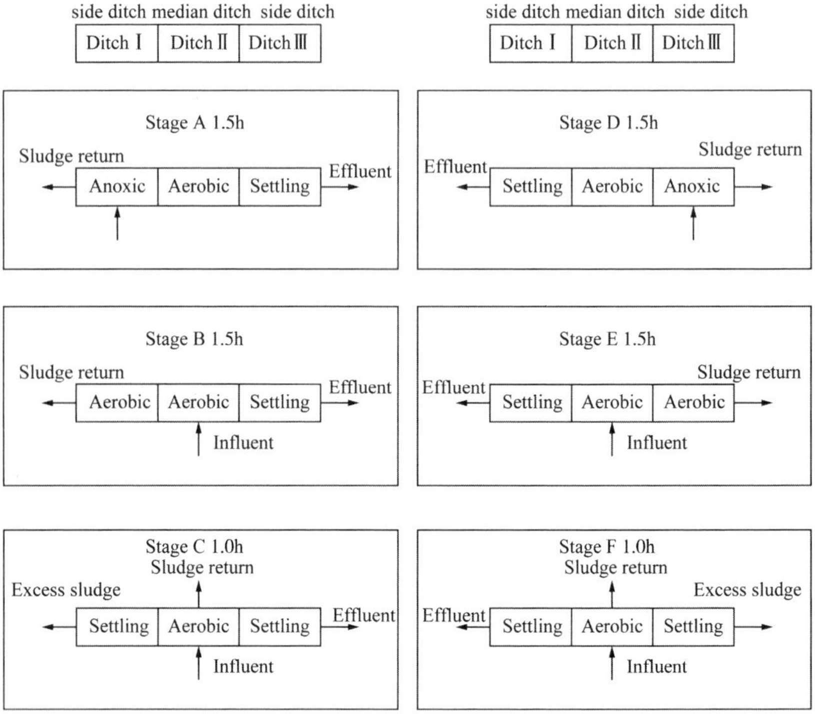

Figure 7-7 The running of Luofang waste water treatment plant

For the designment, the HRT in the anaerobic tank and the oxidation ditch is 0.65 h and 15.2 h, respectively. The removal of N and P is good (NH_3-N<1 mg/L and TP< 0.7 mg/L in the effluent).

7.5 Biological aerated filter

The amount is 8×10^4 m^3/d for the second stage project in Malanhe waste water treatment plant in Dalian city. The main waste water is domestic waste water and a small quantity of industrial waste water. The main process is the high density sedimentation tank and the BIOFOR with pre-denitrification. The figure is as followed (Figure 7-8).

The running is showed in the Table 7-8.

Table 7-8 The water quality in the influent and the effluent

Item/(mg/L)	BOD_5	COD_{Cr}	SS	TN	NH_3-N	TP
Influent	80~110	180~340	100~200	25~35	20~30	2.5~4.5
Effluent	3~10	20~40	2~12	5~15	0.5~7	0.4~1.5

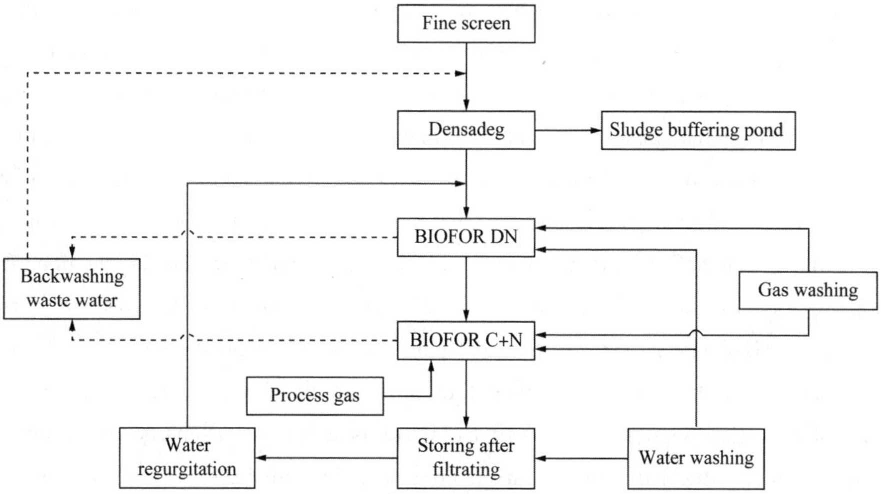

Figure 7-8 The flow chart of Malanhe sewage treatment plant in Dalian

7.6 A domestic example of reconstruction and extension of sewage plant

7.6.1 Introduction

7.6.1.1 Building background

Shanghai Bailonggang Urban Sewage Treatment Plant is located in Heqing, Pudong New area, east of the main waterway of the Yangtze River. It is the terminal sewage treatment plant of the Shanghai Municipal sewage treatment Phase Ⅱ Project, which mainly solves the combined sewage from Huangpu, Luwan and Xuhui, which are the central district of Puxi, as well as the discharge way out of the most of the domestic sewage and industrial waste water in the south of Zhaojiagou and the north of Chuanyang river in Pudong New area.

The main parts of the sewage treatment project, which has been completed, are the eastern section of the middle pipe line, the west section of the middle pipe line in

Longdong Avenue, the connecting pipe in Luoshan Road, the west section of the South pipe line in the west of Luoshan road, the south branch line, the SA, SB, M1, M2 lifting pump station and Bailonggang Urban Sewage Treatment Plant and its discharge pipe. The project also includes Puxi intercepting facility, intercepting branch pipe, intercepting trunk pipe and crossing river pipe of Longhua-airport-south line, etc.

The main sewage pipe in Bailonggang area is composed of the south line and the middle line. The central line includes the eastern part of the central line, the western part of the central line and the south main line. The average sewage volume in the dry season was 172.1×10^4 m^3/d at that time. The south line includes the eastern section of the south line and the west section of the south line. The west section of the south line was built firstly, and transported to Bailonggang sewage plant through the South and middle line connecting pipes of the Luoshan Road extension line and M2 pump station in the middle line. After completing the east section of south line in 2014, the sewage of south line can be sent directly to the Bailonggang sewage treatment plant by the SB pump station .

According to the water environment planning of the Yangtze River Estuary, the estuary of Shanghai section of the Yangtze River is a sewage discharge area, and the sewage is directly drained into the mixed area of 2 km^2 after the pretreatment of dregs and sands.

In 2002, in order to reduce the frequent occurrence of red tide in the Yangtze Estuary, on the basis of the pretreatment plant, the strengthening first grade facility of 120×10^4 m^3/d in Bailonggang was renewed with the world bank loan. It adopts the high efficiency sedimentation tank as the main physicochemical treatment process. It aims at reducing the P, and reduces incidence of red tides by changing the proportion of N and P in addition to P.

After the sewage treatment plant completed, all the sewage treated by the high efficiency sedimentation tank reached the design index, which was closely met the second grade of the national discharge standards, a large number of direct discharge of the Yangtze River pollutants was reduced, the water environment of the Yangtze Estuary was effectively protected, an active role was played in the environmental protection of Shanghai.

With the rapid increase of the volume of sewage, the actual volume of water in 2007 is stable at $160\sim180\times10^4$ m^3/d, the peak flow rate is more than 200×10^4 m^3/d, which is far beyond the design capacity, and the effluent is only for phosphorus removal, which can't meet the second grade of the national discharge standards of *Discharge standard of pollutants for municipal waste-water treatment plant* (*GB* 18918—

2002). Therefore, since 2007, upgrading and extension of the factory was included in the Shanghai third round of three year environmental action plan. The scale of upgrading and modifying project is 120×10^4 m^3/d, and the scale of extension project is 80×10^4 m^3/d, the total scale of the plant reached 200×10^4 m^3/d. At the end of September 2008, the upgrading and extension project of Bailonggang sewage treatment plant was put into operation. The maximum daily handling capacity is more than 200×10^4 m^3/d.

According to the situation of planning and forecasting water consumption growth, in 2015, Bailonggang sewage quantity is 280×10^4 m^3/d, the planned sewage volume was 350×10^4 m^3/d in the long term in 2020. Therefore, after the upgrading and extension project of the Bailonggang sewage treatment plant, Shanghai restarted the construction of the second period of the extension of the Bailonggang sewage treatment plant. The second period of extension project was completed at the end of 2012 and the project was officially opened in February 2013, the scale of the project is 80×10^4 m^3/d.

7.6.1.2 Present situation of sewage treatment Plant before revamping

Shanghai Bailonggang sewage treatment plant was built in 1999. It was originally a pretreatment plant matched with the second period project of sewage treatment. It has thick grid, fine grid, rotational flow grit chamber and other pre-processing facilities, and equipped with 29.67 m^3/s export pump station and deep water drain pipe. Bailonggang sewage treatment plant began to rebuild the pretreatment plant and built a first level intensified plant in 2002. It was put into operation in October 2004. It adopts the first grade of chemical phosphorus removal process to remove phosphorus as the main target and retain the possibility of upgrading the two grade treatment. The design scale was 120×10^4 m^3/d in the dry season, the peak flow rate in dry season 18.06 m^3/s, and the peak flow rate of 21.85 m^3/s in the rainy season.

The design water qualities are shown in Table 7-9.

Table 7-9 The design of treatment plant for influent and effluent qualities

mg/L

Items	COD_{Cr}	BOD_5	SS	NH_3-N	PO_4-P
The design for influent qualities	320	130	170	30	5
The design for effluent qualities	180	70	40	30	1

The sewage treatment plant is filled with the cofferdam on Xiangyangyu. The design ground elevation is 4.40 m, and the surrounding is separated by two embankments from inside and outside and other beaches, and the top of the dyke is about 8.00 m. The whole plant can be divided into three parts: pretreatment plant, physicochemical treatment area and export pump station.

The pretreatment plant consists of a total water well, a metering well, a thick grid, a fine grid, a rotational flow grit chamber, and the sewage after the sanding is remitted into the water outlet canal and flows to the first grade strengthening treatment structure.

The first grade enhanced physicochemical treatment area, that is, the construction of the first phase of the Bailonggang sewage treatment plant, mainly consists of 3 high efficient sedimentation tanks, a dosing rooms and pharmacy store, a sludge storage tank, a sludge dehydrating machine room, a deodorizing room and a group of reclaimed water reuse facilities of 2,500 m^3/d.

The export pump station is located on the northeast side of the pretreatment plant, on the edge of the Yangtze River, covering an area of 3.6 hm^2, separated by an independent wall, and it was connected to the preconditioning plant thorough a special road. There is one storm water pump, one export pump, one high outlet well and one comprehensive building in this area. The sewage from the intersection well is pumped into the Yangtze River through the high position well and the ϕ4,200 mm drain pipe.

Phosphorus removal is the main goal in the first stage of construction. Therefore, it's strict to control the phosphorus in effluent indicators, and it's not strictly required for BOD, SS and ammonia nitrogen.

Sludge is treated by using high centrifugal dewatering machine, until the dehydrated sludge water content was 75%, then it was disposed by sanitary landfill. The factory has a 27 hm^2 sludge landfill, covering a total area of 30 hm^2. The sludge from Bailonggang and Zhuyuan sewage plant was concentrated and disposed for a period of time, and then used as garbage covering soil and urban greening soil. The sewage sludge treatment process is shown in Figure 7-9.

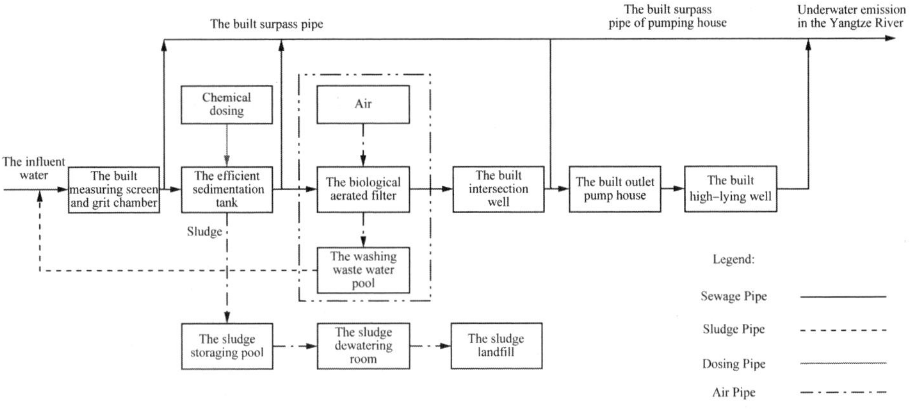

Figure 7-9 The flow chart of the existing process of Bailonggang sewage treatment plant

The influent of the first grade strengthening plant of Bailonggang comes from the middle line and south trunk line of the second period of sewage treatment project, and another part of the sewage is brought into the middle line through the connecting pipe of the west section of south line and the middle line.

The south line and the central line mainly collects the confluent sewage from the central urban area. Because of the relocation of the factories in the central urban area and the long distance transportation of sewage, it has brought more groundwater infiltration along the way, and the water quality is lower than the old sewage plant in the central urban area. The BOD is less than 100 mg/L, NH_3-N is less than 25 mg/L and SS is less than 80 mg/L, and the water quality of initial runoff flow is changed greatly at rainy days. The concentration is even higher than that in the dry season. The south trunk line collected a large volume of Pudong New Area sewage along the way, including a large number of industrial waste water, with low water quality and high chroma, and there were varied colors in different periods. The sewage concentration in the south trunk line is 40% ~ 50% higher than the middle line, NH_3-N even up to 50 mg/L, and sometimes the maximum of COD_{Cr} is more than 1,000 mg/L, The quality of sewage from Wang Gang Industrial Area in Pump Station No. 5 is poor, and the concentration of organic pollutants is high. The variation range of TP is 5 ~ 10 mg/L. According to the operation data of the Bailonggang sewage treatment plant, the actual influent of the former treatment plant is stable at $160 \sim 180 \times 10^4$ m^3/d, and the volume of water is more than 200×10^4 m^3/d at the peak, which has exceeded the design scale of the first grade strengthening treatment plant of Bailonggang.

7.6.2 Renovation and extension scheme of sewage plant

7.6.2.1 The upgrading and modifying of Bailonggang sewage treatment plant

(1) Project scale The target of the upgrading and modifying project of the Bailonggang sewage treatment plant is the first grade strengthening plant with the average flow rate of 120×10^4 m^3/d in the dry season, and the total change coefficient of the peak flow rate in the dry season is 1.3. Upgrading and modifying projects are still designed according to 29.67 m^3/s in the rainy season. The 80×10^4 m^3/d sewage exceeding 120×10^4 m^3/d is incorporated into the extension project, and sludge treatment is included in the sludge treatment and disposal project of Bailonggang waste water treatment plant.

(2) The design for influent and effluent qualities The influent water quality was determined according to the 80% frequency of 1997 ~ 2006 years' water qualities, and

the effluent quality was carried out in the second grade of the national discharge standards of *Discharge standard of pollutants for municipal waste water treatment plant* (*GB* 18918—2002), as shown in Table 7-10.

Table 7-10 The design effluent qualities list of upgrading and modifying project

Items	COD_{Cr}/ (mg/L)	BOD_5/ (mg/L)	SS/ (mg/L)	NH_3-N/ (mg/L)	TP/ (mg/L)	pH	LAS/ (mg/L)	Petroleum/ (mg/L)	Coliform group/ (CFU/L)
Effluent	≤100	≤30	≤30	≤25(30)①	≤3	6~9	≤2	≤5	$\leqslant 10^4$

① the value in brackets is the control index of water temperature <12℃, LAS and oil need to control the source.

(3) The content of modifying project The modifying project mainly includes the upgrading of the existing 120×10^4 m^3/d first grade strengthening treatment process, the transformation of the existing pretreatment plant and the first grade strengthening treatment section, building new biological treatment facilities, the improvement of the public works, the supporting facilities and the automatic control of the whole plant, the water, gas and mud supervision of the whole plant.

When the new construction is designed, it takes into account the large scale of the project. If the traditional design ideas and patterns are adopted, the number of structures is large, the distribution is loose, and the connection pipe between the structures is complex, which is not conducive to the inspection and management of the factory. Therefore, the layout of the group was used, that is, with 30×10^4 m^3/d scale as a group, concentrates on the arrangement of the initial sedimentation tank, the AAO biological reaction tank, the secondary sedimentation tank and the sludge pump house.

(4) Process flow The modifying project makes full use of the existing facilities such as high efficiency sedimentation tank and so on. The pretreating sewage is introduced into the newly built AAO biological reaction sedimentation tank and the first grade enhanced high efficiency sedimentation tank respectively. The effluent will be treated with the ultraviolet ray disinfection. After reaching the national secondary standard, the effluent can be connected to the original intersection well, and discharged from the outlet pump house and the high level well. The sewage treatment processes is shown as Figure 7-10.

The sewage biological treatment adopts the multi-mode AAO process, which can run in the normal or inverted AAO mode according to the needs and seasonal changes, and can also run according to the water quality and treatment requirements according to the nitrogen and phosphorus removal mode or only the phosphorus removal mode. The flow process is shown in Figure 7-11.

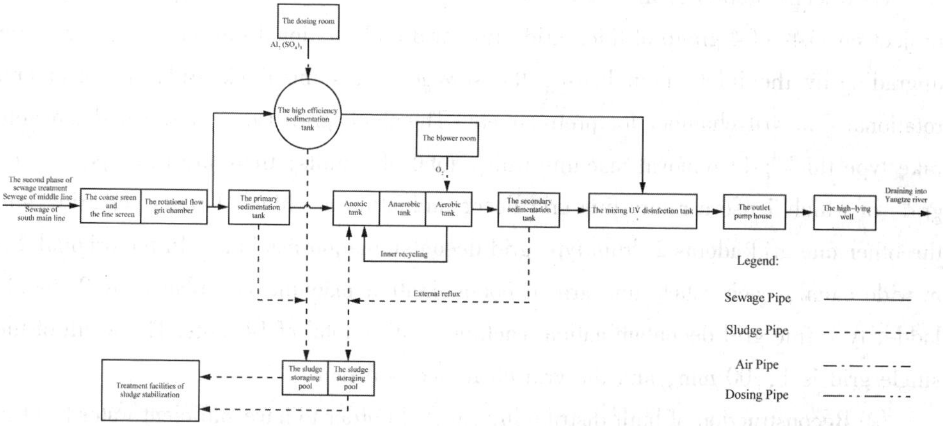

Figure 7-10 The process flow chart of the upgrading and modifying project of Bailonggang sewage treatment plant

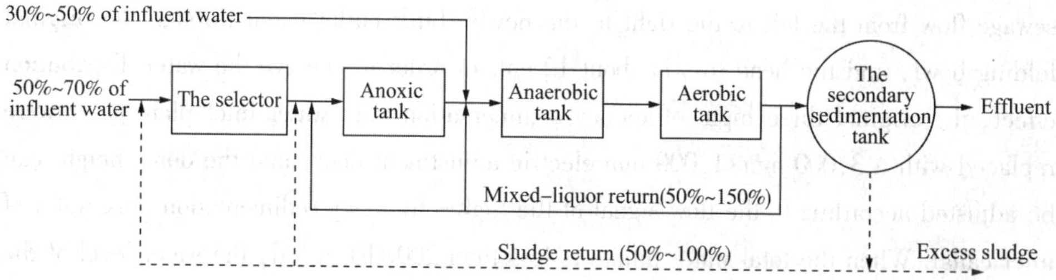

Figure 7-11 The process flow chart of multi-mode AAO process (inverted AAO mode)

Under this operating mode, the influent is divided into two points, the intake of the selected area is 70% ~ 50%, and the intake of the anaerobic area is 30% ~ 50%. The distribution ratio is adjusted according to the phosphorus content of the effluent. If the phosphorus removal efficiency needs to be strengthened, more water will enter the front part of the anaerobic zone, and more organic carbon sources will be given to phosphorus removal bacteria, and vice versa. The inner reflux canal delivers the effluent nitrate to the beginning of the anoxic segment, providing the raw material for denitrification. The external reflow canal sent the activated sludge condensed from the secondary sedimentation tank to the selected area and the anoxic area. The selected area could effectively inhibit the excessive reproduction of filamentous bacteria and prevent the sludge bulking.

(5) The design of main structures.

① Reconstruction of the thick grid, fine grid and rotational flow grit chamber. The project consists of 4 group of thick grid, fine grid and rotational flow grit chamber. After upgrading by the inlet pump house, the sewage enters the thick grid, fine grid and rotational flow grit chamber for pretreatment. The thick grid uses a suspended movable rake type thick grid removal machine with a total of 4 units; there are two types of fine grid, of which 7 use a rotary fine grid decontamination machine, 1 units per grid, and the other one grid adopts 2 drum type grid decontamination machines. In the original 2.8 m wide canal, each rotary fine grid decontamination machine is replaced by 2 double ladder type fine grid decontamination machine, with a total of 14 units. The width of the single grid is 1,300 mm, and the grid clearance is 6 mm.

② Reconstruction of built distributing canal. In order to have sufficient water head in the upgraded biological treatment section, the water level in the total canal is raised to 6.25 m, subtract the 60×10^4 m^3/d of sewage that enters the high-efficiency sedimentation tank from the 200×10^4 m^3/d of sewage into the plant, and the remaining 140×10^4 m^3/d of sewage flow from the left to the right to the newly-built outlet canal through the original folding bowl, and the head loss is about 12 cm. In order to achieve the water distribution effect, the original three high-efficiency sedimentation tank water inlet plate gates were replaced with a 3,000 mm×1,000 mm electric adjustment door, and the dome height can be adjusted according to the flow signal of the high-efficiency sedimentation tank water of inlet canal. When the total water intake is less than 200×10^4 m^3/d, the water level of the canal is kept constant to 6.25 m, and the total water intake is higher than 200×10^4 m^3/d, and the water intake of the high efficient sedimentation tank is kept constant to 60×10^4 m^3/d. Add a 13.0 m×11.5 m outlet well to the south side of the existing water distribution canal, lead 24,000 mm×3,000 mm box culverts to the new deployment wells, and set up two sets of manual-electric dual-use cast iron inlaid copper gates. The process design of distribution canal is shown in Figure 7-12.

③ Reconstruction of built high efficient sedimentation tank. There are 3 high efficient sedimentation tanks, each with a plane size of 71.0 m×36.0 m, and the surface of the high efficient sedimentation tank is open, the odor has a great influence on the environment. Therefore, the deodorization is carried out, the cover area of the single seat is 1,502 m^2, and the steel ribs cover plate is used to cover the surface. The observation parts adopt the fixed cover plate.

④ New bio-reactive sedimentation tank. There were 2 bioreactor sedimentation

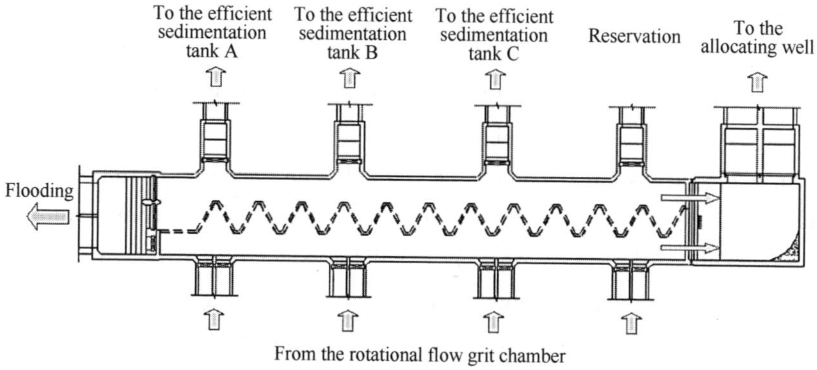

Figure 7-12 The water diversion process design of distribution canal

tanks, each containing 2 groups. The water volume of each group was 15×10^4 m^3/d, and the total treatment volume was 60×10^4 m^3/d. The bio-reaction sedimentation tank is made up of three segments: primary sedimentation section, AAO section and secondary settling section.

Among them, there are 2 primary sedimentation sections, each of which is divided into 2 groups, each group has 8 tanks. The design peak surface load is 3.2 $m^3/(m^2 \cdot h)$, and the average surface load is 2.5 $m^3/(m^2 \cdot h)$. There are 2 A/A/O sections, each of which divided into 2 groups, each group has 2 sections, and single flow of each section is 7.5×10^4 m^3/d, each of section consists of 4 corridors, including 2 corridors in aerobic area, another 2 corridors for the selection area, anaerobic area, anoxic area and alternate area, the effective water depth is 6 m, the residence time of the selected area is 0.45 h, the residence time of the anaerobic area is 1.80 h, the residence time of the anoxic area is 3.15 h, the residence time of the alternate area is 1.8 h, the residence time of the aerobic zone is 8.8 h, the aerobic zone sludge load is 0.11 $kgBOD_5/kgMLSS$, the sludge concentration is 2.5 g/L, the sludge yield is 0.6 kgDS/kgBOD, the aerobic zone mud age is 14.7 d, the gas/water ratio is 4.8 : 1, the total surplus sludge is 37,440 kg/d.

The aeration equipment adopts the microporous membrane aeration pipe, the vertical mixer is installed in the alternate zone, and the high-speed submersible mixer is set up in the anaerobic and anoxic zone. There are 2 secondary settling sections, and each section is divided into 2 groups, each group has 24 tanks. It consists of 96 grids width 9 m, 43.8 m long, and 4 m horizontal sedimentation tank, and the effective water depth is 3.5 m. With the chain plate shaving machine, the slime is first scraped through the chain plate scraper to the slime bucket near the inlet, and then the sludge in the slime bucket is discharged into the sludge canal through the sludge gate. The external reflux sludge pump at the end of the sludge canal transfers the reflux sludge to the AAO section, and the remaining sludge are raised by the 3 residual sludge pumps placed in it, then they are treated by the sludge treatment facilities.

In the secondary settling section, the variable cross section distribution canal and the small resistance corridor are adopted. After the inlet of the two sedimentation tanks is passed through the orifice flow, the plate and the water distribution wall and evenly entered into the tank. So as to improve the separation effect of the sludge and water in the secondary sedimentation tank, the baffle is set up in the water inlet.

As the primary sedimentation tank, the AAO biological reaction tank and the secondary sedimentation tank are constructed, the length of the single tank is close to 300 m and the width is close to 250 m. In order to facilitate the daily inspection and equipment hoisting maintenance, the electric bicycle lane is set up nearby the tank and connected to the ground road.

⑤ New hybrid ultraviolet disinfection tank. A new hybrid ultraviolet disinfection tank was built in this project, with a plane size of 42 m×31.6 m. The design scale is 120×10^4 m^3/d, including the effluent of high efficiency sedimentation tank with 60×10^4 m^3/d and the effluent of bio-reactive sedimentation tank with 60×10^4 m^3/d.

There are 8 sets of UV disinfection devices, including 200 modules and 1,600 low-voltage high-intensity lamps. The ultraviolet light transmittance was more than 55%, the lamp tube life was >12,000 h, the number of fecal coliform group in the effluent was less than 10^4/L, the contact time in peak flow was 2.5 s, and the effective UV dose was 18.3 mW/cm^2.

⑥ New blower room The new blower room is 54.24 m×26.34 m, with a total of 6 single stage centrifugal blower, 4 for use and 2 for spare. The single air volume is 625 m^3/min, the wind pressure is 0.07 MPa, the power is 970 kW, the blower configuration is covered with a sound insulation cover, and the noise outside is less than 80 dB.

The blowers are settled according to the most unfavorable case, the design oxygen demand is 6,482 kgO^2/h, and the actual oxygen demand is 5,151 kgO^2/h in the near future.

7.6.2.2 The extension project

(1) Project scale The design scale of the extension project of the Bailonggang sewage treatment plant are as follows:

Average flow rate in dry season: 80×10^4 m^3/d;

Peak flow rate in dry season: 12.03 m^3/s;

The total variation coefficient of peak flow rate in dry season: 1.3;

(2) Water quality of influent and effluent The effluent quality of the extension project reaches the second grade of the national discharge standards, and is rechecked according to the calibration index, and the designs water qualities of influent and effluent as shown in Table 7-11.

Table 7-11 Water qualities of the extension project mg/L

Items	COD_{Cr}	BOD_5	SS	NH_3-N	TP
Influent water quality	340	150	170	30	6
Design of the scope of the check	250~400	100~200	120~200	25~35	4~7
Effluent water quality	100	30	30	25	3

(3) The process flow The extension project also adopts the multi-mode inverted A/A/O nitrogen and phosphorus removal process.

The sewage from the south line first enters the thick grid and the inlet pump house, and then flows through the fine grid, the rotational flow grit chamber, the water distribution well and the AAO biological reaction tank. After the separation in the secondary sedimentation tank, it flows into the ultraviolet disinfection tank, and is drained into the Yangtze river through the outlet pump house, the high position well and the deep water discharge tube. The process flow is shown in Figure 7-13.

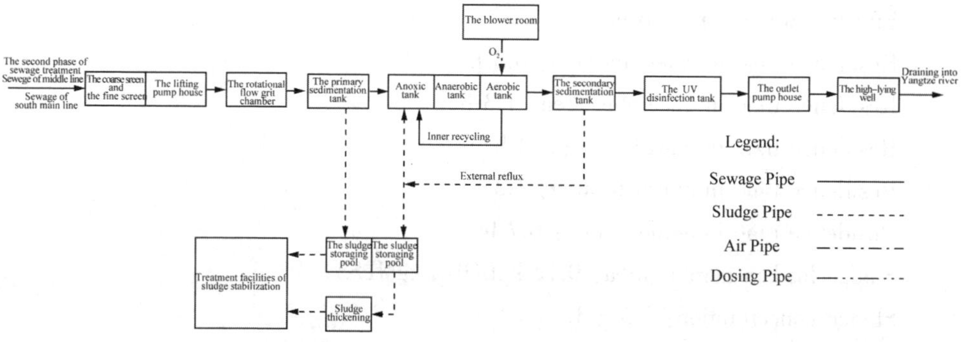

Figure 7-13 The flow chart of the extension project

(4) Main processing structure design.

① Biological reaction sedimentation tank. This project adopts the multi-mode AAO process, which can run in the normal or inverted AAO mode according to the needs and seasonal changes and can also run according to the water quality and treatment requirements according to the nitrogen and phosphorus removal mode or only the phosphorus removal mode.

There are two biological reaction sedimentation tanks, each of which is divided into 2 groups, each with an external dimension of 303.1 m×246.9 m, a treatment volume of 20×10^4 m^3/d, and a total treated water volume of 80×10^4 m^3/d. The biological reaction sedimentation tank consists of three sections: primary sedimentary section, A/A/O section, and second sedimentary section. The main design parameters are as follows:

Design scale (average flow in dry season): 80×10^4 m^3/d

(peak flow in dry season): 43,333 m^3/h

The coefficient of variation by hours: 1.3

A: primary sedimentary section (2 seats, 2 groups per seat, 8 tanks per group)

The scale of tank: 1,042 m^3/h

Single tank size: 53.3 m×6 m×3.5 m

Design surface load for peak: 4.3 $m^3/(m^2 \cdot h)$

Design surface load for average: 3.3 $m^3/(m^2 \cdot h)$

Residence time: 0.83 h

B: AAO section (2 seats, each group is divided into 2 groups)

The scale of group: 8,333 m^3/h

Effective volume per group: 100,000 m^3

Residence time (handling 200,000 m^3/d): 12 h

Effective water depth: 6 m

Residence time in selecting area: 0.3 h

Residence time in anaerobic area: 1.3 h

Residence time in anoxic area: 2.4 h

Residence time in alternate area: 1.3 h

Residence time in aerobic area: 6.7 h

Sludge load in aerobic area: 0.12 $kgBOD_5/kgMLSS$

Sludge concentration: 2.5 g/L

Sludge yield: 0.6 kgDS/kgBOD

Aerobic sludge age: 13.4 d

Oxygen demand: 6,482 kgO_2/h

Air-water ratio: 4.5 : 1

Total surplus sludge volume: 49,920 kg/d

C: Secondary sedimentation tank (2 seats, 2 groups per seat, 24 tanks per group)

The scale of tank: 348 m^3/h

Single tank size: 43.8 m×9 m×3.5 m

Design surface load for peak: 1.15 $m^3/(m^2 \cdot h)$

Design surface load for average: 0.88 $m^3/(m^2 \cdot h)$

Residence time in peak flow: 3.1 h

The main functions, dimensions and equipment configuration of the biological reaction sedimentation tank are as follows:

A: Primary sedimentary section

The primary sedimentary section of this project consists of 32 of 6 m-width, 53.3 m-length parallel flow primary sedimentation basin with a depth of 3.5 m. A chain scraper is used. Sludge is first scraped by a vertical scraper until the sinker near the inflow end, and then the horizontal scraper scrapes the sludge in the 4-settler bucket to the sludge pool next to the tank. Finally, the sludge was lifted to the sludge treatment facility by three primary sludge pumps placed in the sludge tank.

Major equipment (total quantity):

Chain scraper:

Quantity: 32 sets

Parameters: width 6 m, length 53.3 m (Single)

Equipment power: 0.55 kW

Grooved Chain Scraper:

Quantity: 8 sets

Parameters: width 2.5 m, length 24.6 m

Equipment power: 0.37 kW

Primary sludge pump (submersible pump):

Quantity: 24 sets (16 for use, 8 for spare)

Parameters: flow rate: 35 m^3/h (Run for 4 hours), head: 20 m

Equipment power: 5.5 kW

B: A/A/O section

The A/A/O section of the project is 122.45 m wide, 142.05 m long and 6 m deep. It can be operated on a single strip of 10×10^4 m^3/d, each strip consisting of 4 corridors, including 2 corridors of the aerobic zone, and the other 2 corridors of selecting zones, anaerobic zones, anoxic zones, and alternating zones.

Major equipment (total quantity):

Internal recycling sludge pump (submersible axial flow pump):

Quantity: 32 sets (24 for use, 8 for spare, 8 sets are variable frequency pump)

Parameters: Single flow rate: 2,084 m^3/h, head: 0.8 m

Equipment power: 12 kW

Oxygenating equipment:

Type: Microporous membrane aeration tube

Quantity: 18,750

Parameter: Single gas volume: 8 m^3/h

Mixing equipment:

Type: Vertical Stirrer (alternating area)

Quantity: 32

Equipment power: 5.5 kW

Type: Underwater Stirrer (Anaerobic zone, Anoxic zone)

Quantity: 96

Equipment power: 7.5 kW

C: Secondary sedimentation section

The secondary sedimentation section of this project consists of 96 parallel primary sedimentation tanks of 9 m-wide, 43.8 m-long, with a water depth of 3.5 m. Chain plate scraper is used. Sludge is scraped by the scraper on the chain plate to the sinking bucket near the inlet, and then the sludge in the sinking bucket was discharged through the sludge discharging gate into the sludge canal. The return sludge pump at the end of the sludge canal conveys the return sludge to the inlet of the A/A/O section. The remaining sludge is lifted from the three remaining sludge pumps placed therein to the sludge treatment facility for disposal.

The secondary sedimentation section water distribution adopts variable cross-section distribution canals and low head loss corridors. After passing through the orifices to the inlet of each secondary sedimentation tank, they are evenly distributed into the tank through baffles and water distribution grid walls. The advantage of this water distribution method is that the head loss is low, but the investment is large. Due to the low flow rate in the small resistance corridor, some activated sludge will be deposited in the tank, and there will be floating sludge on the surface. Therefore, a non-metal chain scraper is added to the corridor to scrape the floating sludge on the surface and settling sludge in the bottom of tank.

Major equipment (total quantity):

Horizontal Chain Scraper:

Quantity: 96 sets

Parameters: Single width 8.5 m, length 43.8 m

Equipment power: 0.55 kW

Chain scraper of distribution canal:

Quantity: 16 sets

Parameters: Single width 5 m, length 55 m

Equipment power: 0.55 kW

② UV disinfection tank. A new ultraviolet disinfection tank has been built in this project with a plane size of 26.70 m×20.08 m and the design scale are 80×10^4 m^3/d. Eight sets of ultraviolet disinfection devices were installed in this structure, a total of 132 modules and 1,056 lamps.

A: Main design parameters:

Daily processing capacity: 800,000 m^3/d

Peak flow: 43,333 m^3/h

BOD5: 30 mg/L

SS: 30 mg/L

Influent fecal coliform count: >10^7 CFU/L

Ultraviolet transmittance @ 253.7 nm: >55%

Lamp life: >12,000 h

Effluent fecal coliform count: ≤10^4 CFU/L

Lamp type: low voltage and high strength

Effective UV dose: 18.3 mW/cm^2

B: Major equipment (total quantity):

Number of lamps: 1,056

Single lamp output power: 125 W

③ Blower room. The extension project includes a new blower room with an overall size of 54.24 m ×26.34 m. Six single-stage centrifugal blowers are set, 4 for use and 2 for spare. The single air volume is 625 m^3/min, the air pressure is 0.07 MPa, and the motor power is 930 kW. The blower is equipped with a soundproof cover, and the noise outside the machine is less than 80 dB. To adapt to the high temperature environment in summer, water cooling is used, and cooling towers are added on the roof for cooling water recycling.

To reduce the cleaning frequency of the inlet filter of the fan and reduce the inlet air temperature, the inlet gallery, the peripheral filter and the inlet louver are set. To reduce the humidity of the inlet air, the inlet walls and the water-resisting layers are set to prevent the moisture.

Sound-absorbing materials are used on the walls and roof of the blower room, and the glass windows are double-decked and vacuum-insulated.

④ Outlet project. The new design of the discharge outlet is planned for the South Line, and the volume of waste water intercepted during the rainy season is 31.98 m^3/s. The designed discharge water level is 5.18 m for the five-year-cycle tide level. The construction method of the discharge outlet adopts the pipe jacking method. The project includes one high-lying well, two 3.5 m deep discharge pipes and one near-shore emergency discharge pipe.

7.6.2.3 The second period of extension project

(1) Design scale The construction of the second period of extension project will be carried out by the use the 2 groups of long range land of the upgrading and modifying project, which is located on the south of the extension project. The scale of the construction in the dry season is 80×10^4 m^3/d. Constructions of some facilities such as the inlet pump house are matching with the south line, will be built at a long-term of 43.71 m^3/s, and the equipment will be installed at a flow rate of 32 m^3/s of the rainy season.

(2) Water quality of influent and effluent The influent water quality of the second period of extension project is determined according to the cumulative frequency of 80% of the statistical value of the previous year combined with the characteristics of the water quality of the south line. The effluent quality of the extension project reaches the First-B grade waste water emission standard of the *Discharge standard of pollutants for municipal waste water treatment plant* (*GB* 18918—2002). As shown in Table 7-12.

Table 7-12 The water qualities of the second period of extension project

Items	COD_{Cr}/ (mg/L)	BOD_5/ (mg/L)	SS/ (mg/ L)	NH_3-N/ (mg/ L)	TN/ (mg/L)	TP/ (mg/L)	pH	LAS/ (mg/L)	Oils/ (mg/L)	Coliform group/ (CFU/L)
Influent	350	150	170	30	40	5.0	6~9	—	15	$10^6 \sim 10^7$
Effluent	≤60	≤20	≤20	≤8(15)	≤20	≤1.0	6~9	≤1.0	≤3	≤10,000

(3) The process flow The operation of completed upgrading and modifying project and extension project shows that the total nitrogen and total phosphorus in the existing multi-model A/A/O nitrogen and phosphorus removal process are still not able to meet the First-B grade emission standard and it is necessary to enhance denitrifying nitrogen and phosphorus removal and control of TN and TP. Although there are few high-quality

carbon sources for influent water, carbon and nitrogen are relatively poor, so the adopted process should firstly meet the requirements of denitrification.

The operation of completed upgrading and modifying project and extension project that have been completed are all arranged in large groups. The multi-mode AAO treatment process is stable in operation. The second period of extension project can be extended to maintain the unity of the whole plant process, and it is also convenient for future operations and management.

Because there are few carbon sources in the influent, the concentration of TP exceeds the limit of biological phosphorus removal, and the TP compliance rate is lower.

Therefore, the second period of extension process uses A/A/O denitrification and auxiliary chemical phosphorus removal process, and the coagulant is $Al_2(SO_4)_3 \cdot 18H_2O$, with a molar ratio of 1.5∶1, and a dosage of 75 m^3/d. The auxiliary chemical dephosphorization process is added based on the first-level enhanced dosing facilities, and the coagulant is directly added to the end of the biological reaction tank. Part of the phosphorus is removed by the biosynthetic sludge, and the remaining phosphorus is removed by simultaneous chemical precipitation in the secondary sedimentation tank. At the same time, the design parameters of the multi-mode AAO treatment process are adjusted, and the denitrification time is prolonged to strengthen the denitrification. The treated tail water is disinfected and discharged into the Yangtze River through the deep-water pipe from the pump house.

Sewage from the south line flows along two 4 m-diameter pipes into the coarse screen. After being pumped into the pump house, it flows through the fine-grid, the horizontal flow grit chamber, the diversion gate shaft, the primary sedimentation tank, and the AAO biological reaction tank, the primary sedimentation tank, the UV disinfection tank, the outlet pump house, the high-lying well and then is discharged through deep-water pipe into the Yangtze River. The process of the second period of extension project is shown in Figure 7-14.

(4) Main processing structure design As the treatment standard is improved but the site and floor layout are not changed, the residence time in reaction tank is increased by deepening the pool depth to ensure the treatment effect. Two biological reaction sedimentation tanks are arranged on the south side of the extension project, in addition to the thick grids and the inlet pump houses, the fine grids, the horizontal flow grit

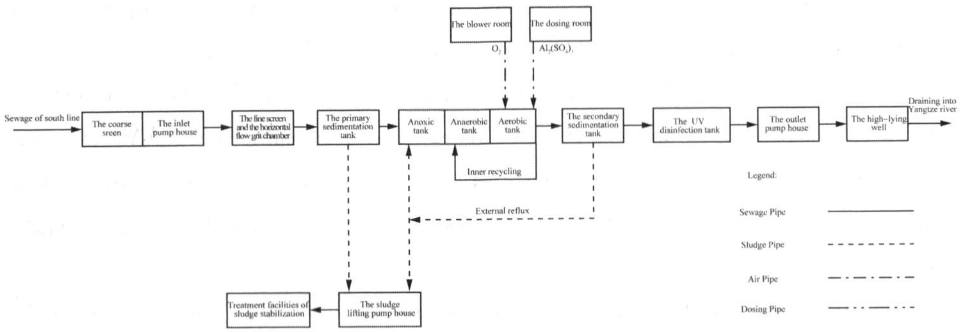

Figure 7-14 The flow chart of the second period of extension project

chamber, the blower rooms, the ultraviolet disinfection tanks, the sludge thickening tanks, the transformer room and other buildings.

① Thick grid and inlet pump house. The second period of extension project involves the construction of a new thick grid and water inlet pump house located on the west of the land occupied by the extension project and on the south of the central control building.

The thick grid and the inlet pump house are separated and connected by the middle pipes. The inlet channel is divided from the pipeline and is directly connected to the pump suction inlet. Because there is no large front pool, the accumulation of gravel in the process of long-distance transportation is not easy to deposit, and the thick grid wells and the inlet pump house can be used for the construction of caisson and underground continuous wall excavation, respectively, to reduce investment. The connecting pipe between the thick grid well and the pumping station adopts a pipe with a diameter of 3.5 m, the thick grid well also serves as a pipe receiving well, and the inflow pump house serves as a pipe working well.

The thick grid and the intake pump house include a grid well, a main pump house, a pump-back sink pool, a control room, a duty room, and a power distribution room.

The civil design scale of the pump station is 43.71 m^3/s, and the equipment is configured for the short-term rainy season flow of 32 m^3/s.

The sewage from the South Line was pumped by the SB pump station and sent to the Bailonggang Sewage Treatment Plant through the pressure pipe. The elevation of the influent water level of the incoming water is related to the volume of water passing through each stage. The relationship between the elevation of the influent water level and the volume of water entering the plant is shown in Table 7-13.

Table 7-13 The influent level of sewage of south line

Stages	Average flow long term in dry season	Peak flow of long term in dry season	Maximum flow of long term in rainy season	Accident maintenance of partial pipe section
Influent water level/m	1.30	-0.46	-2.11	-4.40

Each pump is equipped with an independent inlet pipe and outlet pipe. The loss of the pipe head is only related to the flow rate of the single pump, and it is not necessary to consider the parallel operation conditions. In normal operation, the water level behind the pump is 8.0 m.

The grid well dimensions are 45.0 m×25.0 m. The two DN4,000 mm reinforced concrete pipes in the south line enter the grid well through the pipe-jacking. The front pool in the grid well is partitioned into two parts through the partition wall. The 4,000 mm×3,000 mm manual-electric dual-use cast iron gates are installed and connected to each part in the partition wall, and every part can operate independently. Eight grab-type thick grids are set in the grid wells. In front and behind of grids, there are provided with gates and stacked beam doors, which are used for overhauling thick grids and gates, respectively.

A stainless steel perforated aeration pipe is installed in the well to prevent the sedimentation of gravel, and H_2S generated by confluent sewage can be blown off to protect subsequent equipment. There is a deodorizing hood on the well, which collects and deodorizes the odor and blowing air.

After removing the slags, the waste water from the thick grid enters the pump house through four DN3,500 pipelines, each DN3,500 pipe is divided into two DN2,000 sewer pipes into the pump suction ports, and eight DN2,000 inlet pipes correspond to eight pumps respectively. The pump house has a circular structure with an internal diameter of 59.2 m and a burial depth of 20.1 m. A total of 8 pumps are set in the diameter direction. 6 units are installed in the future and 2 units are installed in the future. According to the characteristics of the water volume and head, the inlet pump adopts a vertical volute-type mixed-flow sewage pump. The double-base motor is adopted. The pump house is designed into three layers: ground floor, motor floor and water pump floor.

Ground floor lays a control room, a power distribution room; motor layer sets hydraulic system, a ventilation motor and maintenance facilities; pump layer sets a maintenance platform and pipe gates.

The height of the pump layer and the motor layer is 7.2 m, so the control shaft length is less than 15 m, and universal joint is used between the motor and the pump.

In the pump house, there is a building area of 900 m^2 for on-duty room, controlling room and distribution room. The other living room, office and auxiliary facilities are considered together in the sewage treatment plant design. The pumping station outdoor design floor elevation is 4.40 m.

② Fine grid horizontal flow grit chamber. The purpose of removing in the grit chamber of this project is:

Particle diameter> 0.21 mm, specific gravity> 2,600 kg/m^3, removal rate> 100%

Particle diameter > 0.15 mm, specific gravity> 2,600 kg/m^3, removal rate > 90%

Therefore, the sand settling speed is designed for 0.85 m/min.

The short-term peak flow during the rainy season is 32 m^3/s, and the long-term peak flow during the rainy season is 43.7 m^3/s. The civil works are designed according to the long-term scale and checked according to the short-term flow. Therefore, there are 2 fine grids and horizontal flow grit chambers, each of them with 2 pools. The single scale is 21.85 m^3/s, and only one pool is being implemented recently.

The fine grid uses a type of rotary drum, with a total of 4 pools, each pool sets 2 sets of rotary drum fine grid of 3,000 mm diameter and 8 mm grid gaps, with a total of 16 sets. The installation angle is 35 degrees and the motor power is 3 kW.

Each horizontal flow grit chamber has 16 corridors. The corridor is 36 m long, 4 m wide and 1.8 m deep.

Water is distributed in a symmetrical manner to achieve a uniform distribution of water, with a total of four, each channel is used in the form of one match with two, and finally evenly distributed to 16 corridors. The design horizontal speed is 0.28 m/s in the future and 0.19 m/s in the long-term. The designed flow of sand is 0.03 kg/m^3, then in the dry season the volume of sand is 54 m^3, the bulk density is 1,500 kg/m^3 and the moisture content is 60%.

The horizontal flow grit chamber is more susceptible to odor, 4 sets of sand-washing devices are installed. The sand-pump pressure is used to feed materials on the two sides. There are 8 sets of pumps, 4 for use and 4 for spare. After sand washing, the organics are separated by sand and returned to the water, the sand was disposed out of plant.

The adjusted outlets for each channel of the grit chamber are set to suit different flow changes in the near-to-long term.

The grit chamber is equipped with a deodorizing system, and after the odor is collected, biological deodorization is performed.

All the metal materials on the pool are made of stainless steel, and the protection class of electrical equipment adopts IP65.

③ Biological reaction sedimentation tank. There are two biological reaction sedimentation tanks, each of which is divided into 2 pools, each with an external dimension of 300.15 m×249.90 m, a treatment volume of 40×10^4 m^3/d, and a total treated water volume of 80×10^4 m^3/d.

The influent of the second period of extension project′ s biological reaction sedimentation tank comes from the newly-built water distribution wells in this project.

The biological reaction sedimentation tank consists of three sections: the primary sedimentation section, the A/A/O section, and the secondary sedimentation section. The influent from the newly-built water distribution well first enters the primary sedimentation section, and the large-particle suspended solids in the sewage of the primary sedimentation section are removed and the effluent flows into the selective tank of the A/A/O section. In this tank, the sewage and the sludge from the secondary sedimentation tank are fully mixed. In the first and last sections of the tank, there is both an external return sludge inlet, the sludge returned from different periods can be fully mixed, so that the dominant bacteria in the external return sludge can be further strengthened.

A: primary sedimentary section

The primary sedimentary section of this project consists of 32 of 6 m-wide, 53.6 m-length parallel flow primary sedimentation basin with a depth of 3.5 m. A chain scraper is used. Sludge is first scraped by a vertical scraper until the sinker near the inflow end, and then the horizontal scraper scrapes the sludge in the 2-settler bucket to the sludge pool next to the tank. Finally, the sludge was lifted to the sludge treatment facility by three primary sludge pumps placed in the sludge tank.

a) Main design parameters:

Design surface load for peak: 4.2 $m^3/(m^2 \cdot h)$

Design surface load for average: 3.2 $m^3/(m^2 \cdot h)$

Residence time: 0.83 h

Sewage volume: 54,400 kg/d

Water content: 97.5%

b) Major equipment (total quantity):

Chain scraper:

Quantity: 32 sets

Parameters: width 6 m, length 53.6 m (Single)

Equipment power: 0.55 kW

Primary sludge pump (submersible pump):

Quantity: 24 sets (16 for use, 8 for spare)

Parameters: flow rate: 35 m^3/h (Run for 4 hours), head: 10 m

Equipment power: 3.7 kW

B: A/A/O section

There are 4 groups of A/A/O sections in this project, each group has a width of 122.15 m, a length of 142.10 m and an effective depth of 7 m. It consists of selecting zones, anaerobic zones, anoxic zones, and aerobic zones. High-speed submersible mixers are installed in the selecting, anoxic and anaerobic zones. The aerobic section installs a tubular microporous aerator. A vertical turbine stirrer is installed in the alternating area.

a) Main design parameters:

Total residence time: 14.12 h

Residence time in anaerobic selecting area: 0.39 h

Residence time in anaerobic area: 1.17 h

Residence time in anoxic area: 2.4 h

Residence time in aerobic area: 7.9 h

Residence time in alternate area: 1.56 h

Sludge concentration: 2.8 g/L

Sludge load in aerobic area: 0.122 $kgBOD_5/kgMLSS$

Sludge yield: 0.55 kgDS/kgBOD

Aerobic Sludge Age: 14.7 d

Air-water ratio: 5 : 1

b) Major equipment (total quantity):

External return sludge pump (submersible axial flow pump):

Quantity: 48 sets (32 for use, 16 for spare, 16 sets are variable frequency pump)

Parameters: Single flow rate is 1,050 m^3/h, head is 3 m

Equipment power: 16 kW

Internal return sludge pump (submersible axial flow pump):

Quantity: 32 sets (24 for use, 8 for spare, 8 sets are variable frequency pump)

Parameters: Single flow rate: 2,084 m^3/h, head: 0.8 m

Equipment power: 14 kW

Excess sludge pump (submersible centrifugal pump):

Quantity: 24 sets (16 for use, 8 for spare)

Parameters: Single flow rate: 150 m^3/h, head 20 m

Equipment power: 18.5 kW

Oxygenating equipment:

Type: Microporous aeration tube

Quantity: 14,000

Parameter: Single length is 1 m

Mixing equipment:

Type: Submersible mixer

Quantity: 96 sets

Equipment power: 9 kW

Type: Vertical Turbine Stirrer

Quantity: 32 sets

Equipment power: 6.5 kW

C: Secondary sedimentation section

The secondary sedimentation section of this project consists of 96 parallel secondary sedimentation tanks of 9 m-wide, 45.8 m-long, with a water depth of 4 m. Chain plate scraper is used. Sludge is scraped by the scraper on the chain plate to the sinking bucket near the inlet, and then the sludge in the sinking bucket was discharged through the sludge discharging gate into the sludge canal. The return sludge pump at the end of the sludge canal conveys the return sludge to the inlet of the AAO section. The remaining sludge is lifted from the three remaining sludge pumps placed therein to the sludge treatment facility for disposal.

a) Main design parameters:

Design surface load for peak: 1.1 $m^3/(m^2 \cdot h)$

Design surface load for average: 0.84 $m^3/(m^2 \cdot h)$

Residence time in peak flow: 3.5 h

Sewage volume: 49,920 kg/d

Water content: 99.5%

b) Major equipment (total quantity):

Chain Scraper:

Quantity: 96 sets

Parameters: Single width 8.5 m, length 43.8 m

Equipment power: 0.55 kW

④ Flow-adjustable well. In order to achieve a flexible deployment of water in the south and middle lines, a new flow-adjustable well is built in the project, which is divided into 2 parts. The external dimensions of the well is 20 m×20 m.

The flow-adjustable well connects with the pretreatment box culvert. With the flow-adjustable well as the boundary, the south side is the second period of extension project and the long-term water distribution, and the north is the upgrading and modifying project and water transmission and distribution of the pretreatment plant.

The flow-adjustable well has 4 interfaces. Each interface is equipped with a 4 m×3 m spherical iron gate and a stop log gate for overhauling. Multi-channel ultrasonic flowmeters are installed on box culverts on both sides of the north and south, and an on-line flow adjustment system is available to regulate the flow of the middle line and south line.

⑤ Allocating well. In order to ensure the uniform distribution of the effluent after adjustment to the biological reaction sedimentation tank, a new allocating well is construed. There are 2 cells in each well, each cell has 4 distribution channels. The external dimension of allocating well is 35.80 m×16.0 m.

Sewage from the flow-adjustable wells enters the allocating wells and is distributed through eight separate channels. Each channel is provided with a fixed crucible with a length of 50 m. The water depth during the maximum flow is 0.12 m. Then the sewage was distributed through a manual-electric dual-use regulating gate to the biological reaction sedimentation tank. Eight sets of ultrasonic flowmeters are installed at the eight box culverts where there the distribution wells and the biological reaction sedimentation tanks was connected, and the water distribution is controlled according to the opening degree of the gates which related to the flow signal.

A slot of stop log gate is provided in front of each inlet gate to shut off the flow when the gate is overhauled.

The anti-corrosive paint is applied to the shaft of the well and the ventilation and deodorization facilities are also set.

⑥ Blower room. The second period of extension project includes a new blower room with an overall size of 54.24 m ×27.24 m. Six single-stage centrifugal blowers are set, 4 for use and 2 for spare. The single air volume is 700 m^3/min, the air pressure is 0.08

MPa, and the motor power is 1,350 kW. The blower is equipped with a soundproof cover and the forced cooling is used. Each blower is equipped with electric butterfly valves, air filters, electrostatic filters, electric vent valves and silencers.

Main equipment parameters of blower room:

A: performance parameters of centrifugal blower

Wind volume: 700 m^3/min

Wind pressure: 0.08 MPa

Power: 1,350 kW

Quantity: 6 sets (4 for use, 2 for spare)

B: performance parameters of electric butterfly valve

Diameter: 1,000 mm

Power: 2.2 kW

Quantity: 6 sets (combined with blower)

(5) Reclaimed water system The second period of extension project is expanded to build a new 2,500 m^3/d reclaimed water reuse facility. The reclaimed water will be used for flushing, greening, and dissolving medicinal purposes. The quality of the reclaimed water determined in the second period of extension project is shown in Table 7-14.

Table 7-14 The reclaimed water effluent index

Items	TP	BOD_5	TSS	NH_3-N	pH	Smell	Escherichia coli
Value/(mg/L)	1	10	1,000	10	6.5~9.0	No unpleasant smell	2,000 CFU/L

The new reclaimed water treatment system uses a membrane treatment process. The process flow is shown in Figure 7-15.

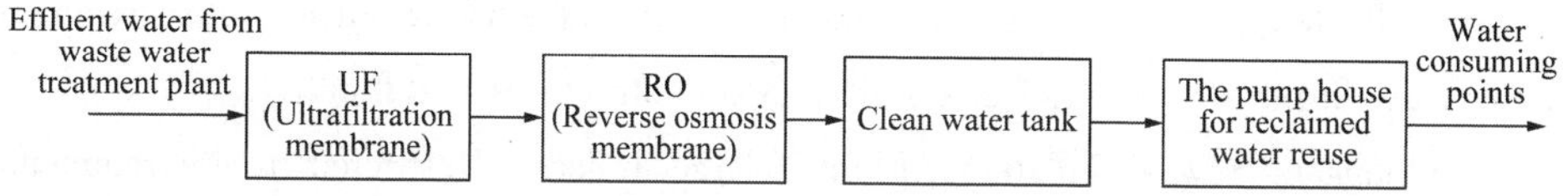

Figure 7-15 The flow chart of reclaimed water treatment

7.6.2.4 Characteristics and key points of reconstruction and extension of Bailonggang Waste Water Treatment Plant

(1) Characteristics of Bailonggang Waste Water Treatment Plant As a large-scale waste water treatment plant, Bailonggang Urban Sewage Treatment Plant has continuously improved according to the external pipe network and the improving standard of water

discharge. The requirements for water quality judgment, plan layout, pipeline arrangement and overall planning for the entire plant are very high.

① The scale is large, the water qualities and quantities change greatly, and the system is complex.

The treated object of Bailonggang Waste Water Treatment Plant is the combined sewage, and the qualities and quantities of water changes greatly. There are three water inlet systems for the middle line, south main line and south line. The water quality from different influent systems is also very different. How to deal with changes in water quantity and quality is the most important issue to be solved. Next, the large scale of the Bailonggang Sewage Treatment Plant puts forward higher requirements on hydraulic flow patterns, tank layouts, and equipment selection. There is almost no previous experience to draw upon.

② The high - quality carbon source for confluent sewage transported over long distances is insufficient, so it's not conducive to denitrification and phosphorus removal.

During the long-distance transport of confluent waste water, some of the rapidly biodegradable organics have been consumed. Therefore, the rate of nitrification and denitrification of sewage and the carbon-nitrogen ratio of influent water are much lower than those of other waste water treatment plants. The denitrification reaction requires a large volume of carbon source. When the high-quality carbon source is insufficient, it can only be solved by adding the carbon source, but this greatly increases the operating cost. In addition, the long inlet water pipeline of Bailonggang Sewage Treatment Plant makes the inlet water's temperature too low, which has a great impact on the NH_3-N of effluent.

③ Affected by the layout and land occupation of existing waste water treatment plants, the layout designing of new buildings and structures is difficult.

Bailonggang Sewage Treatment Plant is large in scale. Restricted by environmental conditions, the land using for upgrading is very insufficient. As the whole plant processing process adopts a parallel arrangement from west to east, and eventually the sewage will be pumped into the Yangtze River through the outlet pump house, it will be very difficult to add processing structures into the existing process. In addition, large-diameter box culverts are used for the connection between the structures, it is difficult to reform the pipelines.

(2) Ideas of reform and extension.

① Use frequency statistics methods to determine the influent water quality, and set the verification water quality to create conditions for raising the bid. After statistical comparison, from 1997 to 2014 as a result of changes and popularity of pipe standards, the water quality of the Bailonggang Waste Water Treatment Plant has increased year by year, and gradually deviated from the original design water quality indicators. Therefore, the extension and reconstruction projects use the frequency statistics methods to determine the quality of incoming water. According to the characteristics of service scope, the changing trend of citizens' water habits, and situation of networks, a comprehensive analysis will be conducted to set the checking water quality and leave sufficient margins to meet the requirements for future re-raising.

② Adopt a safe and reliable mature process to ensure the stable operation of large-scale waste water treatment plants. The scheme of the large-scale waste water treatment plants must consider the safety and stability of operations first and also consider the economics of operation, When the process plan is determined, a layered comparison and selection method is used to optimize the recommended program, and the short-and-long-term integration is achieved. The objective of all of sewage in the dry season was treated in biological treatment process and rainwater was treated by first-level strengthen process is achieved. The project adopts a multi-mode inverted AAO activated sludge process. It can be run flexibly according to different external environmental conditions and processing requirements at different stages to achieve optimal economic and efficiency.

③ According to the previous stage of operation, the existing facility processing capacity is verified, and the transformation parameters in next stage will be determined. The water quality and volume of water in the Bailonggang Waste Water Treatment Plant changes greatly, causing uncertainty in the value of design parameters. For this purpose, in-plant simulators are set up to simulate the effects of different operating conditions under the conditions of actual water intake. An intelligent calculation system based on activated sludge No. 2 model was used to predict the operational control parameters. At the same time, it analyzes a large volume of operational data and compares design parameters to provide scientific basis for the next stage of optimization and save project investments.

④ Implement in phases according to one plan. One-time planning and phased implementation of the construction strategy is Adopted, and access points for the long-term are reserved. At present, the access points are temporary sealed by bricks and reinforced concrete and separated by a gate. Therefore, the inserting of long-term access can simply be modified through closing the gate, without causing water cut.

⑤ Rational use of facilities to ensure all confluent sewage was treated and discharged in the river. The drainage system in Shanghai is co-existence of combined and separated. Although the government has invested a large amount of funds to intercept sewage in most of the pumping stations along the Huangpu River and the Suzhou River, due to the imperfections of the drainage system and the mixing of rain and sewage, there is still a large amount of pollutants discharged into the river with the pump station on rainy days, causing greater pollution to the inland water systems such as the Huangpu River and Suzhou River. In order to prevent the contaminants of rainy overflowing sewage that exceed the biological treatment capacity to be discharged in the Yangtze River, the reconstruction and extension project uses the margin peak value of the bioreactor and the existing high-efficiency sedimentation tanks to treat the overflowing sewage in rainy days. The sewage in dry season was treated by the biological treatment process, and the overflowing in rainy days was treated by first-stage strengthen treatment process.

⑥ Fully consider the continuous water implementation plan during the design. When the sewage treatment plant is reformed, how to choose a reasonable plan and take effective measures to prevent the transformation process from causing the existing facilities to stop operation, or to minimize water stoppage is the key to the feasibility of the reconstruction project. The Bailonggang Sewage Treatment Plant is the final export of the entire sewage treatment second period project and sewage from the south main line. Once the water is stopped due to renovation, it will have a great impact on the production and living of residents and enterprises and institutions in the upstream service area. Therefore, when new and old cast-in-place facilities are connected, all of them are aimed at the corresponding situation and taking into account the uneven settlement of the old and new structures. The "new built-connect" approach is used to make the old and new structures connect and shorten the time for water cut-off. Water stop and expansion joints are installed at the interface.

(3) Design features The extension project of Bailonggang Sewage Treatment Plant takes the principle of "ecological, low-carbon, green, and environmental protection" as its guiding ideology. The concept of energy-saving, environmental protection, and scientific and technological innovation is widely adopted in the project, and the technical system formed by integrated technological innovation results is fully

applied. An organic technology system has been formed to bring the overall technology to the international advanced level.

① The design of the Bailonggang Waste Water Treatment Plant makes full use of existing facilities on the basis of meeting the pollutant reduction and emission standards, and saves investment and lowers the cost. Safe, reliable, mature, reliable and low-cost operation process was chosen, and the matching of the main line and the sewage treatment plant are comprehensively considered. Full consideration is given to the land use situation of the last piece of a sewage treatment plant that can be built in the center of the city, it's planned once and implemented in phases to achieve the aim of that sewage in dry season was treated by the secondary biological treatment process, and achieve the aim gradually that the initial rainwater was treated by first-stage strengthen treatment process and then discharged into the river.

② The short-term primary sedimentation tank and bio-selection tank can be set to stabilize water quality during sunny days, eliminate grease, ensure the stability and reliability of the biological treatment, and also handle large volumes of sediments and retain carbon sources.

③ Intensive layout was adopted; more land was reserved for further development. Due to the large scale of the project, if the traditional design ideas and models are adopted, the number of structures was large, the distribution was loose. There are many pipelines connecting the structures, which is not conducive to the inspection and management of the plant too. Therefore, the cluster intensive distribution method is adopted, a group of 40×10^4 m^3/d is used as a group, and primary sedimentation tanks, AAO biological reaction tanks, secondary sedimentation tanks, and return sludge and surplus sludge pump houses are centrally arranged. This will save space and allow the land for further development.

④ Since the factory area is wide and the size of the structure is large, the whole factory maneuver inspection channels are set up for easy management, and the managers can drive the motor vehicle to go directly to the pool for inspection.

⑤ Based on the characteristics of the project, a large number of scientific research topics such as "stable diversion", "proportional water distribution" and "uniform mixing" were launched. The production equipment was set up in the factory for continuous tracking and debugging and lots of obtained data support the optimization designs. At the same time, senior foreign companies are required to do computer simulations of the design scheme and selected parameters and check the processing effects under different design conditions simultaneously to provide a reliable scientific basis for design.

7.6.2.5 The situation after the completion of reconstruction and extension

Shanghai Bailonggang Sewage Treatment Plant has undergone several reconstructions and extensions, and it has built a 120×10^4 m^3/d first-class strengthen treatment facility in 2004, a $(160\sim200)\times10^4$ m^3/d biochemical treatment facilities which can meet the second grade of the national discharge standards in 2008, the second period sludge treatment and disposal facilities in 2010, sludge emergency treatment and disposal facilities and newly constructed water treatment facilities for 80×10^4 m^3/d which can meet the First-B grade of the national discharge standards in 2012. So far, the biochemical treatment scale of Bailonggang Sewage Treatment Plant is 280×10^4 m^3/d. It is the fourth in the world and the first in Asia.

The second period of the extension project, which has a processing capacity of 80×10^4 m^3/d, has been commissioned since April 2013. Except for the inlet pumping station and pretreatment combined with the south line, the remaining systems have been operating stably. Its effluent quality reaches the First-B grade waste water emission standard of the *Discharge standard of pollutants for municipal waste water treatment plant* (*GB* 18918—2002).

The completion time, scale, construction contents and main processes of the completed project of Bailonggang Sewage Treatment Plant are shown in Table 7-15.

Table 7-15 A brief introduction of the Bailonggang Sewage Treatment Plant

No.	The abbreviation of projects	Time of completion	Scale	Construction standards	Adopted process
1	Preprocessing	1999	$172\times10^4 m^3/d$	Remove dregs andgravel	Thick & fine grid+Grit chamber+outlet pump house+deep sea discharge
2	Enhanced primary treatment	2004	$120\times10^4 m^3/d$	Better than First-B grade	The efficient sedimentation tank
3	Upgrading and modification	2008	$120\times10^4 m^3/d$	Better than Second class(biochemical treatment)	A/A/O process
4	Extension	2008	$80\times10^4 m^3/d$	Second class	A/A/O process

续表

No.	The abbreviation of projects	Time of completion	Scale	Construction standards	Adopted process
5	Sludge treatment	2010	204 tDs/d (digestion); 70 tDs/d(drying)	The water content of sludge is less than 80% after thickening, digestion and drying	gravity/mechanical thickening, anaerobic digestion, drying by fluidized bed
6	The second period of extension	2012	80×10^4 m^3/d	First-B grade	A/A/O process+Auxiliary chemical phosphorus removal
7	Emergency treatment of sludge	2012	300 tDs/d	The water content of sludge is less than 60%	dilution+conditioning+plate-frame pressure filtration+outward transport

(1) General layout The Bailonggang Sewage Treatment Plant has basically formed four major regions following the construction of the upgrading and modifying project, the extension project, the second period of extension project and the sludge treatment and disposal project. Including sludge landfill, the whole plant covers about 2 km^2.

The first area is the northernmost sludge treatment and disposal plant with an area of about 39.05 hm^2. The second area is an upgrading and modifying area in the middle of the plant, covering about 29.36 hm^2. There is pretreatment area, first-level strengthen treatment area, two new built upgrading biological reaction sedimentation tanks, reclaimed water reuse area, living area, and sludge landfill. The third area is the extension project on the south side. It consists of new built biological reaction sedimentation area and discharge area. The fourth area is the southernmost second period of extension project, which consists of the south line pretreatment area, biological reaction sedimentation area, and supporting facilities area, as shown in Figure 7-16.

(2) Structure grouping Because the scale of a waste water treatment plant is very large, reasonable grouping must be carried out to reduce the number and investment of equipment. It is also easy to overhaul and maintain without affecting the normal operation of the entire plant. The Bailonggang Sewage Treatment Plant adopts a intensive arrangement of clusters and is arranged at a size of 40×10^4 m^3/d per block. Each block is divided into two separate groups of 20×10^4 m^3/d each, which can be operated independently at 10×10^4 m^3/d. Once a single tank need to be overhauled, it would have little effect on whole processing capacity. The upgrading and modifying project has a total of 4 groups, and the second period of extension project has 2 groups.

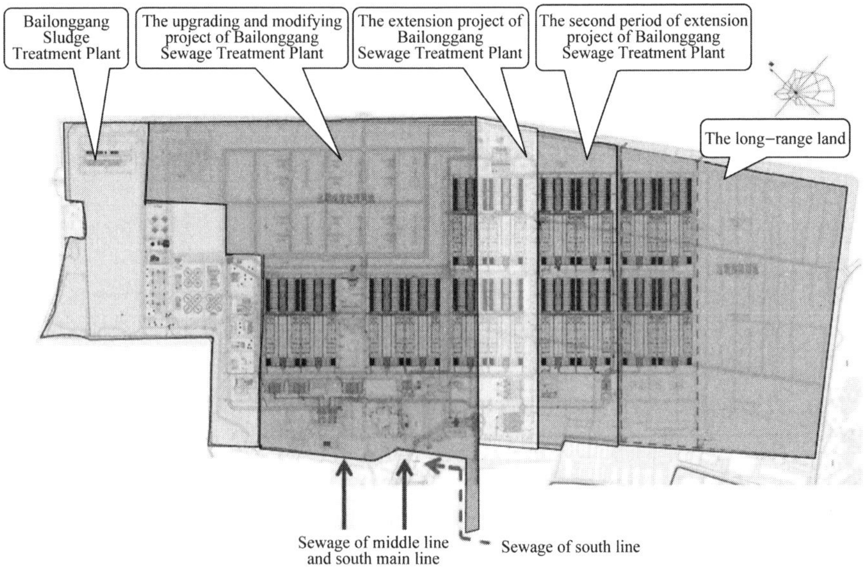

Figure 7-16 Plan layout of Bailonggang Sewage Treatment Plant

(3) In-plant flow distribution The water distribution of large-scale sewage treatment plants is very critical. The water volume of the combined-flow sewage treatment plant changes greatly, and the stable water volume can reduce the impact on the biochemical treatment process and ensure the effluent can reach the discharge standards.

Due to the differences in the volume of dry season among the middle line, south main line and south line, and incompatibilities of the capability of six biological sedimentation tanks and the draining capability of two pumping stations, the in-plant flow distribution is required. Both the before-pretreat distributing and after-pretreat distributing can be adopted. The flow can also be distributed according to different seasons or different times includes rainy season and dry season, or short-term and long-term.

The short-term flow distribution in dry season is shown in Figure 7-17, and the short-term flow distribution in rainy season is shown in Figure 7-18.

To stabilize the volume of influent of the biological treatment, three high-efficiency sedimentation tanks are used as auxiliary flow adjustment measures. After adjusting the opening degree of the gate well, it's not frequent to adjust in other flow rates to ensure that the biological treatment flow is relatively stable, thus reduce the impact on biological treatment.

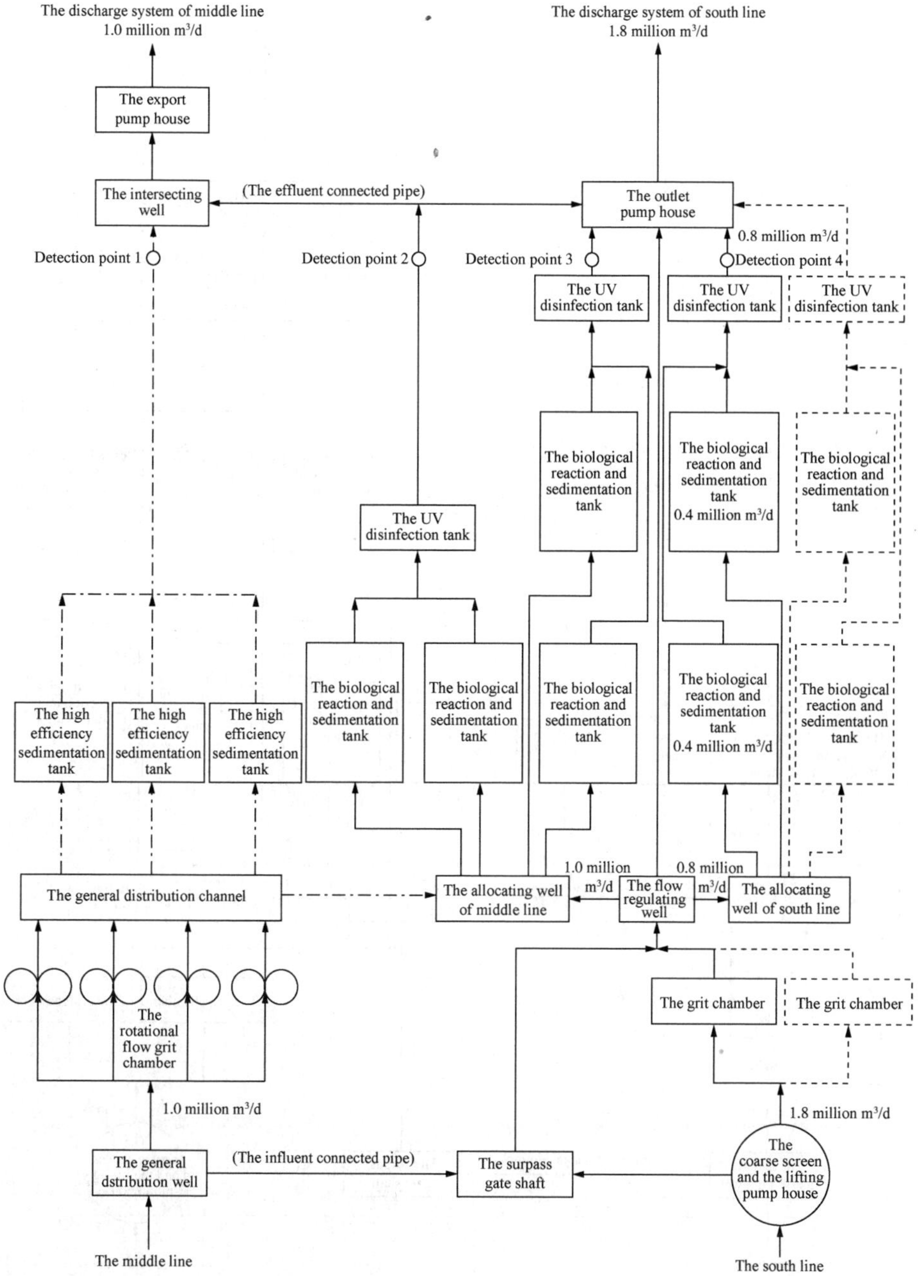

Figure 7-17 The flow chart of short-term flow distribution in dry season

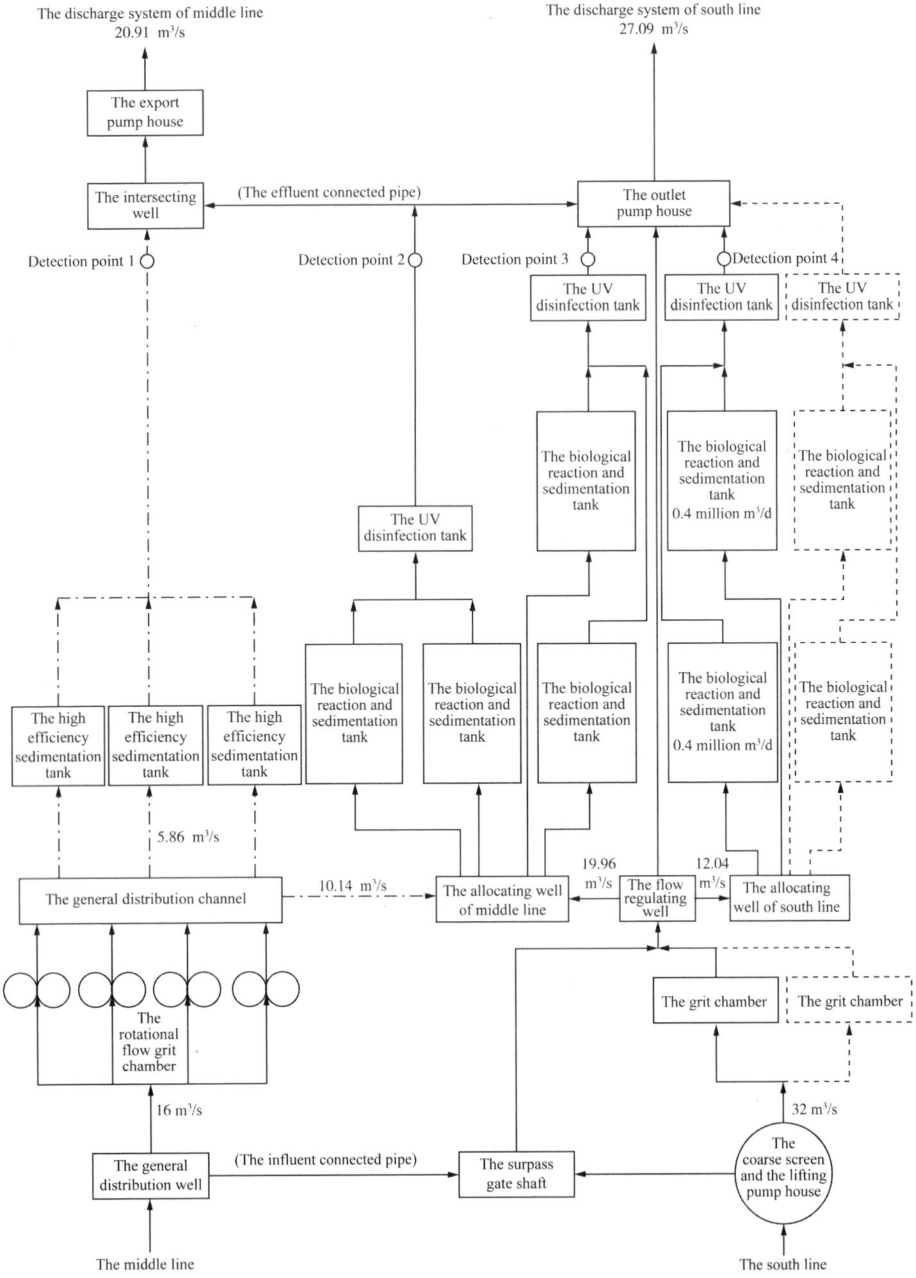

Figure 7-18 The flow chart of short-term flow distribution in dry season

7.6.3 Operation effect

The upgrading and modifying project of Bailonggang Sewage Treatment Plant was completed and put into operation in September 2008. The second period of extension project was also put into trial operation in April 2013. Due to the lack of water on the South Line, the second period of the extension received the sewage from the middle line system. The volume of sewage treated every day is about $200\times10^4 \sim 240\times10^4$ m^3/d. After the biological reaction system was put into trial operation for more than one year, it has all achieved reaches the First-B grade of the national discharge standards and achieved good social and environmental benefits.

Questions

1. What factors should be considered when design a municipal waste water treatment plant?
2. How many anaerobic-aerobic activated sludge processes explained in the chapter? Please clarify the characteristics of these processes. In your opinion, which one is your favorite when you are engineer to design a municipal waste water treatment plant?
3. Please make a comparison between different oxidation ditch and present the differences. Please be sure which oxidation ditch is suitable to remove N.
4. Please explain characteristics and key points of renovation and extension scheme of sewage plant simply.

Chapter 8　Water reclamation and reuse

Emphasis: What is water reclamation. Importance of water reclamation and reuse. The process of water reclamation and reuse. How to meet the demand of water reclamation and reuse standard.

Nut: Water reclamation and reuse processes.

The social economic, and environmental impacts of past water resources development and inevitable prospects of water scarcity are driving the shift to a new paradigm in water resources management. New approaches now incorporate the principles of sustainability environmental ethics, and public participation in project development. With many communities approaching the limits of their available water supplies, water reclamation and reuse have become an attractive option for conserving and extending available water supply by potentially.

① Substituting reclaimed water for applications that do not require high-quality drinking water.

② Augmenting water sources and providing an alternative source of supply to assist in meeting both present an future water needs.

③ Protecting aquatic ecosystems by decreasing the diversion of freshwater, reducing the quantity of nutrients and other toxic contaminants entering waterways.

Reducing the need for water control structures such as dams and reservoirs.

④ Complying with environmental regulations by better managing water consumption and waste water discharges.

⑤ Water reuse is particularly attractive in the situation where available water supply does not meet expanding water demands in a growing community. Increasingly, society no longer has the luxury of using water only once. Typical reclaimed water reuse applications (see Figure 8-1).

(a) agricultural irrigation by means of subsurface drip

(b) landscape irrigation (highway median strip)

(c)industrial reuse for cooling and process water

(d) commercial reuse for car washing

(e) recreational lake

(f) empty rapid infiltration basin used for groundwater recharge

Figure 8-1 Typical reclaimed water reuse applications

8.1 Water reclamation

Water reclamation is the treatment or processing of waste water to make it reusable with definable treatment reliability and meeting water quality criteria. Water reuse is the use of treated waste water for beneficial uses, such as agricultural irrigation and industrial cooling. Treated municipal waste water represents a more reliable and significant source for reclaimed water as compared to waste waters coming from agricultural return flows, storm water runoff, and industrial discharges. Centralized waste water treatment has become commonplace in urban areas of China, as a result of the water shortage and related waste water treatment regulations. New technologies in decentralized and satellite waste water treatment have also been developed.

The benefits of water reclamation and reuse and factors driving its future are summarized as followed.

- Water is a limited resource. Increasingly, society no longer has the luxury of water only once.
- Acknowledge that water recycling is already happening and do it more and better, the quality of reclaimed water is appropriate for many nonportable applications such as irrigation and industrial cooling and cleaning water thus providing a supplemental water source that can result in more effective and efficient use of water.
- To meet the goal of water resource sustainability it is necessary to ensure that water is used efficiently water reclamation and reuse allows for more efficient use of energy and resources by tailoring treatment requirements to serve the end-users of the water.
- Conservation of fresh water supplies.
- Management of nutrients that may lead to environmental degradation.
- Improved protection of sensitive aquatic environments by reducing effluent discharges.
- Economic advantages by reducing the need for supplemental water sources and associated infrastructure. Reclaimed water is available near urban development where water supply reliability is most crucial and water is priced the highest.
- Nutrients in reclaimed water may offset the need for supplemental fertilizers, thereby conserving resources. Reclaimed water originating from treated effluent contains nutrients; if this water is used to irrigate agricultural lend, less fertilizer is required for crop growth. By reducing nutrient (and resulting pollution) flows into waterways, tourism and fishing industries are also helped.
- Reclaimed water is readily available in the vicinity of the urban environment, where water resources are most needed and are highly priced.
- Reclaimed water provides a reliable water source, even in drought years, as production of urban waste water remains nearly constant.
- Technically and economically proven waste water treatment processes are available now that can provide water for nonportable applications and can produce water

of a quality that meets drinking water requirements.

• Growing recognition among water and waste water managers of the economic and environmental benefits of using reclaimed water.

• Increasing awareness of the environmental impacts associated with overuse of water supplies, and community enthusiasm for the concept of water reclamation and reuse.

• Greater recognition of the environmental and economic costs of water storage facilities such as dams and reservoirs.

• The growing numbers of successful water reclamation reuse projects throughout the world.

• The introduction of new water charging arrangements (such as full cost pricing) that more accurately reflect the full cost of delivering water to consumers, and the growing use of these charging arrangements.

• Increased costs associated with upgrading waste water treatment facilities to meet higher water quality requirements for effluent disposal.

• Necessity and opportunity: Motivating factors for development of water reclamation and reuse projects such as droughts, water shortages, prevention of seawater intrusion and restrictions on waste water effluent discharges, plus economic, political, and technical conditions favorable to water reclamation and reuse.

Water reuse offers an alternative water supply that is consistently available in urban Areas, even during drought years, for various beneficial uses. However, because of its genesis from municipal waste water (traditionally known as sewage), acceptance of reclaimed water as an alternative water source has to overcome unique hurdles. In the developed countries, reclaimed water is treated using strict water quality control measures to ensure that it is nontoxic and free from disease causing microorganisms, but it does carry potential risks inherent in the use of any resource exposed to human waste. Concerns for health and safety must be addressed in the planning and implementation of water reclamation and reuse. It has been found that the success of water reclamation and reuse projects in many parts of the world has hinged on the pressures associated with the urgent necessity for water coupled with the opportunity to develop water reuse systems.

Environmental ethics involves the application of moral responsibility in relation to management of the natural environment. Similar as the principle of sustainability, environmental ethics has emerged in response to serious environmental degradation resulting from societal activities such as over-allocation of natural resources. There are several theories of environmental ethics that are used to describe human obligations in the protection of natural systems. The anthropocentric (human-centered) perspective emphasizes environmental protection for the survival and well-being of humans alone. The egocentric (nature-centered) perspective regards humans as only one element of the broader natural community and bases moral responsibility on the intrinsic value and rights of nature.

8.2 Current and potential future global water shortages

The total volume of renewable freshwater in the global hydrologic cycle is several times more than is needed to sustain the current world population. However, only about 31 percent of the annual renewable water is accessible for human uses due to geographical and seasonal variations associated with the renewable water. On a global scale, annual withdrawals for irrigation are over 65 percent of the total with- drawn for human uses; 2,500 out of a total of 3,800 km^3. Withdrawals for industry are about 20 percent, and those for municipal use are about 10 percent. The implementation of water reclamation and reuse projects is driven mainly by existing and projected water shortages in specific water-poor countries. Other factors such as preventing saltwater intrusion into freshwater resources in coastal areas and prohibition of waste water effluent disposal into sensitive environments will certainly influence water reuse decisions. The impacts associated with current and projected world population, water requirements, and potential global and regional water scarcity are considered briefly in the following discussion. The rate of growth in industrialized countries is well under one percent per year. In developing countries, however, the growth rate exceeds two percent per year, and in some parts of Africa, Asia, and the Middle East it exceeds three percent per year. As a result, over 90 percent of all future population increases will occur in the developing world. Six countries currently account for half of the annual population growth: India, China, Pakistan, Nigeria, Bangladesh, and Indonesia. The population in the United States was estimated at about 285 million in 2001 and was growing at an annual rate of about one percent.

8.2.1 Irrigation water use

The expansion of the aerial extent of irrigated land-use due to population growth is one of the most important contributors to the increase of total water use in the world. In 1995, over 65 percent of the total global water withdrawal for human uses was for irrigation, which includes both agricultural and nonresidential landscape applications. Irrigation consumes a large volume of water through evaporation from reservoirs, canals, and soil and through incorporation into and transpiration by crops. Consumptive use is the portion crops, consumed by humans or livestock, or otherwise removed from the immediate water environment. Depending on the technology and management, consumptive use associated with irrigation can range from 30 to 90 percent of the total water withdrawn.

8.2.2 Domestic and industrial water uses

Conversion of farmhand into residential and industrial areas results in a decrease in agricultural water use and a concurrent increase in domestic and industrial water uses. A large share of the water used by households, services, and industry–up to 90 percent in areas where total water use is high–is returned as waste water. While a large proportion of the water used in domestic and industrial water is collected as waste water, water is such a degraded state that treatment is required before it can be discharged or reused.

Globally, the water resources in various regions and countries are expected to face unprecedented pressures in the coming decades as a result of continuing population growth and uneven distributions of population and water. Although the number of persons served has increased, about 1.1 billion people, or about 18 percent of the world population lacked access to clean drinking water, and 2.4 billion did not have adequate sanitation services in 2000. Surging populations throughout the developing world are intensifying the pressures on limited water supplies. The concentration of populations within urban areas further exacerbates the disparity between water demand and regional water availability.

8.3 The important role of the water reclamation and reuse

Water reclamation and reuse involves considerations of public health and also requires close examinations of infrastructure and facilities planning, waste water treatment plant siting, treatment process reliability, economic and financial analyses, and water utility management involving effective integration of water resources and reclaimed water. Whether water reuse will be appropriate depends upon careful economic considerations, potential uses for the reclaimed water, public health protection, stringency of waste discharge requirements, and public policy where the desire to conserve rather than develop available water resources may override other obstacles. In addition, the varied interests of many stakeholders, including those representing the environment, must be considered. The principal categories of water reuse applications for reclaimed water originating from treated municipal waste water are shown in Table 8-1, in descending order of volume of use. The majority of water reuse projects is for nonportable applications such as agricultural and landscape irrigation and industrial uses. Groundwater recharge can be designed for indirect potable reuse where groundwater is recharged with reclaimed water and replenishes portions of potable groundwater.

Table 8-1 Water reuse categories and typical applications

Category	Typical application
Agricultural irrigation	Crop irrigation
	Commercial nurseries
Landscape irrigation	Parks
	School yards
	Freeway medians
	Golf courses
	Cemeteries
	Greenbelts
	Residential
Industrial recycling and reuse	Cooling water
	Boiler feed
	Process water
	Heavy construction
Groundwater recharge	Groundwater replenishment
	Salt water intrusion control
	Subsurface control
Recreational/environmental uses	Lakes and ponds
	Marsh enhancement
	Stream flow augmentation
	Fisheries
	Snowmaking
Nonportable urban areas	Fire protection
	Air conditioning
	Toilet flushing
Potable reuse	Blending in water supply reservoirs
	Blending in groundwater
	Direct pipe to water supply

The standards for the reused water are as Table 8-2~Table 8-5.

Table 8-2 Standards for the reused water

Items	Cooling water CECS61:94		River landscape CECS61:94	Non-drinking water CJ 25.1—89		Agricultural irrigation water GB 5084—92		Discharge standard of pollutants for municipal waste water (GB 18918—2002)
	Once-through cooling water	Circulating cooling water		Toilet flushing	Washing car	Water crop	Rainfed crop	Standard A
pH	6.0~9.0	6.5~9.0	6.5~9.0	6.5~9.0	6.5~9.0	5.5~8.5		6~9
SS/(mg/L)	30	—	30	10	5	≤150	≤200	10
Turbidity/(NTU)	—	5	—	10	5			
Odour	—	—	All pleasure	All pleasure	All pleasure			
Color	—	—	—	30	30			30
BOD_5/(mg/L)	30	10	20	10	10	≤80	≤80	10
COD_{Cr}/(mg/L)	—	75	75	50	50	≤200	≤300	50
TP/(mg/L)	—	—	<2 (summer)			≤2	≤2	0.5
TN/(mg/L)	—	—	—					15
NH_4^+-N/(mg/L)	—	—	<10 (summer), <20(other season)	20	10	≤12	≤30	5
Fe/(mg/L)	—	0.3	0.4	0.4	0.4			
Mn/(mg/L)	—	0.2	—	0.1	0.1			
Chloride/(mg/L)	300	300	350	350	300	200	200	
Total hardness ($CaCO_3$)/(mg/L)	850	450	—	450	450			

续表

Items	Cooling water CECS61:94		River landscape CECS61:94	Non-drinking water CJ 25.1—89		Agricultural irrigation water GB 5084—92		Discharge standard of pollutants for municipal waste water (GB 18918—2002)
	Once-through cooling water	Circulating cooling water		Toilet flushing	Washing car	Water crop	Rainfed crop	Standard A
Total alkalinity ($CaCO_3$)/(mg/L)	500	450	—	—	—			
All salts/(mg/L)	—	—	—	1,200	1,000			
residual chlorine/(mg/L)	—	0.1~0.2	—	0.2	0.2			
Bacteria/(count/L)	—	5×10^5	—					
Coli group/(count/L)	—	—	10,000	3	3			1,000
Volatile penol/(mg/L)	—	—	—					
Petroleum/(mg/L)			—					1
Anionic surfactant/(mg/L)			—	1.0	0.5	≤5	≤8	0.5

Table 8-3 Standards for non-drinking water (GB/T 18920—2002)

Items	Toilet flushing	Washing road, fire-fighting	City virescence	Washing car	Construction
pH	6.0~9.0				
Color	30				
Odor	All pleasure				
Turbidity (NTU)	5	10	10	5	20
Total soluble solids/(mg/L)	1,500	1,500	1,000	1,000	
BOD_5/(mg/L)	10	10	20	10	15
NH_4^+-N/(mg/L)	10	10	20	10	20
Anionic surfactant/(mg/L)	1.0	1.0	1.0	0.5	1.0
Fe/(mg/L)	0.3	—	—	0.3	—
Mn/(mg/L)	0.1	—	—	0.1	—
DO/(mg/L)	1.0				

Table 8-4 Standards for landscape water

Items	Landscape water for ornamental value			Landscape water for entertainment value		
	River	Lake	Landscape	River	Lake	Landscape
Basic requirements	No floating matters and no odor					
pH	6.0~9.0					
BOD_5/(mg/L)	10	6	6	6	6	6
SS/(mg/L)	20	10	10	—		
Turbidity(NTU)	—			5	5	5
DO/(mg/L)	1.5	2.0	—	—	—	—
TP/(mg/L)	1.0	0.5	1.0	0.5	—	—
TN/(mg/L)	15					
NH_4^+ N/(mg/L)	5					
Coliform (count/L)	10,000	2,000		500		not detectable
Residual chlorine/(mg/L)	0.05					
Color	30					
Petroleum/(mg/L)	1.0					
Anionic surfactant/(mg/L)	0.5					

Table 8-5 Standards for recharge of ground water (Shanghai)

Item	Standards	Item	Standards
Temperature/℃	<15, winter; >30, summer	As (mg/L)	<0.02
Taste and odor	No taste and no odor	Hg (mg/L)	<0.001
Color	No color, <20	Cr^{6+} (mg/L)	<0.05
Turbidity (NTU)	<10	Pb (mg/L)	<0.01
pH	6.5~8.0	Cd (mg/L)	<0.01
Chloride/(mg/L)	0.5~1.0	Se (mg/L)	<0.01
DO/(mg/L)	<7	Cyanide (mg/L)	<0.01
Oxygen consumption/(mg/L)	<5	Volatile penol (mg/L)	<0.002
Iron/(mg/L)	<0.5	Bacteria (CFU/L)	<100
Manganese/(mg/L)	<0.1	Coliform (CFU/L)	<3
Cu, Zn/(mg/L)	<1	Others	No radioactive substance

Integrated water resources planning is a process that promotes the coordinated development and management of water, land, and related resources to maximize the resultant economic and social welfare in an equitable and sustainable manner. A framework to compare competing interests, including those of future generations, does not currently exist in water management and planning. A new definition of sustainable water development is also needed that expands the traditional supply and demand approach and encompasses environmental and social issues.

8.4 Technology development

Water reuse has been practiced for over 5,000 years; however, during the last 100 years efforts have been made in many regions of the world, for the production of high quality reused water, following strict quality guidelines. Water reuse can be practiced at domestic (grey water-waste water from clothes washers, bath tubs, showers and sinks) levels, and in municipal, industrial, and agricultural sectors.

Technological approaches include innovation in the present technology based on scientific discovery and practical feasibility. Many technologies are available that produce high quality and safe drinking water or water suitable for other beneficial uses. Technological approaches depend on the type of source water as well as the purpose of the end product. Surface and ground water sources with less than 3,000 mg/L total dissolved solids (TDS) can be treated with conventional technologies such as coagulation, flocculation, lime-softening and sedimentation and filtration processes. Membrane technologies can also be used to remove the TDS effectively. Ground waters and seawaters having TDS higher than 3,000 mg/L should be treated using membrane or thermal desalination technologies.

Natural systems are considered in treatment schemes where the water quality objective is a potable water reuse. These involve "natural barriers" such as soil aquifer treatment, wetlands and other land application for soil infiltration to provide natural treatment of organic and nutrient compounds and microbial contaminants. This is achieved by providing a retention time suitable for natural attenuation of contaminants followed by blending with the water sources. Application of natural systems is subject to climate and hydrogeological conditions. These systems are not the preferred method currently due to lack of standardized guidelines for design and operation.

Engineered solutions are designed by taking the advantages of chemical, physical and characteristics of the contaminants in the water sources. Conventional sand filtration is commonly used to remove the suspended solids but membrane filtration processes such as microfiltration and ultrafiltration have become a more common practice in recent years. Chemical coagulation, precipitation, neutralization, oxidation, hydrolysis and other reactions are considered for organic contaminant removal. Trace organics including volatile contaminants can be removed using granulated organic carbon (GAC) filters and ion exchange columns. Nutrients and other microbial contaminants can be removed using natural systems such as wetlands, soil application, and air-stripping and reverse osmosis processes. Disinfection is performed using chlorination, and or ozonation, and or ultraviolet exposure (UV) depending on the specific objective of the water reuse application. Table 8-6 shows the different membrane technologies and their specific characteristics in water treatment and desalination applications. Microfiltration and ultrafiltration are used in lieu of conventional sand filtration process due to higher efficiency and smaller footprint. Nanofiltration can be used to remove some specific contaminants at a molecular weight cutoff with much lower specific energy consumption compared to reverse osmosis. Salts and other dissolved and micro- or trace contaminants can be removed using reverse osmosis membrane filtration.

Table 8-6 General characteristics of membrane technologies for water treatment and desalination

Membrane type	Particle capture size	Typical contaminants removed	Typical operati on pressure ranges	Key applications
Microfiltration	0.1~10 μm	suspended solids, bacteria, protozoa	0.1~2 bar (1~30 psi)	Water treatment plants, pre-treatment in desalination plants, the preparation of sterile water for industries, such as pharmaceuticals, etc.
Ultrafiltration	ca. 0.003~0.1 μm	colloids, proteins, polysaccharides, most bacteria, viruses (partially)	1~5 bar (cross-flow) 0.2~0.3 bar (dead-end and submerged) (3~80 psi)	Drinking water treatment, the pre-treatment process in desalination, and membrane bioreactors
Nanofiltration	ca. 0.001 μm	viruses, natural organic matter, multivalent ions (including hardness in water)	5~20 bar (70~220 psi)	Treatment of fresh, process and waste waters
Reverse osmosis	ca. 0.001 μm	almost all impurities, including monovalent ions	10~100 bar (800~1,200 psi)	Treatment of fresh, process and waste waters, desalination of sea water

8.4.1 Typical water reclamation processes

In the section, the typical water reclamation processes based on the tertiary treatment are introduced.

(1) Coagulation, clarification and filtration system(Figure 8-2).

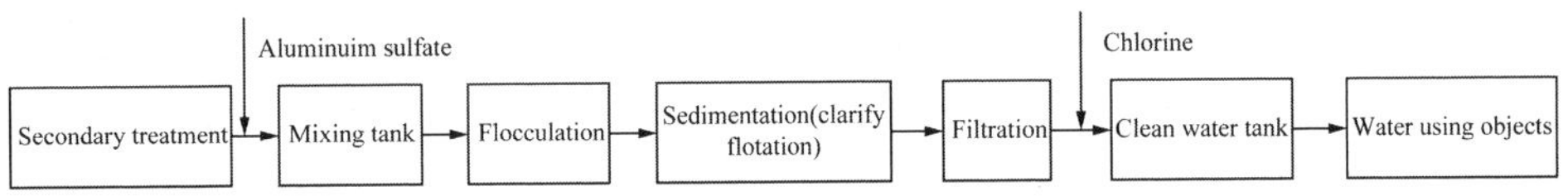

Figure 8-2 The flow diagram of coagulation, clarification and filtration system

(2) Direct filtration system(Figure 8-3).

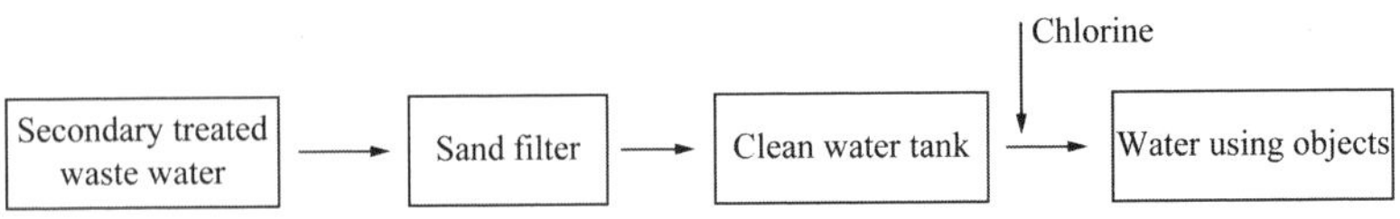

Figure 8-3 The flow diagram of direct filtration system

(3) Micro-flocculation system (Figure 8-4).

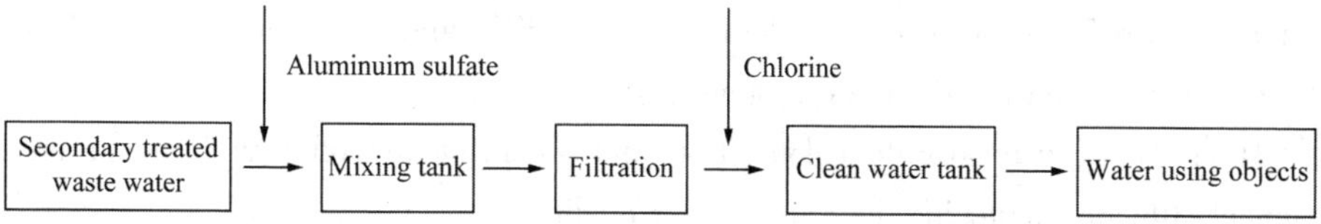

Figure 8-4 The flow diagram of micro-flocculation system

The three systems mainly remove the suspended solids in the secondary treatment effluent. For the removal of COD and BOD is more than 60%. At the same time, a part of colour, taste and odour is removed.

(4) Biofilm filtration system (Figure 8-5).

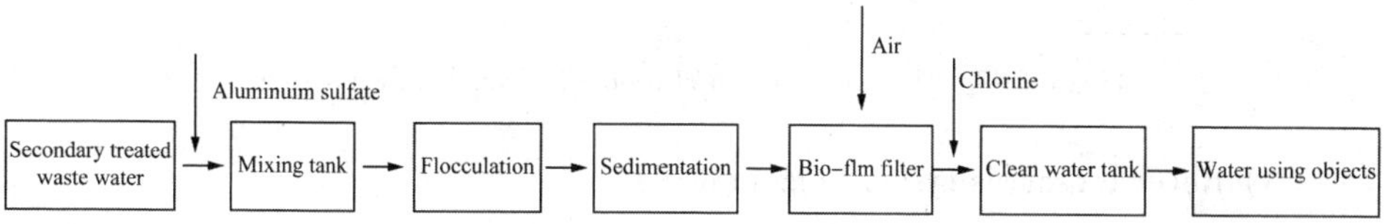

Figure 8-5 The flow diagram of biofilm filtration system

(5) Air flotation filtration system (Figure 8-6).

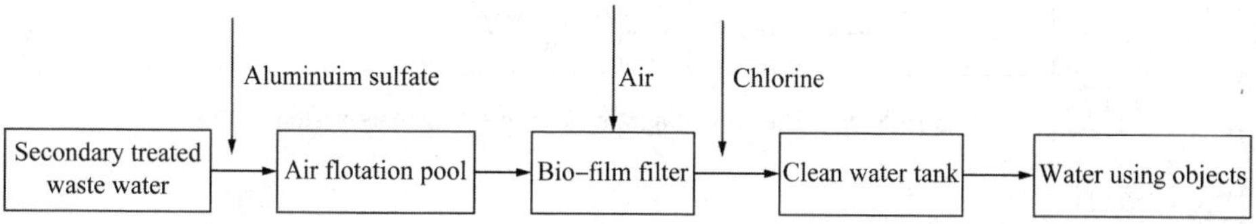

Figure 8-6 The flow diagram of air flotation filtration system

The two filtration systems are to create the aerobic conditions to promote the biofilm growth on the surface of the filters, resulting in the filtration tank with the function of biodegradation of the soluble organic matter. Therefore, the common filtration has double function of the biological oxidation and physical filtration. The systems not only remove the soluble organic matter and suspended solids but also improve the nitrification, improving the reused water quality. The removal of COD and BOD is higher 20% ~ 30%, and the bacteria is higher 85% ~ 90%, compared with the common filtration.

(6) Ozone system (Figure 8-7).

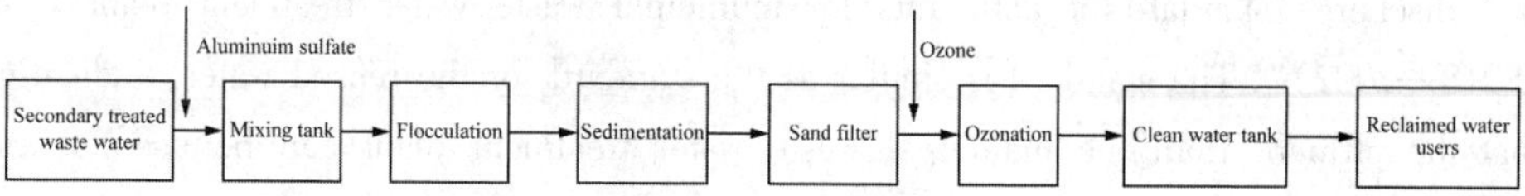

Figure 8-7 The flow diagram of Ozone system

The system can remove the colour and taste and odour in the secondary treatment effluent. The removal of taste and odour is near to 30% and the colour is up to 70%.

(7) Ozone-activated carbon system(Figure 8-8).

The system can remove the colour, taste and odour, micro-pollutants in the secondary treatment effluent, improving the reused water quality.

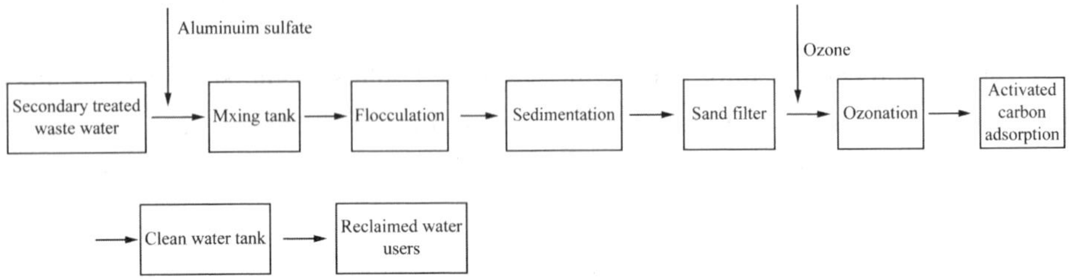

Figure 8-8 The flow diagram of ozone-activated carbon system

(8) Reverse osmosis system(Figure 8-9).

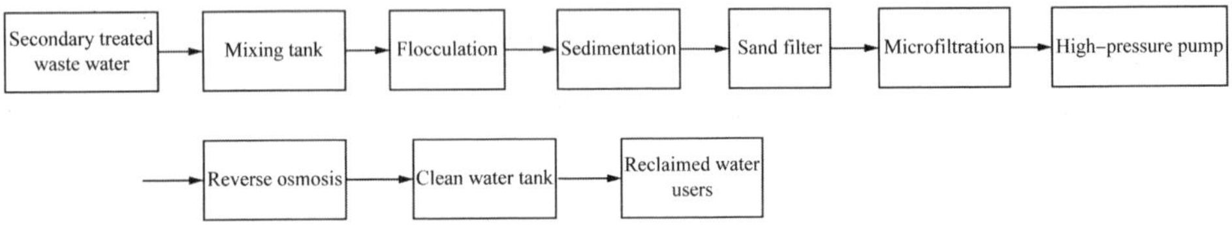

Figure 8-9 The flow diagram of reverse osmosis system

8.4.2 Complete flow scheme optimizing for water reuse

(1) The principle of complete flow scheme At the early 20th century, the activated sludge was created, but the key of the technology is to remove COD. Therefore, the quality of BOD_5 and SS in the effluent are only 20 mg/L and 20 mg/L, respectively. For the N and P in the waste water, the removal is little. To control the pollution in the natural water body, the tertiary treatment was proposed. On the base of the secondary treatment water, the physical, chemical, physical-chemical, biochemical technologies were used to improve the quality in the effluent. The water quality is similar as the reused water quality. Nowadays, in China, the discharge standard for the municipal waste water is " discharge standard of pollutants for municipal waste water treatment plant (GB 18918—2002)". The standard is similar as the standard for the reused water, indicating that the effluent from the municipal waste water treatment plant can be used as the reclaimed water.

In China, the reclaimed water plant can be built on the base of the municipal waste water treatment plant. After the second treatment in the municipal waste water treatment plant, the tertiary treatment is used to improve the micro-pollutants removal to meet the demand of the requirement of the reclaimed water. The widely used processes are as followed.

① The combination of traditional activated sludge and traditional water treatment process.

It is well known that the coagulation, sedimentation, filtration is the key part for the traditional water treatment process. The function of the traditional water treatment process is mainly to remove the debris of the activated sludge in the second sedimentation tank that do not settle. For the secondary treatment effluent, the BOD_5, COD, SS and bacteria can be efficiently removed. Another function is the chemical removal phosphorus by the coagulation. The process is as followed (Figure 8-10).

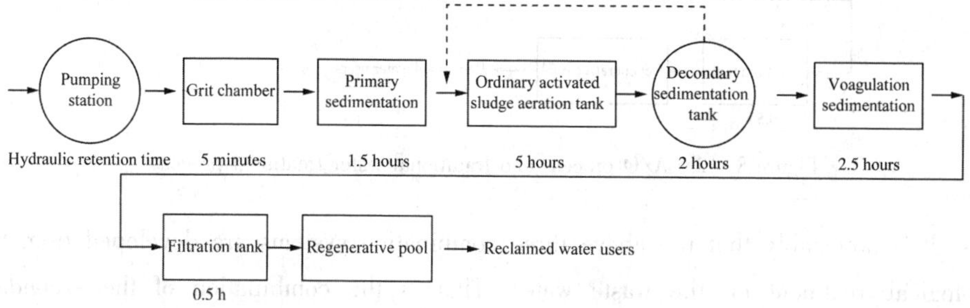

Figure 8-10 Traditional activated sludge and traditional water treatment process

② The combination of A/A/O process and traditional water treatment process. The combination of the systems is to strengthen the removal of organic matter and the N and P. However, the TN, in general, is difficult to meet the demand of the reclaimed water requirement. The P can be removed by the chemical removal. The process is as followed (Figure 8-11).

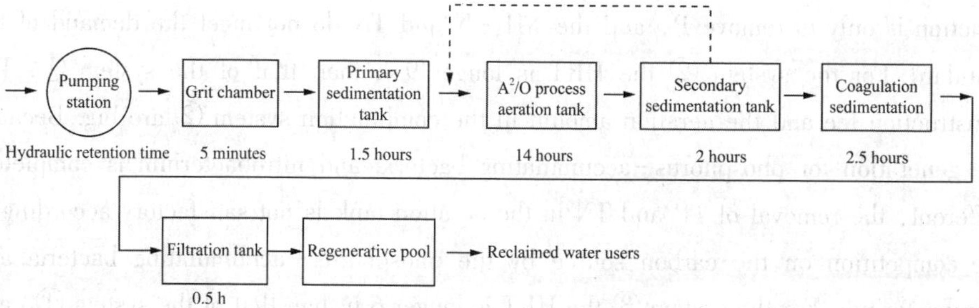

Figure 8-11 A/A/O process and traditional water treatment process

③ The combination of A/O process and traditional water treatment process. According to the function of the above process, the A/O process is selected to replace the A/A/O process, because the A/O process can sufficiently meet the demand on the metabolism conditions of the nitrobacterium and denitrifying bacteria, respectively. While the P can be removed by the coagulation in the advanced treatment processes. After the good designment and running, the combination process is in favor of the demand of the standard A in the discharge pollutants (GB 18918—2002) and the demand of most reclaimed water standard. The process is as followed (Figure 8-12).

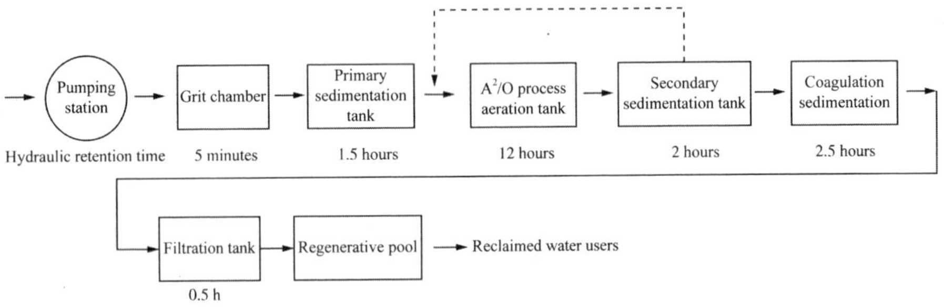

Figure 8-12 A/O process and traditional water treatment process

It is noticeably that the above three combination systems are developed from the biological treatment on the waste water. That is the combination of the secondary treatment of waste water and the tertiary treatment. The pollutant types and pollutant load maybe unreasonable, uneconomical, even mistake, because the designers do not consider the whole processes at the beginning of the designment.

The HRT is 11 h 35 min, 20 h 35 min and 18 h 35 min in the above three systems, respectively. However, in the system① the HRT is so small that it does not meet the demand of nitrification. Although the construction fee and the aeration amount in the combination system is economic, compared with the other two systems, its function is only to remove P, and the NH_3-N and TN do not meet the demand of the standard. For the system ② the HRT is longer 9 h than that of the system ①. The construction fee and the aeration amount in the combination system ② are big. Because the generation for phosphorus-accumulating bacteria and nitrobacterium is completely different, the removal of TP and TN in the aeration tank is not satisfactory according to the competition on the carbon source by the phosphorus-accumulating bacteria and nitrobacterium. For the system ③ the HRT is longer 6 h than that of the system ①, and the aeration amount is bigger. The removal of TP and TN in the aeration tank is satisfactory and the effluent quality meets the demand of the reclaimed water.

According to the analysis of the above three combination systems, the designment on the metabolism of the different bacteria is very important. Therefore, the principle of complete flow scheme is proposed. The key of the principle is to consider the primary treatment, secondary treatment and tertiary treatment according to waste water quality. In the combination system, the removal of pollutant type and load is reasonably distributed to the unit and tank in the whole system. The aim is to get the good quality of reclaimed water by the economic energy and consumption on the resource.

(2) The designment of complete flow scheme The treatment process is regarded as the organic and systematic scheme from raw waste water to the reclaimed water. The municipal waste water is the object and the reclaimed water is the aim, it is to optimize the unit treatment process based on the complete flow scheme. Improving the highly efficient unit and combining the different treatment process, it is to form the complete flow scheme with the characteristic of economy, to reduce the cost of the reclaimed water and improve the water quality, to create the second water supply resource for the city. This is favor of the decrease of the natural water resource.

According to the idea, it is to combine the traditional activated sludge, the anaerobic and aerobic biological treatment and the other kinds of nitrification, denitrification and biological removal of P for the designment of the complete flow scheme. The systems are as followed.

① The combination of A/O process for biological removal P and Canon process for biological removal N.

The process is as followed (Figure 8-13).

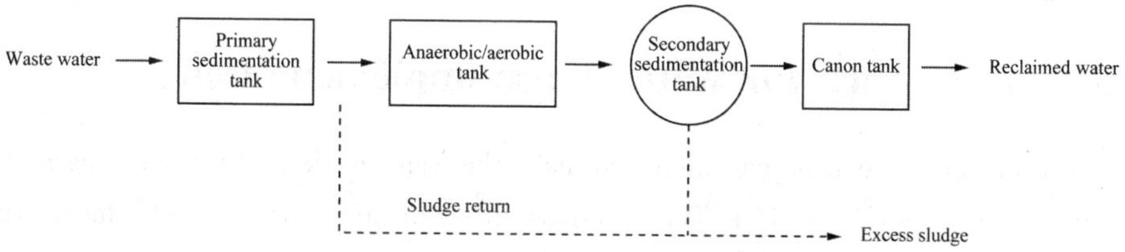

Figure 8-13 A/O process for biological removal P and Canon process

The secondary treatment process is based on the anaerobic-aerobic process, its BOD-SS load is up to 0.2~0.5 kg/(MLSS · d), similar as the traditional activated sludge. At the anaerobic stage, P is released from phosphorus-accumulating bacteria sufficiently. At the same time, the organic matter is reduced by the assimilation of the bacteria, the PHB is formed and stored in the bacteria. At the anaerobic stage, PHB in the body is decomposed by phosphorus-accumulating bacteria, and the organic matter in the water is degraded and the energy is released under the bacteria for the breeding of the bacteria. The most important is to adsorb P excessively and P as the polyphosphates is stored in the body of by the phosphorus-accumulating bacteria. It is noticeably that the tank is not used for the ammoniation and nitrification. The HRT and BOD-SS load of the process is similar as the traditional activated sludge. The advantage of the process is to remove organic matter and P under the conditions of the investment of capital construction and consumption of power with light increase. In the Canon tank, the partial nitrification and anaerobic ammonia oxidation are the same reacted to remove N. Compared with the traditional nitrification and denitrification, the consumption on the aeration is saved half.

② The combination of biological removal P and anaerobic ammonia oxidation.

The process is as followed (Figure 8-14).

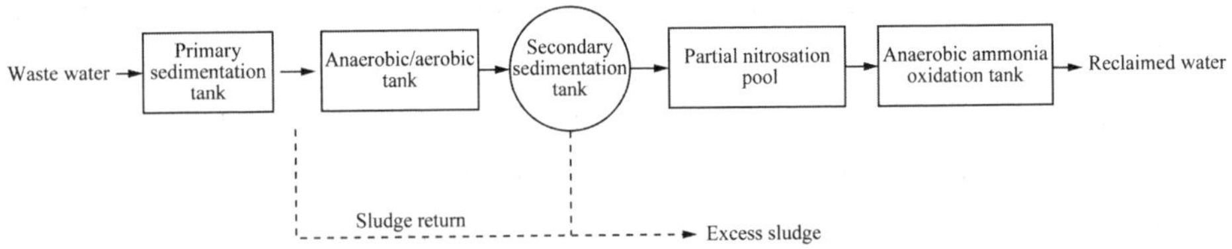

Figure 8-14 Biological removal P and anaerobic ammonia oxidation

In the Canon reactor, half of nitrifying bacteria and anaerobic ammonium oxidation bacteria coexist. Nitrite is the intermediate products. Under the anaerobic ammonium oxidation bacteria, the nitrite reacts with the ammonia nitrogen to produce nitrogen.

8.5 Framework for water reuse implementation

Several options can be considered to make the water reuse and recycle programs to be effective (see Figure 8-15). These options are: education and outreach, incentives, removing barriers, and mandates and regulations.

Education and outreach may include programs that promote certification programs and recognition awards. Programs to disseminate the scientific information and general benefits and essential facts about the water reuse by education and outreach efforts. Educating the users on water withdrawal trends, consumption, discharge and reuse data can help create interest in water reuse implementation.

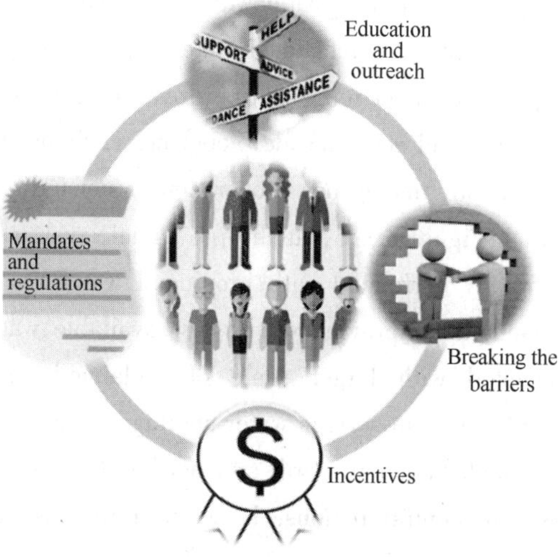

Figure 8-15 Framework for water reuse implementation

An incentive program including direct subsidies, reductions in payments to the government, payments for water reuse, adjusting pricing mechanisms, regulatory relief for recycled water users, government procurement of water recycling/reuse equipment and structuring of water rights to reduce the use of potable water can be beneficial in supporting the water reuse implementation.

Regulatory reliefs may include modification of local regulations depending on the application especially in terms of water quality standards, revising plumbing codes to allow dual piping, and monitoring permitting and inspection requirements for recycled water.

Creating mandatory requirements should consider utility planning and management to develop plans for recycled water, restricting potable water to human or food-related uses, and requiring the use of recycled water for agriculture and other non-potable uses and requiring water recovery systems.

8.6 Cases

8.6.1 Water reuse in China

8.6.1.1 The patterns of the water reuse

Water reuse is among the fastest growing utility sectors in China and reclaimed water is being widely applied to satisfy a significant proportion of water demands. In 2013, there were

571 water reclamation plants (WRP) nationwide equipped to further treat secondary effluents for planned reuse and the reclaimed water pipeline length is 8,300 km. The exploitation and development of water reuse can be driven by a variety of issues, including water shortage, technological and economic development, environmental protection, public awareness, effective regulation, interagency cooperation, water tariff, etc.

As can be seen from Figure 8-16, at the provincial level, the promotion of water reuse activities was strongly correlated with provincial water resource availability and regional GDP levels. Generally, provinces with low available water resource and high GDP levels are associated with larger scale of reclaimed water construction and utilization. Geographically, due to serious water shortage situations in northern regions, the water reuse rate is much higher in north and northeast China, followed by east and south China, but lower in central regions. In western regions, although drought and water deficit issues occur frequently, a low level of economic development has hindered the local water reuse progress to some extent.

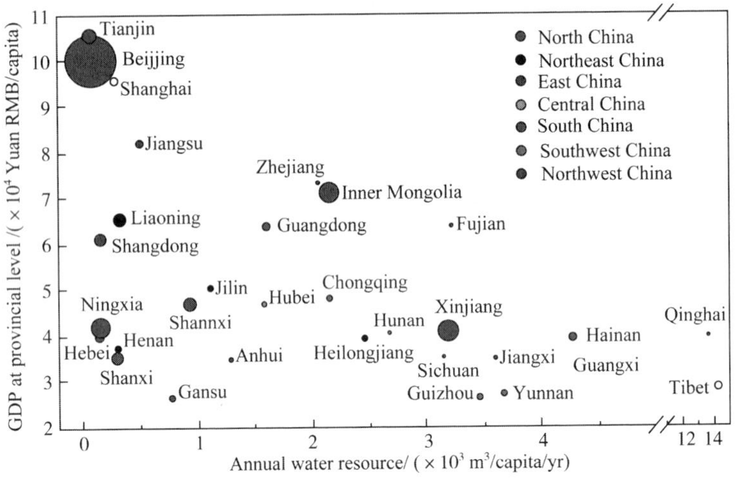

Figure 8-16 Correlations of the scale of reclaimed water facilities (m^3 per capita per year) in urban areas with water resource quantity and GDP level in 31 provinces of mainland China in 2015

Pattern Ⅰ: reclaimed water for single application

In a simple pattern, a centralized water reuse system serves only one type of water quality application (e.g. Intended use A in Figure 8-17). This is probably more likely on a smaller scale. This pattern is employed when 1) reclaimed water is being provided to a singer user, such as one industrial plant or an individual building, or 2) reclaimed water is being provided for a standard user or similar users, such as a community-based residential area where the quality of the reclaimed water needs to meet all requirements specified for water reuse applications.

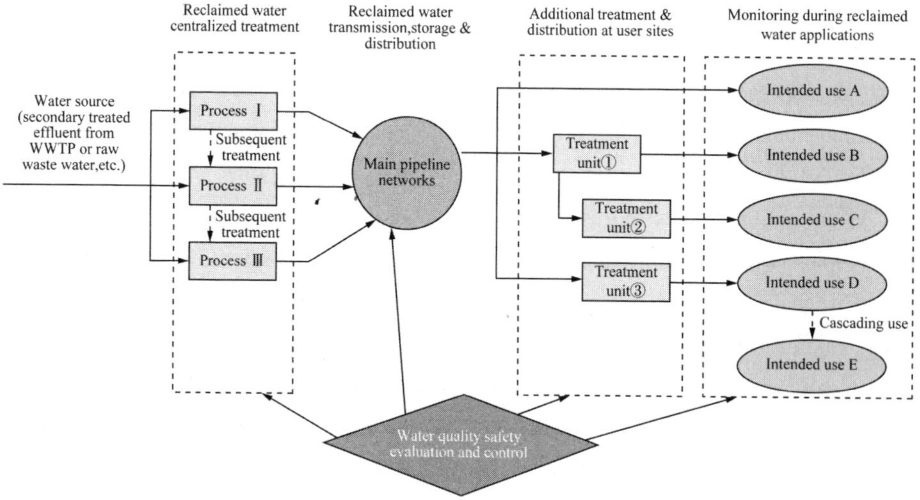

Figure 8-17 Generic framework of a centralized water reuse system

Pattern Ⅱ: reclaimed water for multiple applications

Pattern Ⅱ is being widely applied in large-scale centralized water reuse systems where the reclaimed water is for multiple users with different water quality requirements, such as in an industrial park with several industries (e.g. Beijing Yizhuang Industrial Development Zone, 30,000 m^3/d of reclaimed water with 46 different users), or a region with miscellaneous industrial and domestic uses (e.g. Tianjin Economic-Technological Development Area). Compared to Pattern I, a reuse project with multiple user agglomeration can effectively reduce the carbon footprint of land use, improve use efficiency and reduce pollution and waste. However, because of different user requirements, Pattern Ⅱ is complex in system design and operation. One of the key issues in system design is to select an appropriate treatment technology or technology combinations. For safety considerations, minimum water quality requirements should be ensured for all applications. Basically, the main treatment unit needs to satisfy the water quantity and quality requirements of large and high priority users (e.g. Intended use A in Figure 8-17). Small users with higher quality requirements can connect to the main treatment unit, and either provide an additional treatment unit before distribution or at the point-of-use (e.g. Intended use B-D in Figure 8-17). Alternatively, small users with lower quality requirements can connect directly to the main treatment unit without further treatment. In addition, a cascading use of reclaimed water for different reuse applications (e.g. Intended uses D to E in Figure 8-17) is also a common practice. For instance, industrial uses are large consumers of reclaimed water in China (Figure 8-17), a cascading use can be broadly introduced to industrial parks where waste water from an industrial process can be applied to subsequent reuse applications such as cleaning or ornamental landscape irrigation with/without additional specific water quality improvement.

Pattern Ⅲ: combination of treatment, environmental storage and use

As environment uses represent a large user of reclaimed water in China, Pattern III gives emphasis on the environmental functions of reclaimed water (Figure 8-18). This pattern is mainly considered when water bodies such as natural/artificial wetlands, ponds, lakes, rivers and streams located in close to the treatment system and can be used as storage reservoirs and/or treatment units in which reclaimed water can undergo further purification. For instance, in Beijing, some water bodies such as wetlands and lakes in Olympic Forest Park and Summer Palace receive reclaimed water for scenic environmental uses and are also used as storage reservoirs and treatment units. Specifically, for the Beijing Olympic Park project, reclaimed water from the Qinghe WRP (60,000 m^3/d) is processed by ultrafiltration, ozone oxidation and chlorine disinfection. Of the total water supply capacity, 3,200 m^3/d of reclaimed water is further treated by an eco-biological system including a constructed wetland and an indoor ecosystem (e.g. Environmental storage in Figure 8-18). A proportion of the water is subsequently polished by an aquatic plant pond and a hydraulic circulation system (e.g. Treatment unit 2 in Figure 8-18). The remaining water is supplied for road cleaning landscape irrigation, toilet flushing and other purposes where additional treatment might be needed (e.g. Treatment unit 3 in Figure 8-18). A detailed example is described by Sun et al., 2013. This project provides a good example for the establishment of modern environmental purification and maintenance system by urban water bodies.

8.6.1.2 The water reuse plant in Zaozhuang

(1) Overview The city Zaozhuang lies in Shandong province, at the south of Shandong province. Zaozhuang is a city in shortage of water. Although the ground water resource can meet the demand of the city, with the development, the water quality and quantity will affect the development of the city. Therefore, Zaozhuang city decided to build the water reuse plant based on the Zaozhuang Waste Water Treatment Plant.

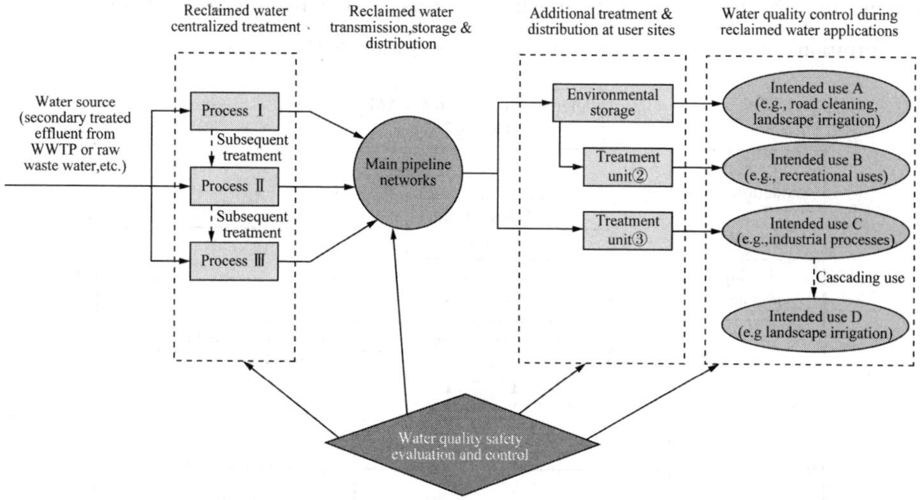

Figure 8-18 Typical example of Pattern Ⅲ for combined treatment, environmental storage and reuse.

(2) The water quality and quantity.

① Influent water quantity. The waste water treatment plant was designed at 1997. The effluent is 70,000 m^3/d.

② Influent water quality. The water quality is descried in the Table 8-7.

Table 8-7 The waste water quality

pH	SS/(mg/L)	Hardness/(mg/L)	Cl^-/(mg/L)	SO_4^{2-}/(mg/L)	NH_3-N/(mg/L)	COD/(mg/L)	BOD/(mg/L)	Colour
6.0~9.0	20	678	61	544	2	100	20	80

③ effluent water quality. The water quality is descried in the Table 8-8.

Table 8-8 The reclaimed water quality

Item	Value	Item	Value
pH	7.0~9.2	NH_3-N/(mg/L)	≤3
SS/(mg/L)	≤5	Fe^{2+}(mg/L)	≤0.5
Alkalinity/(mg/L)	≤500	TP/(mg/L)	≤3
Total hardness/(mg/L)	<600	COD/(mg/L)	≤50
Cl^-/(mg/L)	≤120	Free chlorine/(mg/L)	0.5~1.0
SO_4^{2-}/(mg/L)	≤1,500	Silicic acid/(mg/L)	≤175

(3) The water treatment process Based on the analysis of several schemes, the

effluent from the waste water treatment plant was treated by biological ceramsite, fiber ball filtration, disinfection. It is showed in the Figure 8-19.

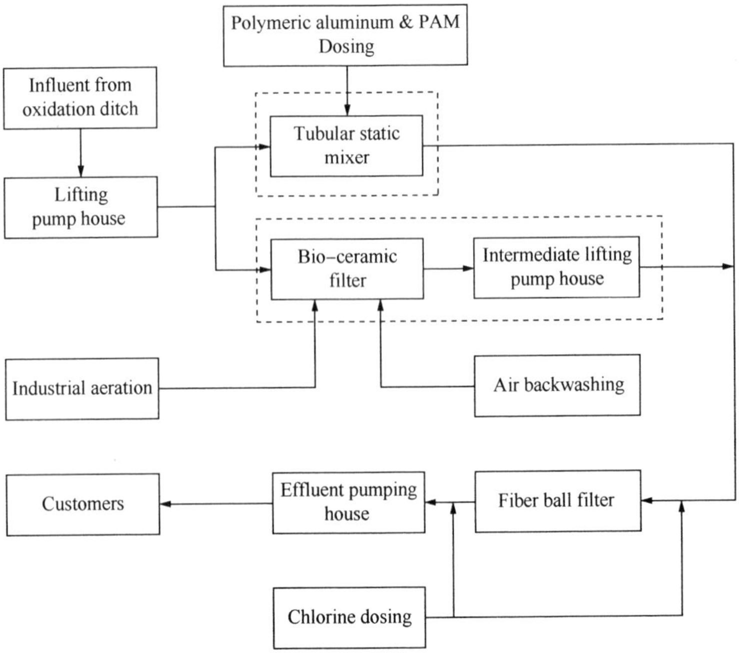

Figure 8-19 The flow chart of the water reuse process

The advantages of the biological ceramsite are as followed.

① The specific surface area of the ceramsite is big.

② The air passes upflow the ceramsite. The distribution of the air is even and the oxygen transfer efficiency is high.

③ The contact time is long for the ceramsite with the waste water. Therefore, the efficiency is high.

④ The sludge age is long, and the residual sludge is little.

⑤ The biological ceramsite filter has the function of filtration.

(4) The designment of the main tank.

① The first pump station.

Submerged pump, $Q=800\ m^3/h$; $H=12$ m. The quantity of the submerged pump is 4.

② Biological ceramsite system.

Design flow: total, 2,083 m^3/h; One room, 173.6 m^3/h.

Filtration speed: 4 m/h.

Aeration amount: 2.71 L/(m^2 · s).

Backwash air amount: 21.6 L/(m^2 · s).

Backwash water amount: 108.5 m^3/h.

Tank quantity: 1.

Room quantity: 12.

Effective water depth: 5.6 m.

The size of the room: 4.8 m×9.6 m×6.1 m

Aeration area: 553.0 m^2(total), 46.1 m^2(room).

Effective volume: 3,096.6 m^3(total), 258 m^3(room).

Materials: reinforced concrete.

Biological ceramsite filter: total depth 3,000 mm; the size 2~5 mm.

Broken stone filter: total depth 200 mm; granular size 4~8 mm, 8~16 mm; each layer 100 mm.

Distribution system of filters: low head loss distribution system of filters.

The main equipment: backwash pump (ZHWD400-400), Q=1,300 m^3/h, H=11 m. The quantity of the pump is 3. Aerators for backwash.

③ The middle pump station.

It is underground pump station. The area is 52.8 m^2, and in the room there are 6 horizontal pumps.

Device type: horizontal low-speed centrifugal pump.

Model: ZHWD400-400

Technical parameters: Q=1,080 m^3/h, H=12.5 m.

Installed power: 55 kW.

④ Coagulant addition equipment.

It is only used when the water quality is abnormal.

The area of the coagulant addition room is 102.6 m^2. The model of the coagulant addition equipment is WA-0.6/1.44B2.

Coagulant addition: rotor flow meter.

Dose of the coagulant: 10 mg/L.

Total dose of the coagulant: 20.83 kg/h.

Model: WA-0.6/1.44B2.

Effective volume of the dissolved coagulant slot: 1.44 m^3.

Installed power: 3.7 kW.

The number: 3.

⑤ Fiber ball filtration.

Fiber ball filtration room occupied 465 m^2. 14 fiber ball filtrations with diameter 3 m were chosen.

Design flow: 2,083 m^3/h (total), 148.8 m^3/h (one pot).

Model: QLJ-3000.

Filtration speed: 21 m/h.

Diameter: 3 m.

Mixing power during backwashing: 22 kW.

Backwash intensity: 30 $m^3/(m^2 \cdot h)$.

Backwash time: 20 min.

Backwash amount: 42 m^3/h.

The backwash pump is horizontal centrifugal pump. The model is ZHWD200-200A. $Q=216\ m^3/h$, $H=8.9$ m. Installed power is 11 kW.

⑥ Chlorine is used for the disinfectant. The dose of the chlorine is 10 mg/L, and the total dose is 20.83 kg/h.

Equipment: vacuum chlorinator.

Model: JHX40.

Chlorine addition amount: 2~40 kg/h.

⑦ The second pump station.

It is semi-underground pump station. The effective volume is 500 m^3, and the area of structure is 51.4 m^2.

Device type: horizontal low-speed centrifugal pump.

Model: ZHWD400-400

Technical parameters: $Q=1,080\ m^3/h$, $H=12.5$ m.

Installed power: 55 kW.

The number: 3.

(5) The project investment and economic analysis.

The investment: 27.23 million Yuan (RMB).

Running cost (no depreciation): 0.28 Yuan/m^3 water.

Consumption on the power: 0.257 kW · h/ m^3 water.

(6) The running Table 8-9 is the effluent of the whole process. Table 8-10 is the effluent without biological ceramsite filtration.

Table 8-9 The effluent of the whole process

Item	Unit	Influent	Biological ceramsite filtration	Fiber ball filtration	Disinfection	Effluent
pH	mg/L	7.23	7.35	7.20	7.34	7.41
COD	mg/L	2.1	1.8	1.6	1.8	2.0
SS	mg/L	1.1	0.9	0.5	1.2	0.8
Turbidity	NTU	0.9	1.6	1.5	0.7	0.6
NH_3-N	mg/L	25.72	22.97	23.93	23.93	19.80

Table 8-10 The effluent without biological ceramsite filtration

Item	Unit	I	II	III
Appearance		Clear	Clear	Clear
pH		7.84	7.51	7.86
Conductivity	μS/cm	1,126	1,367	1,386
Total solid	mg/L	920	960	900
Soluble solid	mg/L	916	924	897
SS	mg/L	3.0	2.0	3.0
Loss on ignition	mg/L	219.6	—	—
SiO_2	mg/L	14.8	5.9	11.6
SO_4^{2-}	mg/L	272	298	268
OH^-	mg/L	—	—	—
CO_3^{2-}	mg/L	0	0	0
HCO_3^-	mg/L	174.67	314.8	248.5
Total hardness	mmol/L	4.15	4.26	4.86
Ca hardness	mmol/L	1.33	2.12	3.62
Mg hardness	mmol/L	2.04	1.15	1.24
Cl^-	mg/L	88.58	98.69	110.86
NO_3^-	mg/L	8.71	8.25	5.70
COD	mg/L	27.1	40.1	44.8
Cd^{2+}	mg/L	—	—	—
Fe^{3+}	mg/L	<0.01	<0.01	<0.01
Turbidity	NTU	1.8	2.1	1.1

8.6.1.3 The water reuse plant in Dalian city

Dalian city is shortage of water and the average amount of water is less than 200

m^3, only the 7% of the average in China. The water has been the limit factor of the economy development and this affects the living. The government paid attention to water reuse. At 1991, the city has built the water reuse demonstration project with the scale of 1×10^4 m^3/d. And then, some waste water treatment plants have been transformed into the water reuse plant. The reused water was mainly used for the landscaping, building operations, municipal water, toilet and urinal flushing et al and the cooling water in several factories.

The Chunliuhe Waste Water Treatment Plant in Dalian is the first water reuse demonstration project in China. The amount of reused water is 1×10^4 m^3/d. The process is coagulation-filtration-disinfection (Figure 8-20).

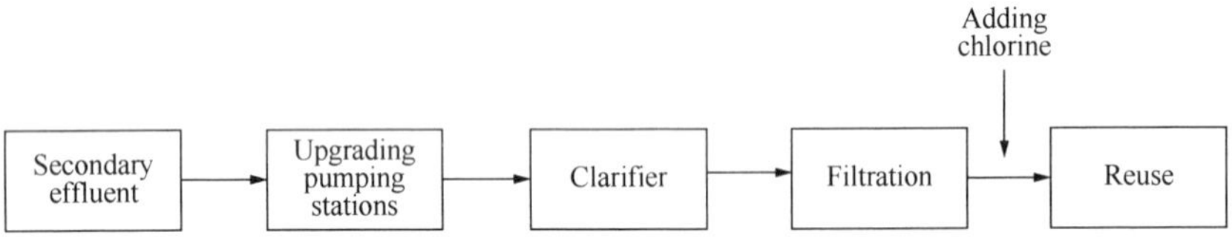

Figure 8-20 The main process for water reuse in Dalian

For the demonstration project, the clarification tank replaced the sedimentation tank. There were 2 circular clarification tanks. In running, it is necessary to add some loess to keep the sludge layer. If the running is improper, adding polyacrylamide as the coagulant aids is to improve the coagulation. There are 4 double valve siphon filters. The filter speed is lower, compared to the portable water. The removal of the pollutants is showed in Table 8-11.

Table 8-11 The removal on the pollutants

Item	Removal/%		Quality/(mg/L)	Item	Removal/%		Quality/(mg/L)
	clarification	filtration			clarification	filtration	
Turbidity(NTU)	62	33	4	TP	10	28	6
BOD_5	37	48	5	SS	10	80	0.1
COD_{Cr}	25	20	39	Iron	—	62	—

The total investment of the water reuse demonstration project of the Chunliuhe Waste Water Treatment Plant is 3.5 million Yuan (RMB). The unit investment is 350 Yuan/m^3(RMB), while the unit investment is 1,052 Yuan/m^3(RMB) for Yinbi project to provide the potable water. Therefore, the investment of the water reuse is obviously lower than that in the Yinbi project. On the other hand, the running cost (do not include the secondary biological treatment) is 0.25 Yuan/m^3(RMB) for the water reuse project, while the running cost is 1.485 Yuan/m^3(RMB) for Yinbi project. At 1995, the cost on the reused water is only 0.48 Yuan/m^3(RMB), while the portable water cost is 1.5 Yuan/m^3(RMB) for the industrial object. It is indicating that the water reuse project not only removes the problem of shortage of water, but also brings good economic benefits.

8.6.1.4 The water reuse engineering in Harbin refinery

(1) Overview The water amount is 8,000 t/d in Harbin refinery. The project was divided into 2 stages. The first is 4,000 t/d, but the plant area, the tanks and the power is designed according to 8,000 t/d. After the construction is finished, the equipment is installed according to 4,000 t/d. The other 4,000 t/d is booked.

The standard for the reused water is showed in the Table 8-12.

Table 8-12 Standard of the reused water

pH	COD/(mg/L)	BOD/(mg/L)	SS/(mg/L)	Turbidity/(NTU)	Total hydride/(mg/L)	Volatile Penol/(mg/L)	Sulfide/(mg/L)	Petroleum/(mg/L)
6.5~8.5	≤20	≤6	≤5	≤5	≤0.2	≤0.01	≤0.01	≤0.5

(2) The process The process is showed in the Figure 8-21.

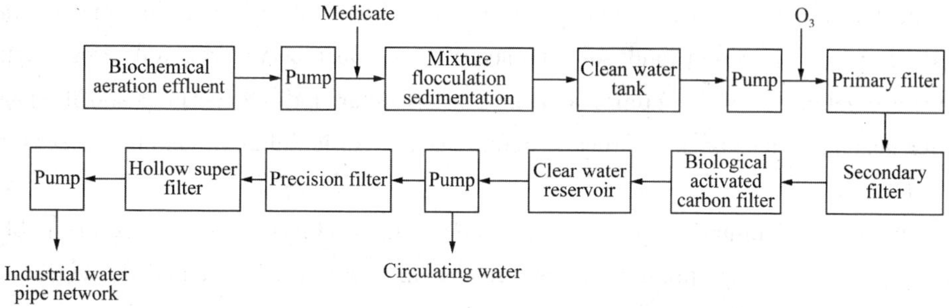

Figure 8-21 The water reuse process in Harbin refinery

The quality of the influent of the water reuse engineering is showed in the Table 8-13.

Table 8-13 The quality of the influent of the water reuse engineering

pH	COD/(mg/L)	BOD/(mg/L)	SS/(mg/L)	Turbidity/(NTU)	Total hydride/(mg/L)	Volatile Penol/(mg/L)	Sulfide/(mg/L)	Petroleum/(mg/L)
7.0~9.0	≤120	≤60	≤2,000	≤30	≤0.5	≤0.5	≤1.0	≤10

(3) The running efficiency At 1998, the water reuse engineering is in practice. The quality of the effluent meets the demand of the water reuse.

(4) The energy consumption and the economic efficiency The energy consumption is as followed. Power, 0.85 kW · h/t (water). Steam, 1.3 kg/t (water). Air, 0.48 m^3/ t (water). Coagulant, 0.018 kg/t (water).

The running cost is 1.05 Yuan (RMB)/t.

8.6.2 The others

8.6.2.1 Agricultural irrigation

Agricultural irrigation evolved from the early practice of sewage farming when untreated municipal waste water (sewage) was directly applied to crops. In some parts of the world that practice is still used, in spite of adverse health and environmental impacts. Agricultural irrigation is the largest user of reclaimed water in the United States and in most of the world. Depending on the use of the crop for food or nonfood purposes and how the irrigation water is to be applied, i.e., by ridge and furrows, sprinklers, or surface/ subsurface drip irrigation, the degree of the required treatment will vary. In almost all irrigation applications, a minimum of secondary treatment is required. For spray application to food crops, if allowed, higher levels of treatment including disinfection are mandatory. Other considerations in agricultural irrigation include the type of crop; topography and soil characteristics; effect of water quality, especially dissolved solids, on soils, crops, and the underlying groundwater; runoff and drainage water management; and infrastructure requirements. Agricultural irrigation is discussed as followed.

Various agricultural crops can be irrigated with reclaimed water. To ensure public health protection, appropriate water reclamation processes must be adopted, depending on the crop type and irrigation method.

(1) Water quality considerations The sustainability of irrigation schemes depends on the proper management of potential impacts of the available water on soils, crops, and the environment. Knowledge of agronomics is used for all stages of a reclaimed water irrigation project, from the feasibility analysis through the design, construction, and management of the constructed system. Relevant topics in agronomics and water quality for the Legation with reclaimed water are introduced, including① soil characteristics; ② suspended solids; ③ salinity, sodicity, and specific ion toxicity; ④ trace elements and nutrients; and ⑤ considerations for crop selection.

It should be noted that specific water quality requirements depend on soil characteristics, climate, plants to be irrigated, irrigation method, and other local conditions. The most important characteristics of irrigation water for determining its quality are: ① concentration of soluble salts, ② relative proportion of sodium to other cations (magnesium, calcium, and potassium), and ③ concentration of boron and other elements that may be toxic to plants.

Soil Texture:

Soil textural classes, used to classify the physical properties of soil, are based on the relative percentages of sand, silt, and clay. The three classes of soil, i.e., sand, silt, and clay, are determined by the fraction of particles that are smaller than 2 mm in diameter. For example, if the soil contains 60 percent silt, 30 percent sand, and 10 percent clay, the soil is classified as silt loam. The class of the soil can be determined from the particle size distribution or estimated by soil scientists in the field. Soils that exhibit the best balance for reclaimed water irrigation are medium textured; very fine sandy loam, loam, silt loam, and silt.

Physical Structure:

Soil particles are aggregated by chemical and biological processes to form natural structures. The structure of soil affects the rate at which air and water move enough it. It also affects root development and the nutrient supply to plants. Soils with prismatic, blocky, and granular structures have better water permeability than the platy or massive (structureless) soils. The prismatic and bloody structures, however, induce peripheral flow of irrigation water around the soil structure, and restrict root development. Therefore, a more granular structure is preferred for growing crops. The structure of soil can be "changed" by specific cultivation practices, including deep tillage, addition of organic matter, and addition of inorganic chemicals such as gypsum calcium sulfate.

Soil chemical characteristics:

Soil is formed primarily from decomposed rocks. The chemical and mineral composition of soil varies with the extent of oxidation, reduction, hydration, hydrolysis, and carbonation that has occurred. Soils composed predominantly of minerals are called mineral soils. In mineral soils, the content of organic matter and oxygen, and the temperature are generally highest at the surface. Microbiological activity is also greatest near the soil surface. The ability of soil to retain cations including nutrients is affected by the soil chemical characteristics and organic contend soils with linked ability to retain cations may require more frequent application of fertilizers.

(2) Elements for the design of reclaimed water irrigation systems A water reuse project for agricultural irrigation may be planned with an existing waste water treatment facility, or with a new facility designed specifically for the intended reclaimed water users. With the existing facilities, the feasibility of water reuse for apicultural irrigation depends on the location of the water reclamation facilities relative to the site of use, the quantity and quality of water required for irrigation, and the cost of providing reclaimed water. Depending on the crops to be irrigated, the existing waste water treatment facilities may have to be upgraded to meet the reclaimed water quality criers'. When planning a new treatment facility to provide reclaimed water for agricultural reuse, it is prudent to locate the facility near reuse opportunities to minimize infrastructure needed to convey reclaimed water.

8.6.2.2 Landscape irrigation

Landscape plants provide various functions such as creating an aesthetically pleasing property; creaking a buffer between streets, parking lots, and noncommercial areas; and providing vertical and horizontal dimensions to a site. Ornamental plants also maintain moisture and may be used to mitigate the effects of heat in urban areas.

Although the water demand for landscape irrigation varies greatly by geographical location, season, and the types of plants, approximately one-dried of residential water use is for landscape irrigation, with significantly higher usage in arid urban areas. For example, the Irvine Ranch Water District in southern California estimates that more than 70 percent of their total water use is for landscape irrigation.

Landscape irrigation with reclaimed water is a viable option to reduce potable water demand, and also as an option to reduce or eliminate waste water discharge to aquatic environment. Factors motivating many local governments to consider the use of reclaimed water include: (1) the high water demand for landscaping, (2) increasing cast of acquiring additional water in urban areas, and (3) stringent waste water discharge requirements.

(1) Water quality If the golf course is the primary user of reclaimed water user, the water reclamation process may be selected and/or adjusted to meet the quality requirements specifically for the golf course. As an alternative, the water quality may be modified by an onsite treatment system, Common onsite conditioning of reclaimed water includes pH adjustment (topically 6.5 to 8.4), corrosion prevention, and SAR adjustment. Fertilizer may be added to reclaimed water at the beginning of the reclaimed

water distribution line, but it is more common to apply fertilizer directly to the turf and other landscape plants as needed. Fertilizer application rates should be adjusted where the reclaimed significance amount of nutrients, but the value of clients in reclaimed water is often ignored, resulting in anoverapplication of fertilizers.

In arid regions, it is sometimes difficult to control leaching fraction to maintain salinity levels in root zone, resulting in a buildup of salts at the end of a dry season. Seasonal rainfall is usually sufficient to lower the salinity level in the root zone; however, additional management measures may be necessary if the salinity level is elevated over years of irrigation with reclaimed water.

(2) Irrigation method selection The irrigation methods used commonly at golf courses include sprinklers, and surface subsurface drip systems. The most common method of turf irrigation with claimed water is by sprinkler application. Subsurface irrigation is popular, especially for tees and greens, because of high irrigation efficiency and very limited human exposure to reclaimed water.

Clogging of sprinkler heads by particles is not a significant issue when the suspended solids are at the levels observed typically in tertiary-treated reclaimed water and the maximum particle size is controlled. Appropriate management measures need to be taken to prevent emitter clogging of surface and subsurface drip irrigation systems.

The entire irrigation system within a golf course is often controlled through a telemetry system a centralized control station. Moisture sensors are placed adjacent to the control valves, and the timing and amount of irrigation are controlled to optimize the irrigation rates. Litigation is typically respected to nighttime and early morning hours when no golfers are present and the turf should be dry by Ale time golfers are ready to play.

The agronomic application rate of reclaimed water is determined by ET and leaching requirements. The capacity of the reclaimed water delivery systems is designed based on the required peak flow. Depending on the relative locations of a water reclamation plant and a golf course, it may be feasible to install a reclaimed water distribution line dedicated to the golf course. Reclaimed water storage facilities may be needed to ensure the reclaimed water supply meets the daily and seasonal uses.

Lakes and ponds can be used for reclaimed water storage and integrated into the golf course to provide challenging obstacles for golfers. If a golf course uses reclaimed water for irrigation and is adjacent to surface waters that cannot be used for the discharge of excess reclaimed water, additional storage may be necessary to ensure all the reclaimed water produced is used by the golf course. In some golf courses, lakes and ponds also collect stormwater runoff. In the event of heavy rain, incidental overflow of the ponds containing reclaimed water may occur. For a reclaimed water storage pond which also serves as a storm water collection system, the golf course may need to obtain a National Pollutant Discharge Elimination System (BODES) permit, and the ponds may have to be designed and/or modified to increase their capacity for storm water.

Control of algae and weed growth is a concern in operation and maintenance of golf course lakes and ponds, and should be considered daring the design phase of a golf course. Measures to control algae growth in the lakes are discussed.

The irrigation distribution system conveys the reclaimed water the transmission system to the application points on the golf course. A distribution system should be designed to minimize dead ends, which can be accomplished by establishing a loop in the distribution system. Valves or hydrants for flushing the pipelines should be provided at all dead ends and low points.

In arid regions, dual distribution and plumbing systems with reclaimed water and another source of water such as potable water may be installed for specific areas such as tees and greens, if the effects of salinity by reclaimed water irrigation need to be mitigated using water of low salinity. A dual distribution system may also be considered upstream of the storage ponds to blend low salinity water and reclaimed water for the entire irrigation system. Where potable water is used as the low salinity water, it is important to take appropriate measures to avoid cross-connection. Backflow preventions; valves and/or air-gap backflow prevention should be installed on the potable water system, and colored pipes or tapes (normally purple) must be used for the reclaimed water distribution lines.

Because water requirements for turfgrass tend to be higher than other types of landscaping plants such as trees and shrubs, and golf courses need to maintain the turf in good condition for golfers, water applications rates for golf courses are generally higher than most other landscape irrigation applications. Intense application of water and the need to maintain a good turf condition necessity careful consideration of leaching and drainage requirements.

8.6.2.3 Urban water use and water reuse applications

(1) Air conditioning Most high-rise commercial and residential buildings have central cooling and heating systems. Water used for air conditioning systems represents one of the major water demands in commercial buildings. Air conditioning systems in

commercial applications considered in this section.

To cool indoor air for buildings and commercial facilities, air conditioning equipment needs a coolant. The coolant can be either water or refrigerants such as liquefied propane gas or hydrofluorocarbons (HFCs). In water-cooled air conditioning systems, a water chiller is used to chill the coolant water (i.e., chilled water cooling system). Water chillers air-cooled or water-cooled in releasing heat from the heat exchanger. For a single building, air-cooled chillers are used most commonly. For lager buildings and commercial sites, water-cooled chillers are used. In a water-cooled chiller system, a cooling tower is used to cool the heat exchanger cooling water. Circulating cooling water can be cooled directly with a cooling tower (open-circuit type), or a separate cooling tower water can be used indirectly for cooling (closed-circuit type).

Cooling towers used for high-rise buildings and commercial sites are usually located on the top of or behind the building where public access is limited. When reclaimed water used for cooling tower systems, dual plumbing systems for potable and reclaimed water supplies must be installed. As noted previously, insulation of a dual plumbing system usually is limited to new construction as it is not feasible moronically to retrofit most existing buildings.

Generally, reclaimed water that meets quality criteria for unrestricted urban uses is acceptable for cooling water, with some specific requirements to prevent scaling, corrosion, and biofouling problems. High ammonia levels in reclaimed water are a special concern as they will cause biologist growth and corrosion; thus, denitrified reclaimed water is prefect for cooling systems. Special precautions must be taken when reclaimed water is used for the cooling towers on buildings in urban areas because of the potential exposure of the public to aerosols emitted from the cooling towers.

Another important water quality issue is prevention of biological growth. Extra caution must be taken to prevent the growth of algae. Cooling water systems are often provided with chemical conditioning systems for pH adjustment and for the addition of other chemicals such as antiscalant, antifoaming agent, and disinfectant.

(2) Fire protection The use of potable water for fire protection requires an appropriately sized distribution system capable of supplying a high flow rate for a limited and irregular time period. When a potable water pipeline is used for multiple purposes including fire protection, the average flowrate is much less than the pipe capacity, resulting in long residence times in the pipeline with accompanying biofilm growth and

water quality degradation. The use of reclaimed water for fire protection allows the potable water pipeline to be designed for much lower peak flowrate and shorter residence times.

While the use of reclaimed water for fire protection can reduce the maximum design flow for a potable water system, additional considerations are required in the design, operation, and maintenance of the reclaimed water distribution system.

The types of fire protection systems are either (1) outdoor systems or (2) indoor systems. Typically, fire hydrants are the principal appurtenances used in outdoor systems. Hydrants are installed directly on the reclaimed enter distribution system. Sprinter systems are used for indoor fire protection.

(3) Toilet and urinal flushing Water demand in high-rise commercial buildings is mostly for toilet and urinal flushing and cooling water for air conditioning. The use of reclaimed water for toilet and urinal flushing in commercial and other buildings can, therefore, reduce the potable water demand. The feasibility of using reclaimed water for toilet and urinal flushing depends primarily on the plumbing and related infrastructure costs. The use of reclaimed water for residential homes and buildings can be more challenging, due to the concerns about potential cross-connections and accidental exposure of reclaimed water, especially to small children.

Reclaimed water toilet and urinal flushing systems may be installed in high-rise commercial and public buildings, residential buildings, and restrooms in public parks. Toilet and urinal flushing systems may be designed as part of a mixed urban water reuse plan with a centralized reclaimed water distribution system, or as part of a small scale, onsite treatment and reuse scheme. With respect to toilets and urinals, the same types of fixtures used with potable water can be used with reclaimed water.

Reclaimed water used for toilet flushing must be odorless and colorless for aesthetic reasons, and highly disinfected for public health protection in case of accidental human contact. Water quality must be considered when using reclaimed water for toilet and urinal flushing. In Japan, where toilet flushing is one of the most common water reuse applications, criteria for odor and color are included in the water quality requirements.

Color and odor are two major water quality issues for toilet and urinal flushing. Organic compounds (mostly humic substances) and inorganic compounds (such as iron) can cause coloration of reclaimed water. Oxidizing agents such as chlorine, ozone, and hydrogen peroxide can be used for the removal of color and odor. In China, to meet the color limit, an ozone treatment unit and an ozone-resistant membrane filtration system were added.

Satellite and decentralized water reclamation and reuse systems are becoming increasingly popular, aided by advances in water reclamation technologies that produce a high quality effluent and have a small footprint. Other types of satellite and decentralized systems utilizing grey water for toilet flushing are found more commonly in European countries and Japan.

Water quality monitoring is important to ensure that public health is protected. Generally, reclaimed water quality monitoring is conducted routinely the water reclamation facility before reclaimed water is distributed.

Questions

1. What is water reclamation?
2. The importance of water reclamation.
3. There are many applications of water reclamation. Presenting some examples.
4. How to reuse the effluent from the waste water treatment plant? What factors should be considered before reuse?
5. If you are the head of the domestic waste water treatment plant, how to save energy?
6. If you are the engineer in the domestic waste water treatment plant, what factors should be focused on for you?

Chapter 9 Waste water recovery

Emphasis: Waste water recovery. The importance of waste water recovery. Waste water recovery method.

9.1 What is waste water recovery

After waste water is treated by physical, chemical or biological methods, the pollutants, such as carbon or phosphorus can be recovered and then the effluent can meet the reuse standard of reclaimed water. This is called waste water recovery. In a word, this is a significant measure to recycle resource and increase water resource efficiency.

9.1.1 Pollutant recovery

Generally, there is lots of pollutants in waste water, such as organic matter (protein, fat, etc.), phosphorus, nitrogen. If the pollutants can be recovered, it is obvious that the waste water will be a significant resource to provide constant matters every time.

It is well known that municipal waste water treatment plant is mainly to remove organic matter. That is to degrade organic matter to CO_2 by biological methods. During the process, it is needed to provide oxygen to oxidize the organic matter by bacterial and this is the dominant for energy use in waste water treatment plant. However. If the organic matter is not degraded into CO_2 by biological methods, the necessary energy consumption can be saved. Therefore, some researchers proposed to recover organic matter while not to degrade organic matter. The accumulated organic matter can be used to produce methane, volatile fatty acid (VFA), polyhydroxyalkanoates (PHA), microbial protein, or alginate. It is obvious that methane production has been the main waste water recovery ways. In some waste water treatment plants, the methane from organic matter can provide parts of power to maintain plant running.

Another important pollutant is phosphorus in municipal waste water. Phosphorus is necessary element for life. In fertilizers, phosphorus can be replaced by other elements because phosphorus is necessary for plant growth. However, phosphorus is non-renewable resource. The storage of phosphate rock is about 71 billions tons for commercial use based on the information from United States Geological Survey. About 70% of phosphate rock is in Morocco. Natural phosphate rock decreased gradually year by year due to mining. It is reported that the discharged phosphorus in waste water can meet 15%~20% of phosphorus demand by people. According to statistic, about 16% of phosphorus is utilized by people and discharged into waste water treatment plant, resulting in that phosphorus content is about 4~16 mg/L in the influent and about 50% of phosphorus in the influent will be utilized and stored by the bacterial. It is indicating that phosphorus recovery from municipal sewage is significant.

9.1.2 Water reuse

The significance of water reuse is as following.

(1) The distribution of water resource is not even Total amount of water resource in China is about 2,812.4 billion m^3, but the water resource per capita is only 2,200 m^3, ranking 110 in the world. In addition, China is one of 13 countries which are abject poverty of water resource.

The distribution of water resource is not even. In China, the difference is very obvious on region and season of water resource. It is more in east and south of China, while it is less in west and north of China. Taking the region at the north of Yellow river as an example, cultivated land is about 64% of China, while water resource is only 19% of total amount. Additionally, in the region, the raining is mainly in July, August and September. Therefore, the shortage of water resource is very obvious.

(2) Water pollution intensifies water resource shortage It is reported that the rivers, lakes and reservoirs have been polluted on varying degrees and about 80% of surface water and 54% of ground water are polluted. Over 90% of city water bodies are polluted heavily.

It is noted that the danger is very nervous due to water pollution. It is reported that about 25% of human disease is directly or indirectly related with water pollution. Drinking worse water can cause some infectious disease, endemic disease, mutagenicity, acute or chronic poisoning.

(3) Water efficiency is low Social water demand increases sharply due to the

increase of population and economy and fast of urbanization process, but the water storage is limited. According to statics, there are 669 countries in China, and water supply is not enough for 400 countries throughout the year, while 110 countries are abject shortage of water resource. There are 32 big countries with millions of people, and 30 countries are troubled due to water shortage. It is reported that the shortage on water resource is about 6 billion m^3, resulting in affecting on 230 billion yuan of total value of industrial output.

It is noticeable that there is heavily waste on water use in industry and agriculture. Although rate of industrial water reuse grows up year by year, up to 83% at 2008, but water requirement per unit of output is still obviously higher than that in developed countries.

In a word, water danger and its water quality problems and ecology problems not only hindered economy development but also caused important social and economic danger. Therefore, it is necessary to save fully water resource and to develop non-traditional water resource. The aim is to alleviate a series of serious problems due to water shortage.

Waste water reuse is an important measure to solve water shortage and it is a effective way which is low cost and fast work. Waste water reuse not only can alleviate water shortage but also can reduce pollutant discharge. It is very significant to better water environment.

9.2 Recovery and energy in municipal waste water

9.2.1 Recovery types and feasibility

There are lots of pollutants in the municipal waste water, which can be recycled and reused.

9.2.1.1 Carbon

Generally, municipal waste water contains lots of organic matter, such as protein, fat, etc. Under bacterial, organic matter can be utilized and then transported into residual sludge. It is reported that the rate of carbon to residual sludge is up to 67%. Therefore, the residual sludge can be recycled, such as methane production, biochar production, biodegradable plastic production.

Another, some researchers pointed that parts of organic matter will be oxidized into CO_2 by bacterial, resulting in carbon waste. Therefore, an idea is proposed that the organic matter can be accumulated by some special materials without biodegradation, decreasing the transformation from organic matter to CO_2. It is obvious that this will avoid of more organic matter waste compared with biological treatment.

It is well known that it is the key for carbon recovery to accumulate organic matter.

9.2.1.2 Phosphorus

In general, the content of phosphorus in municipal waste water is at the range of 4~16 mg/L. About 50% of phosphorus will be transformed into residual sludge, while the other 50% of phosphorus will be discharged into water bodies, resulting in eutrophication.

Based on statics, the demand on fertilizers is about 45 million tons P_2O_5 in the whole world, while about 30 million tons residual sludge is produced. Therefore, if the phosphorus in the residual sludge can be recovered and reused, it can meet the demand on 5% of phosphorus requirement. It is noted that phosphorus recovery from municipal waste water contains huge potentiality.

9.2.1.3 Water

It is reported that the rate of impurities is only 0.1% in municipal waste water, less than sea water. Therefore, most of municipal waste water is clean water which can be reused. About 80% of portable water is discharged as municipal waste water. In fact, at least 70% of municipal waste water can be treated and reused. That is to say, the collection, treatment and recovery of sewage constitute good social water cycle systems, to safeguard natural water cycle systems.

Municipal waste water is the second most important water resource for cities due to its stable amount and without effect by seasons. Nowadays, waste water reuse is easy due to the technology development.

9.2.2 Energy reserves and energy feasibility

Municipal waste water contains lots of energy. Taking activated sludge as an example, energy consumption is about 0.3~0.6 kW · h/m^3, and the average is about 0.45 kW · h/m^3, equal to 1,620 kJ/m^3. If COD content in municipal waste water is 500 mg/L, the energy consumption is about 3.2 kJ/gCOD. It is reported that the energy contained in municipal waste water is up to 14.7~17.8 kJ/gCOD. It is calculated that it is self-sufficient in energy for treatment of municipal waste water.

Anaerobic digestion is used to recover methane from municipal sludge and it is a mature and wide used technology to recover energy. It is reported that the recovered energy by anaerobic digestion is obviously higher than that by statics. It is forecast that the energy provided by anaerobic digestion can meet about 50% of the total demand on power in municipal waste water plant.

It is noticeable that the amount of energy by anaerobic digestion in municipal waste water plant depends on the concentration of organic matter in the influent. That is, COD concentration is higher and the energy provided by anaerobic digestion is higher. According to statics, if COD concentration in the influent is at the average level of 283 mg/L, the energy provided by anaerobic digestion can meet 42% of the demand on total power. On the other hand, if COD concentration in the influent is at the average level of 600 mg/L, the energy provided by anaerobic digestion can meet about 70% of the demand on total power. However, this depends on energy transformation and energy consumption. Taking Haibohe waste water treatment plant as an example, COD concentration in the influent is about 1,500 mg/L, but the energy provided by anaerobic digestion can only meet about 33% of the demand on total power.

Another, hydrothermal process is an effective method to recover energy from municipal waste water. High temperature pyrolysis, incineration can be used to recover energy from municipal sludge and the energy efficiency is higher than that by anaerobic digestion. It is noticeable that the byproduct of high temperature pyrolysis on residual sludge is biochar which can be used as recovery resources.

9.3 Phosphorus recovery

9.3.1 Significance of phosphorus recovery

Phosphorus in municipal waste water is the main promoter for the eutrophication in water bodies. With the development of industry and agriculture, the discharge of phosphorus increased sharply, resulting in the frequency increase of algae bloom in lakes and red tide in offshore waters. This pollution due to phosphorus has threatened seriously the development of agriculture, fishing industry, tourism, etc. At the same time, this will affect negatively the drinking water and food industry. Therefore, phosphorus recovery from municipal waste water is the best way to control eutrophication development and phosphorus resource sustainability.

Phosphorus is very important for life and for food industry, biofuel and chemical industry. Nowadays, China is the second for phosphate rock production, about 3.2

billions tons and about 5% of the total amount in the world. However, according to the mining speed, the mining time may be only 23 years and then there is no phosphate rock to be mined. According to statics, the phosphate rock storage in the world does not meet the demand on phosphorus by people. Therefore, phosphate rock is the same as rare earth and fluorite.

It is well known that natural phosphorus after mining, production and consumption goes to natural water bodies with the discharge of municipal waste water. The value, according to statics, is about 80% going to natural water bodies. Another statics shows that about 90% of phosphorus will be accumulated in the soil and then penetrated into groundwater with the rainwater and irrigation. In these years, developed countries highly focused on phosphorus resource sustainability. Phosphorus recovery from municipal waste water and animal excrement has been the hot point. It is noticeable that phosphorus recovery from municipal waste water is not only to solve phosphorus pollution but also to keep phosphorus resource sustainability.

9.3.2 Basic methods for phosphorus resource

Phosphorus recovery can be divided into chemical methods, such as chemical settling, adsorption/desorption, crystallization and biological methods, such as EBPR.

9.3.2.1 Chemical settling

The key of chemical settling is to add chemical agent (ferric salt, aluminum salt, calcium salt, magnesium salt, etc.) into municipal waste water or liquid with high content of phosphorus, and then the metal ions will react with phosphate to form insoluble phosphate precipitation. Finally, phosphorus can be gotten by separation of insoluble phosphate precipitation and water.

Fe^{2+}, Fe^{3+} and Al^{3+} are used widely to remove phosphorus from municipal waste water. Generally, chloride and sulphide of iron are used to remove phosphorus. The reactions are as followed.

$$FeCl_3+PO_4^{3-} \longrightarrow FePO_4 \downarrow +3Cl^- \tag{9-1}$$

$$3FeCl_3+2PO_4^{3-} \longrightarrow Fe_3(PO_4)_2 \downarrow +6Cl^- \tag{9-2}$$

$$3FeSO_4+2PO_4^{3-} \longrightarrow Fe_3(PO_4)_2 \downarrow +3SO_4^{2-} \tag{9-3}$$

The mechanism of aluminum salt to remove phosphorus is the same as ferric salt. The reactions are as followed.

$$Al^{3+}+(H_nPO_4)^{(3-n)-} \longrightarrow AlPO_4 \downarrow +nH^+ \tag{9-4}$$

Calcium salt is used widely to remove phosphorus due to its cheap and easy management. The key precipitation is HAP. During the reactions, pH and the ratio of Ca to Pis the key factors. The reactions are as followed.

$$5Ca^{2+}+4OH^{-}+3HPO_4^{2-} \longrightarrow Ca_5(OH)(PO_4)_3\downarrow+3H_2O \quad (9-5)$$

Magnesium salt can react simultaneously with nitrogen and phosphorus and form struvite. When the concentration of PO_4^{3-}, NH_4^+ and Mg^{2+} is at some degree and the ion solubility product is larger than solubility product parameter of struvite, the precipitation of struvite will appear. Nowadays, more and more scientists have interest on struvite because the technology can recover simultaneously phosphorus and ammonia nitrogen. According to statics, in the struvite, the ratio of phosphorus as P_2O_5 is up to 51.8%, obviously higher than that in the best phosphate rock (46%). It is indicating that the struvite is same as phosphate rock rich with phosphorus.

In liquid, the formation of struvite can be described as followed.

Main reactions:

$$NH_3+H_3O^{+} \longrightarrow NH_4^{+}+H_2O \quad (9-6)$$

$$Mg^{2+}+NH_4^{+}+HPO_4^{2-}+6H_2O \longrightarrow MgNH_4PO_4\cdot 6H_2O\downarrow+H^{+} \quad (9-7)$$

$$Mg^{2+}+NH_4^{+}+PO_4^{3-}+6H_2O \longrightarrow MgNH_4PO_4\cdot 6H_2O\downarrow \quad (9-8)$$

$$Mg^{2+}+NH_4^{+}+H_2PO_4^{-}+6H_2O \longrightarrow MgNH_4PO_4\cdot 6H_2O\downarrow+2H^{+} \quad (9-9)$$

By-product reactions:

$$Mg^{2+}+4H^{+}+2PO_4^{3-} \longrightarrow Mg(H_2PO_4)_2 \quad (9-10)$$

$$Mg^{2+}+H_2O \longrightarrow Mg(OH)_2+2H^{+} \quad (9-11)$$

$$3Mg^{2+}+2PO_4^{3-} \longrightarrow Mg_3(PO_4)_2 \quad (9-12)$$

$$Mg^{2+}+H^{+}+PO_4^{3-} \longrightarrow MgHPO_4 \quad (9-13)$$

The formation of struvite has close relationship with the activity product (K_{OS}) of PO_4^{3-}, NH_4^+ and Mg^{2+} in the waste water. When the parameter of K_{OS} is larger than solubility product (K_{sp}) of struvite, the precipitation appears spontaneously.

In real waste water, there are lots of coexisting ions, such as Ca^{2+}, K^+ and CO_3^{2-}, which can react with PO_4^{3-}, NH_4^+ and Mg^{2+}, affecting saturation of these ions. Therefore, it is difficult to forecast formation potential of struvite in real waste water. Under the conditions, conditional solubility product is proposed, which can be described as following.

$$P_s = C_{T,Mg^{2+}} \times C_{T,NH_4^+} \times C_{T,PO_4^{3-}} = \frac{K_{OS}}{(a_{Mg^{2+}}a_{NH_4^+}a_{PO_4^{3-}}) \times (\gamma_{Mg^{2+}}\gamma_{NH_4^+}\gamma_{PO_4^{3-}})} = \frac{K_{sp}}{(a_{Mg^{2+}}a_{NH_4^+}a_{PO_4^{3-}})} \quad (9-14)$$

In the equation, $C_{T,i}$ is referred as the total concentration of i component. a_i is referred as ionization degree of i component. P_s is the function of solution pH.

9.3.2.2 Adsorption/desorption

Adsorption/desorption is a physical chemistry process. After the suitable adsorbent is added into the solution, surface complexation or exchange of coordination group in the gap of crystal appears, resulting in the adsorption on phosphorus. Finally, the adsorbent which is used in the solution is dissolved by alkaline solution and then the calcium salt is added to precipitate the phosphate in the solution.

There are many factors affecting adsorption/desorption, such as adsorbent, temperature, time, initial content, pH. In addition, the coexisting ions, such as Cl^-, SO_4^{2-}, will affect negatively the adsorption capacity. For the method, adsorbent is the key. Suitable adsorbent should have the following characteristics, large adsorption capacity on phosphorus, easy production, easy recovery, low price. Traditional adsorbents include hydrotalcite, aluminous quartz, ferric oxide, aluminium oxide, and so on. It is noted that the adsorption capacity on phosphorus is not big compared with the new adsorbents, for example, calcium silicate hydrate, metal hydroxide.

It is noticeable that the efficiency of the method depends on the adsorbent and the initial content of phosphorus. Therefore, for the municipal waste water, it is first to accumulate phosphorus and then to recover by the method.

9.3.2.3 Crystallization

The mechanism of crystallization is the same as chemical settling. After the solid materials with the similar structure and surface characteristics as the materials for chemical settling is added into the solution, the metastable state of phosphate ion is destroyed so that the ion can be accumulated on the surface of crystal seed. With the growth of crystal seed, the phosphate can be separated and recovered. Using struvite to recover phosphorus is the method.

The key factor for the crystallization is the oversaturation of ion. Additionally, pH, reaction time, the ratio of crystalline ion, selection of crystal seed are the important factors affecting on crystallization.

9.3.2.4 Biological methods

Generally, activated sludge to remove nitrogen and phosphorus is the main process to treat waste water. Based on this, EBPR is developed to recover phosphorus and has been focused by more and more people. EBPR process has the function not only to remove phosphorus but also organic matter. In the residual sludge, the phosphorus rate is up to 4% ~ 10%, higher than that in activated sludge, 2%. It is indicting that EBPR can effectively accumulate phosphorus from waste water and then the accumulated phosphorus can be recovered with chemical settling or crystallization. Compared with chemical settling or crystallization, EBPR is suitable for phosphorus recovery from municipal wastewater.

9.3.3 Evaluation on phosphorus resource

Nowadays, it is very clear that the recovery product of nitrogen and phosphorus is more expensive than phosphorus rock. However, the difference between the cost for phosphorus recovery and phosphorus rock mining and making will be shorten, with the increase of degree of maturity on phosphorus recovery and the outstanding of phosphorus rock reduction. It is indicating that the market to recover phosphorus is huge. The most important is that phosphorus recovery can take into account both economy benefits and social benefits.

Struvite is the most widely used method to recover phosphorus from municipal waste water. It is reported that about 1 kg struvite can be recovered from 1,000 m^3 waste water. Therefore, if all countries are going to recover phosphorus, 63 thousand tons phosphorus (P_2O_5) can be recovered. It is similar as the reduction on 1.6% of phosphorus rock mining. If 1 kg struvite can be recovered, the disposal cost on residual sludge will be reduced, suggesting that the reduction on running cost is about 9 RMB. According to abroad statics, the cost on chemical agent is about 0.36 euro, and the production of struvite is about 0.29 ~ 0.63 kg, and the profit is about 0.06 ~ 0.14 euro, if the anaerobic digestion supernatant of 1 m^3 is treated in UK waste water treatment plant. It is noted that the price of $MgCl_2$, NaOH and struvite is 90 euro, 59 euro and 200 euro, respectively. Another, in Australia, the production cost of struvite is about 140 dollar, while the selling price is about 261 dollar. In Japan, the production cost of struvite is about 460 dollar, while the selling price is about 276 dollar.

Questions

1. What is waste water recovery?

2.In your opinion, what is the significance to recover carbon and phosphorus? How to recover carbon and phosphorus from waste water?

3. What is waste water energy, please present its feasibility.

4. There are many kinds of methods to recover phosphorus from domestic waste water, please show the characteristics for different methods.

5. Making a comparison between chemical methods and biological methods to recover phosphorus, please present the advantages and disadvantages.

6. If you are the engineer in the domestic waste water treatment plant, would you like to choose biological methods to recover carbon and phosphorus from domestic waste water, please give the reasons.